Proceedings in Life Sciences

Central Neurone Environment and the Control Systems of Breathing and Circulation

Edited by
M. E. Schläfke H. P. Koepchen W. R. See

With 133 Figures

Springer-Verlag
Berlin Heidelberg New York 1983

Professor Dr. MARIANNE E. SCHLÄFKE
Dr. WOLF R. SEE, M.A.
Ruhr-Universität Bochum
Institut für Physiologie
Universitätsstraße 150
4630 Bochum, FRG

Professor Dr. HANS-PETER KOEPCHEN
Physiologisches Institut
der Freien Universität Berlin
Arminallee 22
1000 Berlin 33, FRG

Cover illustration: Plastogene Cast of Medullary Blood Supply
as seen from the Ventral Aspect.

ISBN 978-3-642-68659-7 ISBN 978-3-642-68657-3 (eBook)
DOI 10.1007/ 978-3-642-68657-3

Softcover reprint of the hardcover 1st edition 1983

2131/3130-543210

Preface

This volume contains the papers presented at the symposion on Central Neurone Environment and the Control Systems of Breathing and Circulation held at Bochum, October 5–7, 1981 in honour of Prof. Dr. Dr. h.c. Hans H. Loeschcke, who retired in March 1981.

His discovery of ventral medullary substrates forming an essential drive for the ventilatory as well as for circulatory control systems, and the elaboration of this concept during the last 25 years have profoundly influenced the concepts in this field. In an age of proceeding specialization on partial mechanisms, his approach has always emphasized the integrative aspects of the control systems, especially the role of the respiratory control system for the ionic homeostasis of the brain and the close interrelation between neuronal respiratory and cardiovascular control. This general intention implies the combination of physicochemical and neurophysiological approaches as well as the study of peripheral function of respiration and circulation, topics which are usually handled separately according to the different methods.

Following this general line, some 40 scientists of various disciplines were invited to contribute to this symposion from their own points of view, all of them more or less concerned with acid-base homeostasis of the body and of the brain tissue, neuronal and overall control of respiration and circulation and the mechanisms of respiratory rhythmogenesis. Particular emphasis was placed on treating in an integrative manner three main topics. These were: (1) Bicarbonate and chloride exchange between extracellular fluid of the brain and glia in comparison to analogous mechanisms in blood (2) the organization of chemical respiratory drives and their role for the maintenance of extracellular ion homeostasis of the CNS and (3) the organization of the central respiratory rhythm generator and its relation to autonomic cardiovascular innervation in dependence on central nervous chemical environment. In addition several pathophysiological phenomena such as respiratory sleep disorders, bronchomotor dispairments, and the Sudden Infant Death Syndrome and their possible connection with an impairment of the described systems were considered.

H.H. Loeschcke gave a review on central chemosensitivity and presented an integrative concept of this mechanism derived from older and recent results of his laboratory. Unfortunately, due to illness he was unable to submit a written account of his lecture. The summarizing scheme of his presentation is given in the round table discussion. This formed the basis of the main objective of this symposion: What is the functional structure of those substrates in the ventral part of the medulla oblongata which provide activation of the respiratory neuronal rhythm generator and of cardiovascular efferent outputs, and how does this system interact with peripheral chemoreceptive and other nonchemical central and peripheral drives to effect the overall control of acid-base homeostasis and cardiorespiratory adaptation?

The manuscripts represent the full or even an extended account of the oral presentations. We decided not to include the individual discussions of each lecture but rather the summarizing round table discussion which took place at the end of the symposium in order not to enlarge this volume unduly.

The symposion was sponsored by the Deutsche Forschungsgemeinschaft (SFB 114, Bionach) and the Gesellschaft der Freunde der Ruhr-Universität Bochum, whom we should like to thank for their support. Additional funding by Drägerwerke, Lübeck, K. Hillerkus, Krefeld, and Carl Zeiss, Oberkochen, is gratefully acknowledged.

We would also like to thank all those who by their willing help have made the organization of the symposion an easy task.

With great pleasure we acknowledge the generous offer of the Springer-Verlag, Heidelberg, to publish the proceedings of the symposion in full length so swiftly and in such an extensive presentation.

Bochum, November 1982

M.E. Schlaefke
H.P. Koepchen
W.R. See

Contents

Contributors

You will find the addresses at the beginning of the respective contribution

Central Neurone Environment

I. LEUSEN and J. WEYNE[1]

"La fixité du milieu intérieur est la condition de la vie libre, indépendante"

Cl. Bernard
Phénomènes de la vie, 1878

It is a fundamental characteristic in biology that cells and cell systems display their normal metabolism and activity only when they are surrounded by an optimal environment. This is particularly the case for nerve cells and muscle cells where the ionic composition of the extracellular fluid is crucial for excitability. In most tissues of the higher organisms environmental composition is largely determined by the composition of blood plasma, and constancy of the extracellular fluid is the result of the circulation with interchange of water, electrolytes and dissolved substances through the walls of the capillary beds, and of drainage by a network of lymphatics.

A particular situation, however, is encountered in the brain. Early in this century Goldman found, in extension of the observations of Ehrlich in 1885, that the brain and the cerebrospinal fluid (CSF) showed no coloration after intravascular injection of dyes such as trypan blue although all other tissues were colored. On the other hand, when the dye was injected into the CSF, the brain became colored. This led to the conclusion that there exists a barrier between blood and brain (CSF) which is impermeable to the dye.

Electronmicroscopic and dynamic (tracer) studies of recent years have given insight into this concept of a blood-brain barrier. The brain capillaries show the particularity that the endothelial cells are joined to each other by tight junctions. Ultrastructural study of these capillaries after injection of low-molecular weight substances such as microperoxidase or lanthanum has disclosed that these intercellular tight junctions are the site of the barrier for these substances (Brightman and Reese 1969). It thus is evident that brain capillaries restrain, in a large way, passive movement of solutes through their intercellular spaces so that active or facilitated (regulated) transport across the endothelial membranes is accepted to play an important role in the exchanges between blood and brain. This is particularly manifest for water soluble substances, and some hexoses and amino acids show facilitated diffusion into the neurone environment (for a review see Pardridge and Oldendorf 1977). Other substances, on the other hand, are prevented from entering brain extracellular fluid.

The blood-brain barrier can thus be considered to be protective in that it controls the rate of exchange between blood and brain. It thus can function as a second line for regulation of the composition of the "milieu interieur" of brain cells. Such is e.g.,

1 Laboratory of Normal and Pathological Physiology, University of Gent, De Pintelaan, 185 B-9000 Gent, Belgium

Central Neurone Environment
(ed. by M.E. Schläfke, H.P. Koepchen and W.R. See)

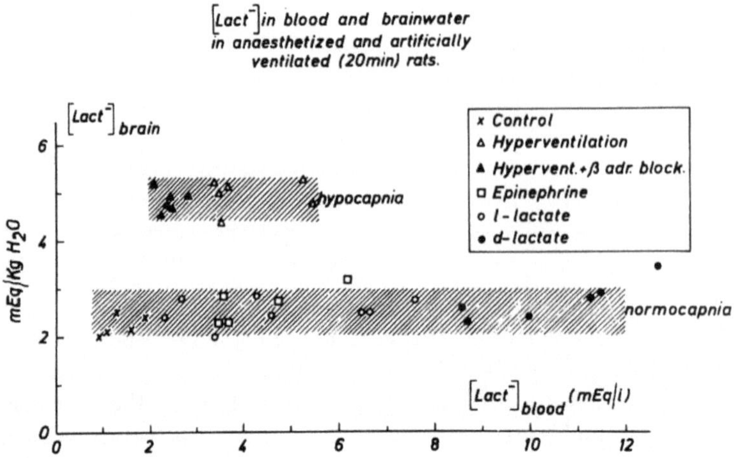

Fig. 1. Lactate concentration in blood and brain in anaesthetized rats which were artificially ventilated for 20 min. In normocapnic rats when blood lactate was increased for 20 min by i.v. infusion of lactic acid or by administration of epinephrine, brain lactate remained very stable. In hypocapnic rats, where there was a simultaneous rise of brain and blood lactate, blood lactate was lowered by administration of propranolol (β adrenergic blockade) and brain lactate remained almost unchanged

illustrated for lactic acid. Lactate concentration in brain appears to be highly independent from the lactate concentration in blood, and it was found to remain unchanged even when the lactate concentration in blood was markedly increased (Fig. 1). Similar results were obtained in CSF (Weyne and Leusen 1975). These observations may seem at variance with observations of Oldendorf, that there is an important uptake by the brain within 5–15 s when labeled lactate is injected into the carotid artery (Oldendorf 1973). The contradiction, however, is only apparent since Oldendorf's experiments also show that lactate uptake decreases with increasing blood concentration, suggesting the involvement of a carrier mechanism which is readily saturated (Pardridge and Oldendorf 1977).

This brings us to questions such as: what is the composition of the brain extracellular fluid (ECF)? what are the factors that have an influence on this composition? Many reviews published in journals and books show the development of research in this field (Siesjö and Sørensen 1971, Cserr et al. 1975, Loeschcke 1976, Bradbury 1979, Wood 1980).

Composition of Brain Extracellular Fluid

Deductions concerning the overall composition of brain ECF have first of all been made from data obtained on CSF, a compartment that is readily accessible and that is in continuity with brain ECF. Electronmicroscopy has indeed shown that the extracellular spaces of the brain are open to the cerebroventricular system and the subarachnoid space, and free penetration of charged particles and non-electrolytes has been demonstrated when they are injected into the cerebroventricular system (Rall et al. 1962, Brightman and Reese 1969, Pollay and Kaplan 1971). Such ready interchange between CSF and brain ECF suggests that both compartments should have a similar composition. However, attention must be given to some characteristics which are of a nature to induce differences between CSF and brain ECF (Wood 1980). CSF can indeed be expected to reflect the compostion of ECF only within the limits of the existence of a diffusion equilibrium between brain ECF and CSF. Or, many factors may intervene to perturb such equilibrium.

The Overall Brain ECF-CSF Equilibrium

It must be taken into consideration that the CSF is not a static compartment but that it is a fluid that is continuously renewed so that there is a bulk flow from the sites of formation to the places of absorption. The main rate of CSF production has been calculated to be approximately 0.4%/min of its total volume (thus total CSF volume is renewed every 6–7 h) (Davson 1967).

The choroid plexus accounts for approximatively 70% of CSF production, whereas the remaining CSF is derived from the capillary bed of the brain and metabolic water production by brain tissue (brain ECF). When the concentration of a substance in the freshly secreted fluid by the choroid plexus would be different from that in the brain ECF, the possibility of a steady difference between CSF and brain ECF can be considered. When the concentration in the secreted fluid is e.g., lower than in the brain ECF, a concentration gradient will exist from brain ECF to CSF. This seems to be the case for several substances and is considered as a mechanism for clearing substances of the brain [(sink action of the CSF; Davson 1967) Fig. 2]. If on the other hand the concentration of a given substance in the plexus secretion would be higher than that

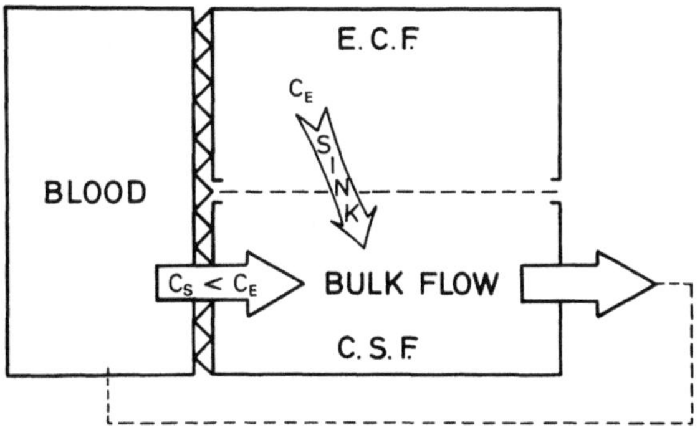

Fig. 2. Relationship between brain extracellular fluid, cerebrospinal fluid and blood. Continuous bulk flow of CSF from the site of secretion (choroid plexus) to sites of absorption (arachnoid villi). The blood-brain barrier separates blood from brain ECF and CSF, limiting transport of solutes, whereas the ependyma separating ECF from CSF does not impede diffusion between these two liquids. Sink-action of CSF: if the concentration of a substance in the freshly secreted fluid (C_s) is less than in the brain ECF (C_e) a concentration gradient develops between brain ECF and CSF and as a result that substance is continuously cleared from the brain ECF

in the brain ECF, the inverse situation will occur. Such exchanges between CSF and brain ECF are responsible for the changes in CSF composition which are observed when the CSF flows through the ventriculocisternal system from the choroid plexus sites to the cisterna magna and the subarachnoid space (Davson 1967, Cserr 1971). It can be accepted that in non-steady state such differences will be more pronounced due to changes in production rate and in bulk flow of CSF or to factors modifying the composition of brain ECF or of CSF. This is clearly demonstrated during perfusion of the ventriculocisternal system with an artificial CSF to which foreign substances are added. Thus, several hours after addition of inuline a concentration gradient was still noted from the lateral ventricle to the depth of brain tissue (Rall et al. 1962). Similar results were obtained with labelled sucrose (Pollay and Kaplan 1971). When, however, the tracer reached the brain via the blood, an inverse concentration gradient between the ventricular system and brain tissue was observed, with the higher concentration in the central tissue blocks.

Local Brain ECF Modifications

It can also be accepted that compostion of brain ECF can show local differences which are not reflected in bulk CSF. Several factors can contribute to this:

– Equilibrium from diffusion and convection will e.g., depend on the distance between the mean CSF compartment and the local brain ECF fraction which is considered. The greater this distance, the more the compostion may be expected to be dissimilar. Differences between local brain ECF and CSF can thus be the result of structural characteristics that impede rapid exchange and equilibration. Indeed, the intercellular spaces are narrow; but on the other hand, it is accepted that the configuration of the ventricular and arachnoid compartments is such that most regions of gray matter of the brain are within a few mm of CSF, so that exchange would occur rather rapidly (Wood 1980).

– It is conceivable that local differences also occur due to the fact that substances, while present in sites of brain ECF, do not have time to diffuse and equilibrate with bulk CSF. Such can e.g., be the case when substances liberated in brain ECF are rapidly metabolized, and enzymatic breakdown of neurotransmitters such as acetylcholine, dopamine, and 5 HT can be cited as typical examples (Wood 1980).

– Finally, changes occurring in some brain ECF sites in relation with neuronal activity may also be very restricted as was shown by the use of ion-selective microelectrodes (Nicholson et al. 1978). In this respect, an important role has been attributed to the glial cells in the regulation and the restoration of the environment of the neurones (Varon and Somjen 1979).

In summary: careful consideration must be given to the possible inhomogeneity of brain ECF and CSF in steady state, and certainly in non-steady state, if correlations are to be made between physiological responses and modifications in the composition of the neurone environment.

As this meeting will be in conjuntion with ventilation phenomena, we will concentrate more particularly on some aspects of the above considerations in relation to acid-base balance.

Acid-base Balance of CSF and Brain ECF

The acid-base balance and the hydrogen ion concentration of CSF are determined, as in other body fluids, by the partial pressure of carbon dioxide (pCO_2) and the bicarbonate concentration ($|HCO_3^-|$). Under normal control conditions the pCO_2 of cisternal CSF is 0.7 to 2 kPa (5–15 mm Hg) higher than in arterial blood while in general CSF $|HCO_3^-|$ is comparable to that in arterial blood plasma. As a result CSF pH is found to be somewhat lower than arterial pH. In many experimental and clinical conditions of acid-base disturbance, important differences are seen between the acid-base composition of CSF and that of blood. This raises the question as to what factors are important for acid-base regulation of CSF (cf. Leusen 1972, Leusen et al. 1981). In view of the direct dependence of CSF pH on the HCO_3^--pCO_2 ratio, two questions are obvious, namely what are the factors which control CSF pCO_2 and CSF $|HCO_3^-|$.

Factors Controlling CSF pCO_2

Carbon dioxide, as non-ionized liposoluble substance, diffuses easily through lipoprotein membranes; so it can be accepted that there is an equilibrium between CSF pCO_2 and its ambient structures. The ambient structures for the CSF are, on the one hand, the static cell structures of the central nervous system (neuronal and glial cells) and on the other hand, the dynamic flowing capillary blood. Nervous tissue cells and blood flow will have important roles in determining the level of CSF pCO_2. The neuronal and glial cells, which continuously produce CO_2 aerobically, form an afferent system adding CO_2 to their immediate environment.

The blood flowing through the cerebral capillaries can be either an afferent or an efferent system depending on the arterial-venous pCO_2 gradient. Under steady state conditions, the blood continuously removes CO_2 and the cerebral venous pCO_2 is higher than the arterial pCO_2. The arterio-venous pCO_2 difference through the brain is about 5–10 mm Hg and is a function of CO_2 production and cerebral blood flow. In steady state a pCO_2 gradient exists between the sites of CO_2 production (cells of the central nervous system) and the sites of CO_2 removal (capillary blood). It is accepted that the pCO_2 of the CSF is normally related to that of the brain ECF.

Whether or not CSF pCO_2 is close to the venous pCO_2 (Pappenheimer et al. 1965) or close to the average between arterial and cerebrovenous pCO_2 [(Krogh model) Caronna et al. 1977, Pontén 1976] is still disputed. The differences are usually small, but in some conditions large pCO_2 gradients between CSF and blood have been obtained. Some years ago Davies, Gurtner, and associates published data showing very high CSF pCO_2 values which largely exceeded cerebrovenous pCO_2. The effect was especially observed during severe blood acidosis and was explained by the so called charged membrane hypothesis (Davies and Gurtner 1973, Razavi et al. 1977). We have also observed high pCO_2 values in CSF which exceeded venous pCO_2 even without blood acidosis. In experiments involving ventriculocisternal perfusion for 45 min, CSF pCO_2 markedly increased with time although arterial and sagittal sinus pCO_2 remained constant (Fig. 3). At the end of the experiment CSF pCO_2 exceeded venous pCO_2 by about 2 kPa (15 mm Hg). Such high CSF pCO_2 values were not present at the outset of the experiment since CSF pCO_2 was between arterial and venous pCO_2. This observation indicates that large pCO_2 gradients in the brain can be observed in some conditions.

Attention has been given to a possible technical origin of these important gradients. Experimental methodology (e.g.; catheterisation of the sagittal sinus) may interfere with local blood flow and in non-homogenously perfused brain, important pCO_2 differences were observed in cerebrovenous blood sampled from different sites (Pontén 1976). Depth of anaesthesia and cerebroventricular perfusion may also be responsible for disturbing the normal relationship between blood and CSF pCO_2 (Berkenbosch 1971). If CSF pCO_2 is indeed susceptible to experimental artefacts, especially in prolonged experiments involving manipulations of the brain (ventriculocisternal perfusion, repetitive puncture of the sinus, and a fortiori its permanent catheterisation) measurements of CSF pCO_2 and pH will be of limited value if one is primarily interested in brain ECF.

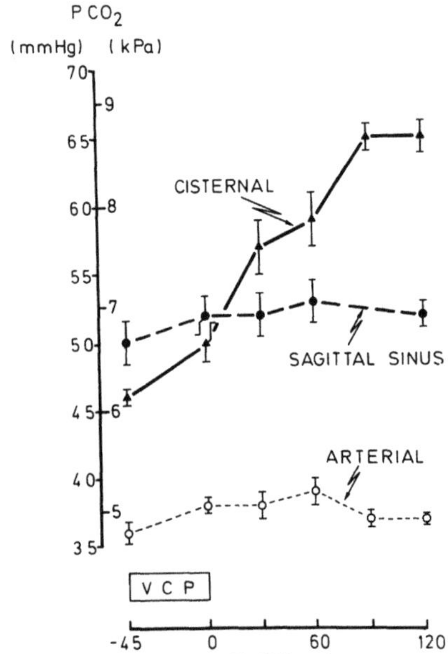

Fig. 3. pCO_2 in arterial and sagittal venous blood and in cisternal CSF in a group of anaesthetized and artificially ventilated dogs (mean ± S.E. n = 6). Between time 0 min and -45 min endogenous CSF was replaced by mock CSF by means of ventriculocisternal perfusion. In spite of a practically constant pCO_2 in blood, CSF pCO_2 gradually increased with time, exceeding sagittal pCO_2 by about 2 kPa at the end of the experiment

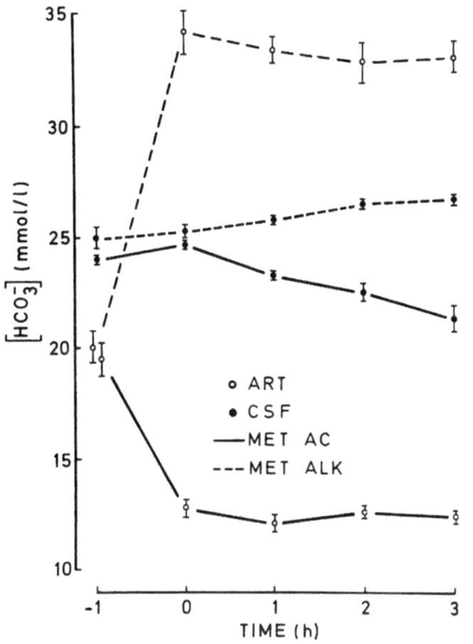

Fig. 4. Bicarbonate concentration in arterial plasma and cisternal CSF in 2 groups of anaesthetized and artificially ventilated normocapnic dogs (mean ± S.E.; n = 6 each). In one group metabolic acidosis was induced for 3 h by i.v. infusion of HCl and in the other group metabolic alkalosis was induced by i.v. infusion of Na_2CO_3. The change in plasma $|HCO_3^-|$ is about 20 mmol/1 and is maintained for 3 h, the change in CSF $|HCO_3^-|$ at the end of the experiment amounts to about 5 mmol/1. The results indicate some dependence of CSF $|HCO_3^-|$ on plasma $|HCO_3^-|$ (in particular on decreasing $|HCO_3^-|$)

Factors Affecting the CSF Bicarbonate Concentration

The HCO_3^- concentration in brain ECF and CSF appears to depend on several factors, and many observations have shown that there is not a simple equilibrium between the HCO_3^- concentration in CSF and that in blood although some dependence is present. This is shown in experiments in which the effect of changes in plasma $|HCO_3^-|$ on CSF $|HCO_3^-|$ was studied in normocapnic conditions. Important modifications of blood $|HCO_3^-|$ are accompanied by only moderate changes in CSF $|HCO_3^-|$ in the same direction, at least in the initial hours (Fig. 4). On the other hand some dependence is evident. Such dependence of CSF $|HCO_3^-|$ on plasma $|HCO_3^-|$ is also manifest in more artificial conditions in which CSF $|HCO_3^-|$ is artificially decreased (Weyne et al. 1979). Figure 5 shows the results from such experiments in which CSF $|HCO_3^-|$ was increased. It appeared that the return of CSF $|HCO_3^-|$ to normal values was importantly impeded when plasma $|HCO_3^-|$ was artificially increased. But, on the other hand, the fall in CSF $|HCO_3^-|$ appeared not to be entirely dependent on plasma $|HCO_3^-|$, and some decrease was still observed even against a positive blood-CSF concentration gradient. Thus, other mechanisms are also involved. This complexity is also observed in experiments in hypercapnia and hypocapnia in which changes in CSF $|HCO_3^-|$ are only partially inhibited when plasma $|HCO_3^-|$ adaptation is artificially antagonized (Pannier et al. 1971, Leusen 1972, Dempsey et al. 1978). To explain the relative independence of CSF $|HCO_3^-|$ from blood $|HCO_3^-|$, factors which may influence the equilibrium between plasma $|HCO_3^-|$ and CSF $|HCO_3^-|$, and mechanisms that may act independently of plasma $|HCO_3^-|$ have been considered.

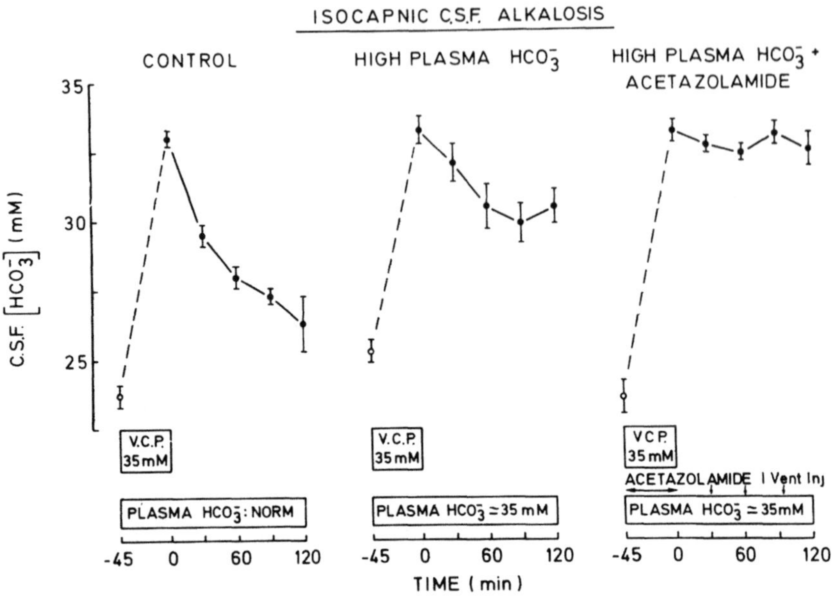

Fig. 5. CSF $|HCO_3^-|$ in 3 groups of anaesthetized and artificially ventilated normocapnic dogs (mean ± S.E.; n = 6 each). After sampling endogenous CSF (*open symbols*) CSF $|HCO_3^-|$ was increased by ventriculocisternal perfusion with mock CSF ($|HCO_3^-|$ = 35 mM), thereafter CSF $|HCO_3^-|$ was monitored for two h. In control dogs (*left panel*) with normal plasma $|HCO_3^-|$ at about 22 mM, CSF $|HCO_3^-|$ fell by about 7 mM in 2 h. In dogs with increased plasma $|HCO_3^-|$ to about 35 mM (*middle panel*) CSF $|HCO_3^-|$ fell by only 3 mM indicating importance of plasma $|HCO_3^-|$ for the fall in CSF $|HCO_3^-|$. The remaining fall in CSF $|HCO_3^-|$ occurred against a concentration gradient with blood. In dogs with increased plasma $|HCO_3^-|$ and in addition with acetazolamide injected intraventricularly (*right panel*) CSF $|HCO_3^-|$ remained unchanged at the high level

Factors Influencing the Plasma / CSF $|HCO_3^-|$ Ratio

Attention has been given to mechanisms which can effect the movement of HCO_3 ions between blood and CSF (through the blood/CSF barrier) and thus can influence the plasma/CSF $|HCO_3^-|$ ratio.

Thus the electrical potential difference which exists between blood and CSF might influence the electrochemical equilibrium and thus the movement and the $|HCO_3^-|$ ratio between blood and CSF, the more that it was found that this potential changes during alteration of blood pH. However, observations which invalidate this possibility were made by Pannier et al. (1971) and also recently by Bledsoe et al. (1981).

Mechanisms Acting Independently from Blood $|HCO_3^-|$

a) *Influence of* pCO_2. Since changes in pCO_2 are reflected in CSF and central nervous system, these pCO_2 alterations could effect the $|HCO_3^-|$ adaptation in CSF. Such an influence was found in experiments in which pCO_2 modifications were induced at constant plasma $|HCO_3^-|$ and can also be deduced from comparative experiments in which changes in plasma $|HCO_3^-|$ were or were not accompanied by pCO_2 modifications. Thus CSF $|HCO_3^-|$ increases in respiratory acidosis even when plasma $|HCO_3^-|$ is maintained at very low levels and it decreases in respiratory alkalosis even when plasma $|HCO_3^-|$ is maintained at high levels (Pannier et al. 1971, Dempsey et al. 1978).

Similarly, in metabolic acidosis the fall in CSF $|HCO_3^-|$ appeared to be more in function of the fall of CSF pCO_2 than of plasma $|HCO_3^-|$ decrease, while the increase in CSF $|HCO_3^-|$ was much more pronounced in metabolic alkalosis when pCO_2 was allowed to increase (Nattie and Romer 1978, Javaheri et al. 1979). Indications of a pCO_2 effect are given by the presence in the central nervous tissue, more particularly in the choroid plexus and in the glial cells, of carbonic anhydrase, an enzyme which catalyzes the hydration of $CO_2 + H_2O \rightarrow H_2CO_3 \rightarrow H^+ + HCO_3^-$ (Giacobini 1962). Very recently it was suggested that carbonic anhydrase is also present in the endothelial cells of the brain capillaries (Laux and Raichle 1981).

Thus the possibility exists of generating HCO_3^- apart from the plasma HCO_3^- concentration. The possible role of carbonic anhydrase (CA) in the production of CSF and in the determination of its HCO_3^- concentration has been studied using CA inhibiting substances.

Under the influence of carbonic anhydrase inhibitors the rate of CSF formation is markedly reduced (Davson 1967). In such studies it was also shown that in respiratory acidosis the increase in CSF $|HCO_3^-|$ was limited after central carbonic anhydrase inhibition. This led to the conclusion that the increase in CSF $|HCO_3^-|$ was due to two factors:

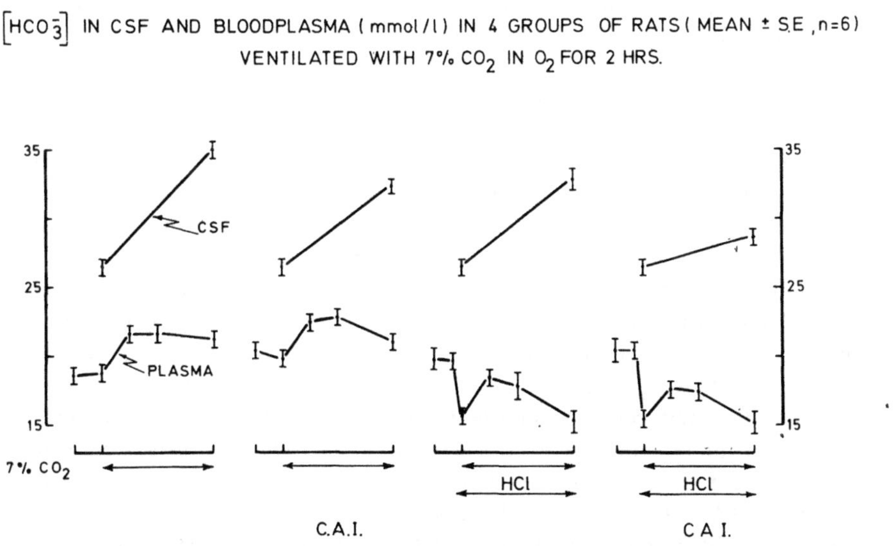

Fig. 6. $|HCO_3^-|$ in CSF and arterial plasma (mM) in 4 groups of rats ventilated with 7% CO_2 in O_2 for 2 h (mean ± S.E.; n = 6 each); before hypercapnia was induced, the animals were normocapnic for 45 min. In control rats (*left panel*) CSF $|HCO_3^-|$ increased by 8.5 mM in 2 h. With intraventricular administration of acetazolamide, resulting in carbonic anhydrase inhibition (CAI) this rise was reduced to 6 mM (*second panel*). When plasma $|HCO_3^-|$ was lowered by i.v. infusion of HCL, the rise in CSF $|HCO_3^-|$ was about 6.5 mM (*third panel*), but when in addition carbonic anhydrase was inhibited the rise in CSF $|HCO_3^-|$ was further reduced to 2.5 mM

one factor related to the hydration of CO_2, a reaction catalyzed by carbonic anhydrase, and the other related to the increase in plasma $|HCO_3^-|$ (Hasan and Kazemi 1976, Maren 1979). Figure 6 represents experiments done in this context on rats (Kazemi et al. 1976). Carbonic anhydrase inhibition also appeared to block the decrease in CSF $|HCO_3^-|$ after its experimental increase, when this fall occurred against a concentration gradient with blood (Leusen et al. 1979) (Fig. 5, right panel).

These studies thus indicate how acetazolamide is able to influence the $|HCO_3^-|$ adaptations occurring in CSF. This effect is most probably due to an influence on carbonic anhydrase containing structures, although an influence of acetazolamide on Cl^- - HCO_3^- exchange mechanisms of the cells has also been presented as a possibility. As carbonic anhydrase and also Cl^- - HCO_3^- exchange mechanisms are not limited to the choroid plexus structures but are present in the nervous tissue.mass (carbonic anhydrase is particularly present in glial cells) this indicates how acid-base equilibrium of brain extracellular fluid can be influenced directly by brain tissue.

b) *Influence of Brain Tissue.* That the brain cells affect the acid-base equilibrium of CSF is shown by the alterations which occur in CSF $|HCO_3^-|$ during perturbations in the nervous tissue metabolism. Thus, in pathological conditions there is lactacidosis in the CSF whenever the brain's oxygen supply is impeded, because lactate diffuses readily from the brain tissue to CSF (cf. Weyne and Leusen 1975). It is therefore evident that the brain cells can take part in the regulation of the acid-base equilibrium of CSF and brain ECF, the more that the cellular compartments of the brain have a high buffer capacity, comparable to that of blood.

During acute pCO_2 alterations, the bicarbonate concentration in brain tissue changes quickly and gives a $|HCO_3^-|$ - pCO_2 dissociation curve comparable to the classical dissociation curve obtained with total blood. The log pCO_2/pH curve derived from the CO_2 dissociation curve is approximately linear for pCO_2 values from about 3.5 to 14 kPa (25–100 mm Hg).

The most significant difference between brain and CSF is seen in the HCO_3^- response to acute pCO_2 changes. While $|HCO_3^-|$ in the brain reacts almost immediately to changes of pCO_2 giving a typical CO_2 dissociation curve, only slow gradual changes are seen in CSF $|HCO_3^-|$. This can be explained by the fact that CSF lacks the non-CO_2/HCO_3^- buffer systems which are present in brain tissue. It is conceivable that this rapid adaptation of intracellular buffer may have more influence on brain ECF than on CSF. Indications are found in the measurements by Ahmad et al. (1976) of cortical surface pH during hypercapnia showing a fast increase in the extracellular $|HCO_3^-|$ which probably finds its origin in the brain cells. Even more important buffer effects were obtained in the intercellular fluid of the nervous tissue (Cragg et al. 1977). In this respect studies concerning possible mechanisms of these exchanges are interesting and are discussed in this symposium (Ahmad and Loeschcke, Javaheri and Weyne). Such local pH measurements indicate that also during (acute) acid-base changes, inhomogenous distribution may exist within the brain. Sometimes the extracellular fluid is better protected than the large cavity fluid. Sometimes, however, the large cavity fluid seems better protected. This is e.g., the case during acute disturbances of plasma HCO_3^- and also during alterations of brain metabolism e.g., hypoxia. In these conditions the changes in brain ECF precede those in the large cavity fluid (Javaheri et al. 1981). This illustrates how in changing acid-base metabolic conditions, the acid-base environment of the brain cells may be different from that of the cisternal CSF.

Because of the significance attributed to acid-base balance of CSF and brain ECF in hypotheses concerning the chemical regulation of ventilation (Loeschcke 1980, Schlaefke 1981) and cerebral blood flow, and in view of the importance of small changes in pH, attention must be paid to potential differences in the chemical composition of brain ECF and CSF. An illustration is found in a recent publication by Fencl and his co-workers (1979). In 1966 this author concluded, from his observations at normal sea level, that the ionic composition of cerebral interstitial fluid and large cavity CSF must be identical (Fencl et al. 1966). Observations made more recently have led to the conclusion that at simulated high altitude $|HCO_3^-|$ in cerebral interstitial fluid is lower than in CSF. This suggests that chemoreceptor cells active in the regulation of ventilation have a more acidic environment in these circumstances (Fencl et al. 1979). Davies (1978) came to a comparable conclusion during hypoxia experiments on anesthetized dogs.

References

Ahmad HR, Berndt J, Loeschcke HH (1976) Bicarbonate exchange between blood, brain extracellular fluid and brain cells at maintained pCO_2. In: Loeschcke HH (ed) Acid base homeostasis of the brain extracellular fluid and the respiratory control system. Thieme, Stuttgart, pp 19–27

Berkenbosch A (1971) CSF-arterial pCO_2 differences with and without ventriculo-cisternal perfusion in the anesthetized cat. Pflügers Arch 330:99–110

Bledsoe SW, Eng DY, Hornbein TF (1981) Evidence of active regulation of cerebrospinal fluid acid-base balance. J Appl Physiol 51:369–375

Bradbury M (1979) The concept of a blood-brain barrier. John Wiley & Sons, Chichester

Brightman MW, Reese TS (1969) Junctions between intimately opposed cell membranes in the vertebrate brain. J Cell Biol 40:648–677

Caronna JJ, Plum F, Siesjö BK (1977) pCO_2 gradients between blood and CSF in rat during alterations of acid-base balance. Am J Physiol 227:1173–1177

Cragg P, Patterson L, Purves MJ (1977) The pH of brain extracellular fluid in the cat. J Physiol (London) 272:137–166

Cserr HF (1971) Physiology of the choroid plexus. Physiol Rev 51:273:311

Cserr HF, Fenstermacher JD, Fencl V (1975) Fluid environment of the brain. Academic Press, New York

Davies DG (1978) Evidence for cerebral extracellular fluid $|H^+|$ as a stimulus during acclimatization to hypoxia. Respir Physiol 32:167–182

Davies DG, Gurtner GH (1973) CSF acid base balance and the Wien effect. J Appl Physiol 34:249–254

Davson H (1967) Physiology of the cerebrospinal fluid. Churchill, London

Dempsey JA, Forster HV, Chosy LW, Hanson PG, Reddan WG (1978) Regulation of CSF $|HCO_3^-|$ during long-term hypoxic hypocapnia in man. J Appl Physiol 44:175–182

Ehrlich P (1885) Das Sauerstoffbedürfnis des Organismus. In: Eine Farbeanalytische Studie, Berlin

Fencl V, Miller TB, Pappenheimer JR (1966) Studies in the respiratory response to disturbance of acid-base balance, with deduction concerning the ionic composition of cerebral interstitial fluid. Am J Physiol 210:459–472

Fencl V, Gabel RA, Wolfe D (1979) Composition of cerebral fluid in goats adapted to high altitude. J Appl Physiol 47:508–513

Giacobini E (1962) A cytochemical study of the localization of carbonic anhydrase in the nervous system. J Neurochem 9:169–177

Hasan FM, Kazemi H (1976) Dual contribution theory of regulation of HCO_3^- in respiratory acidosis. J Appl Physiol 40:559–567

Javaheri S, Nardell EA, Kazemi H (1979) Role of pCO_2 as determinant of CSF HCO_3^- in metabolic acidosis. Respir Physiol 36:155–166

Javaheri S, Clendening A, Papadakis N, Brody JS (1981) Changes in brain surface pH during acute isocapnic metabolic acidosis and alkalosis. J Appl Physiol 51:276–281

Kazemi H, Weyne J, Leuven F Van, Leusen I (1976) The CSF $|HCO_3^-|$ increase in hypercapnia. Relationship to HCO_3^-, glutamate, glutamine and NH_3 in brain. Resp Physiol 28:387–401

Laux BE, Raichle ME (1981) The passage of ^{11}C-bicarbonate ion and $^{11}CO_2$ from blood to brain. J Cerebr Blood Flow Met 1: Suppl 1:383

Leusen I (1972) Regulation of cerebrospinal fluid composition with reference to breathing. Physiol Rev 52:1–56

Leusen I, Nshimyumuremyi JB, Weyne J (1979) Correction of cerebrospinal fluid HCO_3^- after its experimental increase: role of central carbonic anhydrase. J Physiol (London) 295:71P–72

Leusen I, Weyne J, Demeester G (1982) Regulation of acid-base equilibrium of cerebrospinal fluid. In: Wood JH (ed) Neurobiology of cerebrospinal fluid, vol II. Plenum Press, New York, in press

Loeschcke HH (1976) Acid base homeostasis of the brain extracellular fluid and the respiratory control system. Thieme, Stuttgart

Loeschcke HH (1980) Chemical alterations of cerebrospinal fluid acting on respiratory and circulatory control systems. In: Wood JH (ed) Neurobiology of cerebrospinal fluid, vol I. Plenum Press, New York, pp 29–40

Maren TH (1979) Effect of varying CO_2 equilibria on rats of HCO_3^- formation in cerebrospinal fluid. J Appl Physiol 47:471–477

Nattie EE, Romer L (1978) CSF HCO_3^- regulation in isosmotic conditions: the role of brain pCO_2 and plasma HCO_3^-. Resp Physiol 33:177–198

Nicholsen C, Ten Bruggencate G, Stöckle H, Steinberg R (1978) Calcium and potassium changes in extracellular microenvironment of cat cerebellar cortex. J Neurophysiol 41:1026–1039

Oldendorf WH (1973) Carrier mediated blood brain transport of short-chain monocarboxylic organic acids. Am J Physiol 224:1450–1453

Pannier JL, Weyne J, Leusen I (1971) The CSF-blood potential and the regulation of the bicarbonate concentration of CSF during acidosis in the cat. Life Sci 10:Part 1:287–300

Pappenheimer JR, Fencl V, Heisey SR, Held V (1965) Role of cerebral fluids in control of respiration as studied in unanesthetized goats. Am J Physiol 208:436–450

Pardridge WM, Oldendorf WH (1977) Transport of metabolic substances through the blood-brain barrier. J Neurochem 28:5–12

Pollay M, Kaplan RJ (1971) Effect of cerebrospinal fluid sink on sucrose diffusion gradients in brain. Exp Neurol 30:54–56

Pontén U (1976) Carbon dioxide tension relations in the brain in various acid-base conditions. In: Loeschcke HH (ed) Acid base homeostasis of the brain extracellular fluid and the respiratory control system. Thieme, Stuttgart, pp 8–15

Rall DP, Oppelt WW, Patlak CS (1962) Extracellular space of brain as determined by diffusion of inulin from the ventricular system. Life Sci 2:43–48

Razavi AK, Burns B, Scinto AM, Gurtner GH, Davies (1977) The Wien effect in compensated metabolic acidosis. Resp Physiol 29:25–33

Schlaefke ME (1981) Central chemosensitivity: a respiratory drive. Rev Physiol Biochem Pharmacol 90:172–244

Siesjö BK, Sørensen SC (1971) Ion homeostasis of the brain. Alfred Benzon Symposium III. Munksgaard, Copenhagen

Varon SS, Somjen GG (1979) Neuron glia interactions. Neuronscience 17:1–239

Weyne J, Leusen I (1975) Lactate in CSF in relation to brain and blood. In: Cserr HF, Fenstermacher JD, Fencl V (eds) Fluid environment of the brain. Academic Press, London New York

Weyne J, Kazemi H, Leusen I (1979) Relation of CSF HCO_3^- after its experimental lowering in normocapnic conditions. J Appl Physiol 47:369–376

Wood JH (1980) Neurobiology of cerebrospinal fluid. Plenum Press, New York

Evidence for a Carrier Mediated Exchange Diffusion of HCO_3^- Against Cl^- at the Interphases of the Central Nervous System

H.R. AHMAD[1,2] and H.H. LOESCHCKE[1]

The kinetics of the in vivo CO_2 dissociation curve of the brain extracellular fluid (ECF) revealed that as soon as the extracellular PCO_2 was increased, there was a marked and instantaneous increase in the extracellular bicarbonate concentration (Ahmad et al. 1978). Further experiments were designed to find out if there was a counter exchange of HCO_3^- against Cl^- at the interphase of the CNS. In these experiments brain ECF pH, pCO_2, Cl^-, and HCO_3^- changed, following closely mono-exponential time function after a delay of 5–7 s in response to a nearly step increase of end-tidal PCO_2 in artificially ventilated cats. The most important observation from such a preparation was an increase of the extracellular HCO_3^- accompanied by a decrease of Cl^- with a ratio of almost one to one and a time constant of 30 s. A similar rapid HCO_3^--Cl^- exchange has been shown to occur at the blood-brain ECF-barrier in metabolic disturbances while keeping $paCO_2$ constant (Loeschcke and Ahmad, 1980, Ahmad and Loeschcke, 1982). These investigations led to a new concept regarding the exchange process of HCO_3^- and Cl^- between blood plasma and the brain ECF on the one hand and between brain cells and the ECF on the other hand. Since the glial cells provide the bulk of the cell mass, they are the first candidates for the anion exchange. In respiratory acidosis such anion exchange is the leading process because CO_2 from blood enters the extracellular fluid and the brain cells without restriction; then it reacts with intracellular buffers forming HCO_3^- which in turn is exchanged against Cl^-. What is the mechanism of such a HCO_3^- - Cl^- exchange in glial cells under the influence of their high membrane potential?

In red cells a non-electrogenic exchange of HCO_3^- against Cl^- is supposed to be mediated by a carrier protein system, which can be inhibited by stilbene disulfonate derivatives like DIDS, SITS, and salicylate (Gunn et al. 1973, Cabantchik and Rothstein 1974, Passow et al. 1980, Wieth 1979, Wieth and Brahm 1980). The question is whether or not a similar type of mechanism also occurs in the glial cell membrane.

We studied the influence of anion exchange blockers SITS or DIDS on the time courses of the changes in brain extracellular H^+ and Cl^- activity occuring upon

1 Institut für Physiologie I, Ruhr-Universität, D-4630 Bochum, Federal Republic of Germany

2 Max-Planck-Institut für Systemphysiologie, Rheindamm 201, D-4600 Dortmund, Federal Republic of Germany (present address)

Central Neurone Environment
(ed. by M.E. Schläfke, H.P. Koepchen and W.R. See)

alteration of the PCO_2. In this communication the findings from in vivo and in vitro preparations are reported.

Methods

In Vivo Preparation

In chloralose-urethane anaesthetized cats, a hole was drilled in the parietal skull and a pool was constructed using a plexiglass ring of 20 mm height. A pH electrode was balanced on the parietal cortex after fixing the dura at the bone edge. The pool was filled with native cerebrospinal fluid (CSF) maintained at 37°C. The details of the method are described elsewhere (Ahmad et al. 1978. Loeschcke and Ahmad 1980). DIDS or SITS was dissolved in natural CSF and pH was titrated with NaOH to CSF pH equilibrated with 5% CO_2 at 37°C. Before and after incubation of the brain surface (parietal cortex) with 10^{-3} molar SITS or DIDS the time course of the change in pH_e was observed in response to a step increase of end-tidal PCO_2 in spontaneously as well as in artificially ventilated cats.

In Vitro Method

Rat cortical brain slices (thickness: ~0.3–0.5 mm) were prepared immediately after induction of a spinal shock following the method of Fukuda and Loeschcke (1977). The brain slices were cut into two halves. One half was incubated in mock CSF and the other in mock CSF with 10^{-3} molar SITS* or DIDS+. For comparison washed human red cells suspended in Ringer solution were incubated with and without SITS or DIDS (10^{-3} molar) at 37°C. The solutions contained 20 mmol $NaHCO_3$ besides the other electrolytes and were free from protein and phosphate. For microtonometry a CO_2 electrode was modified in such a way that the substrate, for instance a brain slice, was introduced between a silicone membrane and a glass electrode. This device could be exposed to varying gas mixtures of the desired PCO_2 so that pH changes in the *fluid film* between the electrode and the brain slice surface could be compared to those of in vivo studies. The system was kept at 37°C (for details, see Ahmad et al. 1981). Mock CSF, treated or untreated brain slices or red cells (1 µl probe with a cytocrit of 75 or 40%) were separately interposed between the pH electrode and the silicone membrane (Fig. 1 A). The chloride electrode was designed in a similar way using a chlorinated silver plate instead of the glass membrane (Fig. 1 B). The electrode is specific to changes in chloride activity because it is free from cross sensitivity of other ions, especially H^+. The assembled electrodes containing cellular or non-cellular samples were exposed to a rectangular step increase of CO_2 from 4–8% in O_2. The gas perfusion mixture was administered using a 4 way stopcock, so that the time course of the changes in H^+ and Cl^- activity in the fluid film could be evaluated.

---

* (4-acetoamido-4′ isothiocyanostilbene-2,2′ disulphonic acid)

+ (4″-diisothiocyanostilbene-2,2′ disulphonic acid)

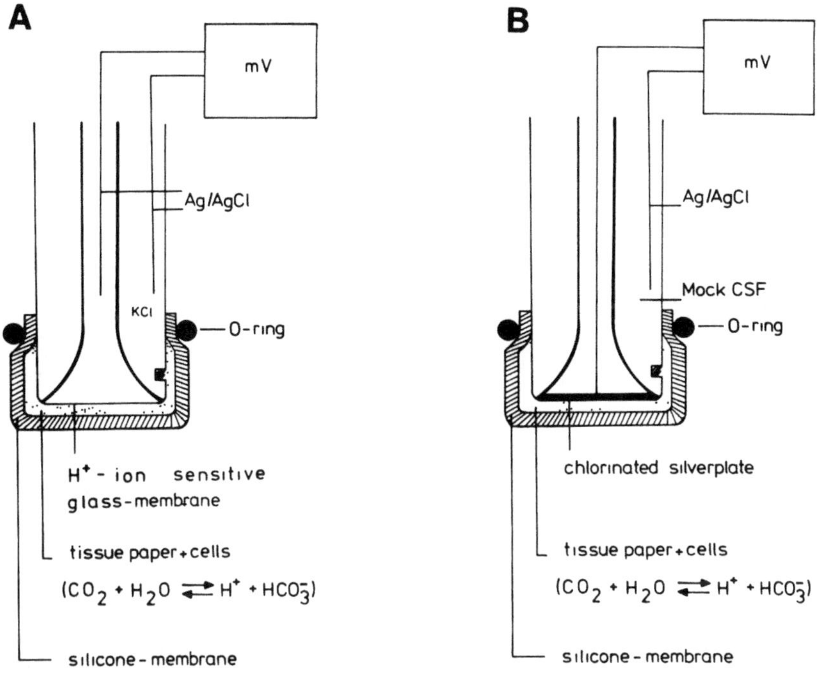

Fig. 1. A Schematic diagram of a microtonometer consisting of a pH-electrode, a thin tissue ("Josef paper") and a silicon membrane. A CO_2 step can be produced in a gas perfusion current at the side of the electrode to study the time course of the change in pH in cellular or non-cellular samples interposed between the membrane and the pH electrode. **B** The chloride electrode was designed in a similar way using a chlorinated silver plate instead of a glass membrane

Results

a) pH Response of the Parietal Cortex to Changes in the End-tidal PCO_2

The arterial blood pressure, tidal volume, pH_e and end-tidal PCO_2 were recorded in a spontaneously breathing anaesthetized cat (Fig. 2). In response to a rectangular increase of end-tidal PCO_2, the extracellular pH decreased by 0.11 units with a time constant of 25 s (Fig. 2A), and 12 s when the brain tissue was treated with 10^{-3} molar DIDS (Fig. 2B). A similar result was obtained, when the brain surface was incubated with SITS or salicylate. A difference in the time constants of the changes in pH_e between treated and untreated brain surfaces was observed although the pH decrements in the steady state were not significantly different. The experiments were repeated in artificially ventilated cats, showing similar results. The time constant of the decrease in pH of the cortical surface pretreated with SITS (10^{-3} molar) was significantly shorter (32.7 ± 1.9 s, n = 14 in 5 cats) than that of controls (48.4 ± 2.3 s, n = 10 in 5 cats). An initial rapid and greater decrease in pH_e in the treated preparation could have increased local cerebral blood flow more than in the control runs. Therefore, enhanced washout of CO_2 would diminish the pH_e decrement in the treated preparation.

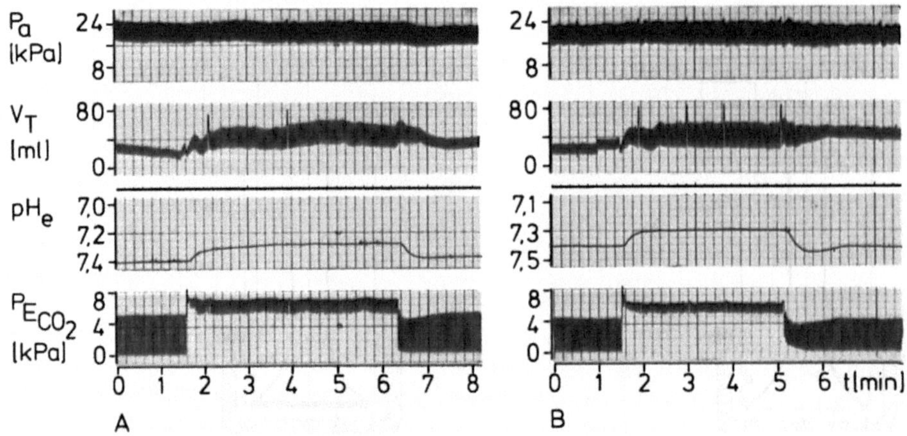

Fig. 2. Continuous record of arterial pressure, tidal volume, extracellular pH (*pH_e*) on the parietal cortex and tidal *PCO_2* in an anaesthetized cat breathing 100% O_2. The end-tidal PCO_2 has been varied in an almost rectangular fashion in order to determine the time course of the change in pH_e before (**A**) and after application of DIDS (**B**). DIDS shortens the time constant of the pH_e response

b) In Vitro pH_e Response at the Rat Parietal Cortex to Changes in PCO_2

Experiments were designed to study the influence of SITS or DIDS on the anion exchange between brain cells and ECF without the influence of changes in local cerebral blood flow.

Three different samples, untreated, or SITS treated cortical brain slices and mock CSF alone were separately examined in the microtonometer (Fig. 3, from top to

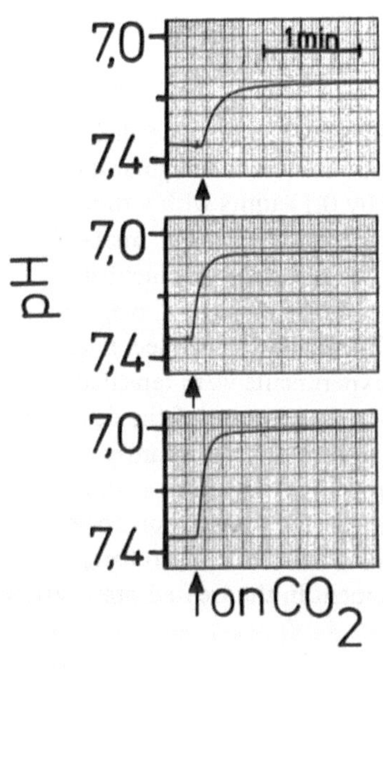

Fig. 3. pH time courses of 3 different samples, following step increases of PCO_2. *From top to bottom* untreated brain slices, SITS treated brain slices and mock CSF. The pattern of response of pH_e in a SITS treated slice approaches that in mock CSF alone

Table 1. Average time constants of the pH response to a step increase of PCO_2 from mock CSF, brain slices, Ringer, and human red cells in the microtonometer. Additionally, the time constant of the pH change in a red cell film was examined without the silicone membrane and the paper. n indicates numbers of slices, red cells or solution samples

Samples in the microtonometer		$\tau\Delta$pH (s) Control	Treated	-Δ pH Control	Treated
With membrane					
Mock CSF	$\bar{x}$	4.5	--	0.312	--
	± SEM	0.13		0.005	
	n	10		10	
Brain slices	$\bar{x}$	9.2	6.0	0.215	0.275
with and without	± SEM	1.8	0.9	0.008	0.005
10^{-3} molar	n	9	7	9	7
SITS	P	> 0.05		< 0.002	
Ringer	$\bar{x}$	4.2	--	0.308	--
	± SEM	0.2		0.0	
	n	10	--	10	
Human red cells	$\bar{x}$	11.5	5	0.115	0.235
cytocrit: 75%	± SEM	1.5	0.5	0.003	0.006
with and without	n	20	17	20	17
10^{-3} molar SITS	P	< 0.0002		< 0.0002	
Without membrane					
Red cell film	$\bar{x}$	0.95	0.6	0.175	0.220
(cytocrit: 40%	± SEM	0.15	0.1	0.005	0.008
with and without	n	28	20	28	20
10^{-3} molar SITS	P	= 0.05		< 0.0002	
Red cell film	$\bar{x}$	0.95	0.45	0.175	0.275
(cytocrit: 40%	± SEM	0.15	0.1	0.005	0.003
with and without	n	28	24	28	24
10^{-3} molar DIDS	P	= 0.005		$\leq$ 0.0002	

bottom). In response to a stepwise increase of PCO_2 the time constants of the pH_e decrease were 9, 6, and 4.5 s, respectively. The amplitude of the pH_e decrease was found to be greater in the SITS treated slice (-0.275) than in the untreated slice (-0.215). Note that the pattern of the pH response in SITS treated brain slices approached that in mock CSF alone. The average time constants from similar experiments for the treated and untreated brain slices, human red cells, Ringer and mock CSF are given in Table 1. The time constants of the change in pH were close to 1 s, when a red cell film (cytocrit 40%) was examined in the microtonometer without the silicone membrane and Josef paper. The significance of the method differences of the group means were tested by an unpaired t-test.

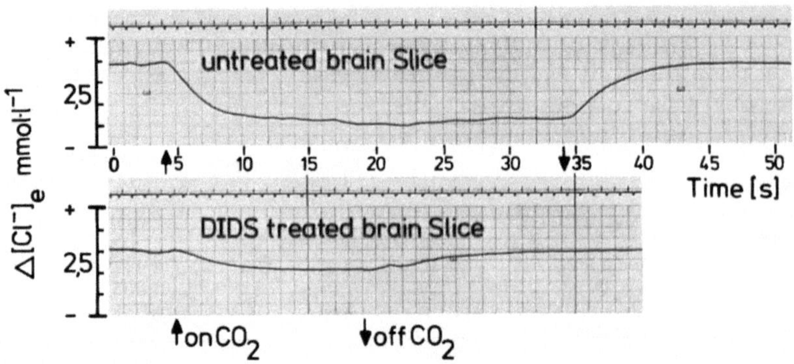

Fig. 4. Time course of the Cl^- response to step changes of PCO_2 at the surface of a treated and untreated rat cortical brain slice in the chloride-microtonometer. Both the uptake and the efflux of the extracellular Cl^- are reduced by DIDS

Table 2. Average time constants of the Cl^- response to a step change of PCO_2 from brain slices and red cells examined in the chloride-microtonometer. n indicates the number of slices and red cell samples

Samples	CO_2 step per cent		$\tau\Delta[Cl^-]$ (s) Control	DIDS treated	$\Delta[Cl^-]$ mmol/l^{-1} Control	DIDS treated
Brain slices	4 to 8	$\bar{x}$	2.9	3.8	−1.42	− 0.52
		± SEM	0.06	0.2	0.023	0.033
		n	9	8	9	8
		P	= 0.0004		< 0.0002	
	8 to 4	$\bar{x}$	3.7	5.5	+ 1.36	+ 0.5
		± SEM	0.18	0.36	0.026	0.016
		n	9	8	9	8
		P	= 0.0005		< 0.0002	
Red cells (cytocrit: 40%)	4 to 8	$\bar{x}$	5.6	10.6	− 0.67	− 0.36
		± SEM	0.38	1.4	0.027	0.064
		n	15	9	15	9
		P	= 0.005		< 0.0002	
	8 to 4	$\bar{x}$	6.1	11.4	+ 0.63	+ 0.16
		± SEM	0.32	1.5	0.025	0.066
		n	13	8		
		P	= 0.0027		< 0.0002	

c) In Vitro Extracellular Cl^--response of the Slice of Rat Parietal Cortex to Changes in PCO_2

A rat cortical brain slice either treated with 10^{-3} molar DIDS or untreated control was interposed between a chloride electrode and a silicone membrane (Fig. 4). The time course of the change in the extracellular [Cl^-] was observed after a rectangular variation of PCO_2 (4–8% and 8–4%). When PCO_2 was increased in the control, the extracellular chloride decreased by 1.12 m mol/l with a time constant of 2.5 s. The DIDS treated slice [Cl^-] showed a decrement of 0.5 m mol/l with a time constant of 3.5 s.

The average time constant and the amplitude of the Cl^- change in treated and untreated brain slices and red cells (1 μl probe, cytocrit 40%) are given in Table 2. Note that both the [Cl^-] uptake during CO_2 and the [Cl^-] efflux during off CO_2 are significantly inhibited by DIDS in brain slices as well as in red cells. This is obvious in the amplitude of the exchange. There is also a prolongation of the exchange as documented in the time constants. The significance of the method differences of the group means were tested by an unpaired *t*-test.

Discussion

The cortical surface of the brain contains mainly glia marginalis, and the interstitial spaces are open to the subarachnoid space as has recently been reviewed (Loeschcke 1980). This offers a good opportunity to study the ion exchange mechanisms, mainly between glial cells and the ECF both in vivo and in vitro. The non invasive application of electrodes with flat ion sensitive areas of about 20 mm^2 onto the brain surface is based on the fact that the fluid film under the electrode is in contact and in exchange with the extracellular fluid of the brain rather than with the cerebrospinal fluid. This technique dramatically shifted the measured time constants of the changes in the acid-base parameters from hours in the cisternal or lumbar CSF to seconds in the brain ECF.

Factors influencing the time constants of the pH or Cl^- change after a step change of PCO_2 in such in vitro preparations are, on the one hand, thickness and diffusion constants of cellular and non cellular layers (fluid film, silicone membrane, paper) and on the other hand cytocrit and buffer capacity of the extracellular and intracellular fluids, if an anion exchange occurs. For the microtonometer technique it should be noted that the fluid film between the brain slice and the H^+-sensitive glass membrane is to a great extent directly accessible to the changes in PCO_2. This diffusion pathway must be considered in addition to the diffusion through the brain slices. Additional factors are the degree of changes in the local cerebral blood flow and the accuracy of the step change in the end-tidal PCO_2 in vivo.

The constancy of the difference between the pH at 4% and 8% CO_2 equilibration in different samples as well as that of the time constants may be taken as evidence of how well the thickness of the cellular and noncellular layer can be reproduced. Leakage of carbonic anhydrase from damaged cells might have played a role in determining the time constant of the pH change. For this reason brain slices as well as red cells were washed before incubation with SITS or DIDS. Since it has recently been shown that SITS and DIDS in 10^{-3} molar concentration inhibit carbonic anhydrase, acetazolamide was not used additionally in the present experiments (Cohen et al. 1978, Edelman et al. 1978, Frömter 1980). It is also known that SITS and DIDS bind covalently with anion exchange protein (cis configuration) only from the exofacial side of the membrane. (Passow et al. 1980). For these reasons, the influence of carbonic anhydrase on the time constant of the pH change in treated samples may be excluded.

The most important observation from the in vitro studies is that the pattern of the pH response in SITS treated brain slices approached that in mock CSF. The question arises whether or not the diminution of the time constant of the pH change is an indicator of the inhibition of the HCO_3^- – Cl^- exchange mechanism. In case of a complete block of the HCO_3^- – Cl^- exchange by stilbene derivatives, both intra- and extracellular compartments would be equilibrated separately in respiratory acidosis. The equilibration of the brain ECF phase would need less CO_2 uptake because of the negligible protein buffer concentration in vivo. In case of the in vitro preparation mock CSF or Ringer without protein and phosphate was used. Consequently, the decrement in pH_e would be expected to be faster than in the presence of cells with intact Hamburger shift. In such a case no HCO_3^- can pass from the cell to the ECF. This would mean that the true ECF of DIDS or SITS treated cells would behave more like isolated ECF, where the anion carrier protein is not available for HCO_3^- and Cl^- ions.

Such a concept is further supported by experiments perfomed in a chloride microtonometer. It has been shown that both uptake and efflux of Cl^- in brain slices as well as in red cells, were significantly inhibited by DIDS. Similar findings were recently reported in a Cl^--tracer study on astroglial cultures (Kimelberg et al. 1979). We conclude that an electrically silent exchange of HCO_3^- against Cl^- occurs between the glial cells and ECF similar to the Hamburger shift in red cells. Such a mechanism may help to understand the in vivo CO_2 dissociation curve of the brain ECF independent of a high glial membrane potential (Wieth et al. 1980, Mückenhoff et al. this symp).

Acknowledgements. The valued technical assistance of Mrs. Renate Hildebrandt and cand. med. Klaus Masthoff, is gratefully acknowledged. Supported by the Deutsche Forschungsgemeinschaft, SFB 114 "Bionach".

References

Ahmad HR, Loeschcke HH, Woidtke HH (1978) Three compartment model for the bicarbonate exchange of the brain extracellular fluid with blood and cells. In: Fitzgerald RS, Gautier H, Lahiri S (eds) Regulation of respiration during sleep and anaesthesia. Plenum Press, New York p 195

Ahmad HR, Luttmann A, Loeschcke HH (1981) Application of a CO_2 electrode as a microtonometer for the study of pH kinetics in the blood. In: Piiper J, Scheid P (eds) Gas exchange function of normal and diseased lung. p 185

Ahmad HR, Loeschcke HH (1982) Fast bicarbonate-chloride exchange between brain cells and brain extracellular fluid. Pflügers Arch (in press)

Ahmad HR, Loeschcke HH (1982) Fast bicarbonate-chloride exchange between plasma and brain extracellular fluid at maintained PCO_2. Pflügers Arch (in press)

Cabantchik ZI, Rothstein A (1974) Membrane protein related to anion permeability of human red blood cells. J Membr Biol 15:207–226

Cohen LH, Mueller A, Steinmetz PR (1978) Inhibition of the bicarbonate exit in urinary acidification by disulfonic stilbene. J Clin Invest 61:981–986

Edelman A, Teulon T, Anagostopoulus T (1978) The effect of a disulfonic acid stilbene on proximal cell membrane potential in necturus kidney. Biochem Biophys Acta 514:137–144

Frömter E (1980) Significance of carbonic anhydrase for HCO_3^- absorption and H^+ secretion in renal tubules. In: Bauer C, Gros G, Bartels H (eds) Springer, Berlin Heidelberg New York, p 419

Fukuda Y, Loeschcke HH (1977) Effect of H^+ on spontaneous neuronal activity in the surface layer of the rat medulla oblongata in vitro. Pflügers Arch 371:125–134

Gunn RB, Dalmark M, Tolsteson CD, Wieth Jo (1973) Characteristics of chloride transport in human red cells. J Genet Physiol 61:185–206

Kimelberg HK, Biddlecome S, Bourke RS (1979) SITS-inhibitable Cl^--transport and Na^+ dependent H^+ production in primary astroglial cultures. Brain Res 173:111–114

Loeschcke HH (1980) Chemical alterations of cerebrospinal fluid on respiratory and circulatory control systems. In: Wood JH (ed) Neurobiology of cerebrospinal fluid, vol I. Plenum Press, New York, p 29

Loeschcke HH, Ahmad HR (1980) Transient adn steady state of chloride-bicarbonate relationship of the brain extracellular fluid. In: Bauer C, Gros G, Bartels H (eds) Biophysics and physiology of CO_2. Springer, Berlin Heidelberg New York, p 439

Passow H, Kampmann L, Fasold H, Jennings M, Lepke S (1980) Mediation of anion transport across the red blood cell membrane by means of conformational changes of the band 3 protein. In: Lassen UV, Ussing HH, Wieth IO (eds) Munksgaard, Copenhagen, p 467

Wieth JO (1979) Bicarbonate exchange through the human red cell membrane determined with ^{14}C bicarbonate. J Physiol (London) 294:521–539

Wieth JO, Brahm J (1980) Kinetics of bicarbonate exchange in human red cells. Physiological implication. In: Lassen UV, Ussing HH, Wieth JO (eds) Munksgaard, Copenhagen, p 467

Effect of the Anion Blocker 'SITS' on Cerebrospinal Fluid [HCO_3^-] in Acute Acid-Base Perturbations

S. JAVAHERI and J. WEYNE[1]

Introduction

The mechanisms involved in regulation of cisternal cerebrospinal fluid (CSF) [HCO_3^-] and H^+ homeostasis during plasma acid-base perturbations are not yet well understood (Leusen 1972, Javaheri and Kazemi 1981). Recent studies on red blood cells have shown that a protein carrier system is involved in transmembrane exchange of certain inorganic anions which can be blocked by a variety of anionic probes (for review, see Deuticke 1977). Disulfonic stilbene derivatives, for example, selectively combine with the protein carrier and diminish the facilitated transport of Cl^- and HCO_3^- (Cabantchik and Rothstein 1972, 1974, Rothstein et al. 1976). In different acid-base disturbances, since changes in cisternal CSF [HCO_3^-] are generally reciprocated by changes in CSF [Cl^-] (Nattie and Romer 1978, Javaheri and Kazemi 1981), we studied the effect of a disulfonic stilbene derivative, SITS (4-acetamido-4'-isothiocyano 2-2'disulfonic stilbene) on the regulation of cisternal CSF [HCO_3^-] during respiratory acidosis and metabolic alkalosis.

Methods

Biological Procedures

Mongrel dogs under anesthesia with Na-pentobarbital and paralysed with gallamine, were intubated after tracheostomy and mechanically ventilated with O_2. Ventilation was adjusted to maintain arterial CO_2 tension (pa_{CO_2}) normal. End-tidal CO_2 concentration was continuously measured by an infrared analyzer. Polyethylene catheters were inserted into femoral arteries and veins for blood sampling, monitoring arterial blood pressure and intravenous injections. Rectal temperature was kept relatively constant by intermittent heating of the thermostated warming table. The head of the animal was fixed in a stereotaxic instrument and two burr holes were drilled into the skull. The cerebral lateral ventricles were punctured simultaneously by

1 Laboratory of Normal and Pathological Physiology, University of Gent, De Pintelaan 185, B-9000 Gent, Belgium

Central Neurone Environment
(ed. by M.E. Schläfke, H.P. Koepchen and W.R. See)

advancing two no. 23 needles, one for each side, secured firmly in the micromanipulator of the stereotaxic instrument (for details of this technique see Weyne et al. 1979). The solution, either mock CSF or mock CSF containing SITS, was then injected intraventricularly at regular time intervals.

Chemical Analysis

Periodically about 0.8 ml of cisternal CSF was obtained anaerobically for immediate measurement of total CO_2 and pH. Cisternal CSF p_{CO_2} was calculated from Henderson-Hasselbalch's equation using published values for CSF pK' and α at 37°C (Mitchell et al. 1965). Cisternal CSF p_{CO_2} and pH were corrected for animal's rectal temperature (Mitchell et al. 1965). Cisternal CSF lactate concentration was determined enzymatically. In experiments with metabolic alkalosis cisternal CSF samples (0.8 ml) were also periodically obtained for measurements of CSF electrolytes. The concentration of SITS in CSF was determined spectrophotometrically. Measurements were done immediately after cisternal CSF samples were obtained and caution was taken to avoid light exposure during the procedure. Periodically arterial blood samples were obtained anaerobically and p_{O_2}, p_{CO_2} and pH were measured by appropriate electrodes. Plasma $[HCO_3^-]$ was calculated using Severinghaus calculator (1966).

Experimental Design

Three series of experiments were done: (a) acute respiratory acidosis; (b) acute respiratory acidosis with HCl infusion and (c) isocapnic metabolic alkalosis. There were two groups in each series. Both groups received an equal volume of intraventricular mock CSF with and without SITS (respectively, the experimental and the control group).

a) Acute respiratory acidosis. During the first hour, normocapnia was maintained and repeated intraventricular injections (0.4 ml into each ventricle) of mock CSF (group I, serving as control, n = 6) or mock CSF containing SITS (group II, n = 6) were given. Arterial p_{CO_2} was then increased to 10–10.7 kPa by ventilating the animals with 7% CO_2 in O_2 for 3 h. Further intraventricular injections were given hourly.

b) Acute respiratory acidosis with HCl infusion. In order to prevent the rise in plasma $[HCO_3^-]$ due to acute CO_2 elevation, 0.16 N HCl was infused intravenously in two groups of animals, one group (group III, n = 6) receiving intraventricular injections of mock CSF, and the other (group IV, n = 6) reveiving mock CSF containing SITS. Experimental design was otherwise similar to that in series (a).

c) Isocapnic metabolic alkalosis. In this series, metabolic alkalosis was induced by intravenous infusion of iso-osmotic Na_2CO_3 and arterial p_{CO_2} was kept constant throughout the experiment. As in the previous series, two groups of animals were studied; in the control group (group V, n = 9), mock CSF and in the experimental group (group VI, n = 7) mock CSF containing SITS was periodically injected into the cerebral ventricles. In this series, 0.2 ml of mock CSF was injected into each cerebral ventricle each time; in addition the duration of observation was extended to 7 h. At the end of the first hour, during which time repeated cerebral intraventricular injections had been given, plasma [HCO_3^-] was increased by about 14 mmol/l within 15 min and was maintained approximately at that level for the next 6 h. Further intraventricular injections of mock CSF or mock CSF containing SITS were given hourly.

Data were analyzed by analysis of variance and Student's *t*-test; *P* values < 0.05 were considered significant.

Results and Discussion

a) Acute respiratory acidosis; groups I and II. Changes in arterial plasma and cisternal CSF acid-base variables in these two groups are depicted in Figs. 1 and 2. Increments in arterial plasma p_{CO_2} and [HCO_3^-] and in CSF p_{CO_2} in the control group (group I, receiving intraventricular mock CSF) were similar to those in the experimental group (group II, receiving mock CSF containing SITS). The rise in CSF [HCO_3^-] in the SITS treated group, however, was significantly lower than that in the control group and this tendency was observed during the 3 h of hypercapnia (Fig. 2). The difference: in cisternal CSF [HCO_3^-] noted in groups I and II were not due to differences in CSF lactate concentration which increased similarly in both groups.

b) Acute respiratory acidosis with metabolic acidosis; groups III and IV. Changes in arterial plasma and cisternal CSF acid-base variables in these groups of animals are shown in Table 1. The increments in arterial and CSF p_{CO_2} were similar in both grou] in this series, the rise in plasma [HCO_3^-] which is associated with acute CO_2 elevation,

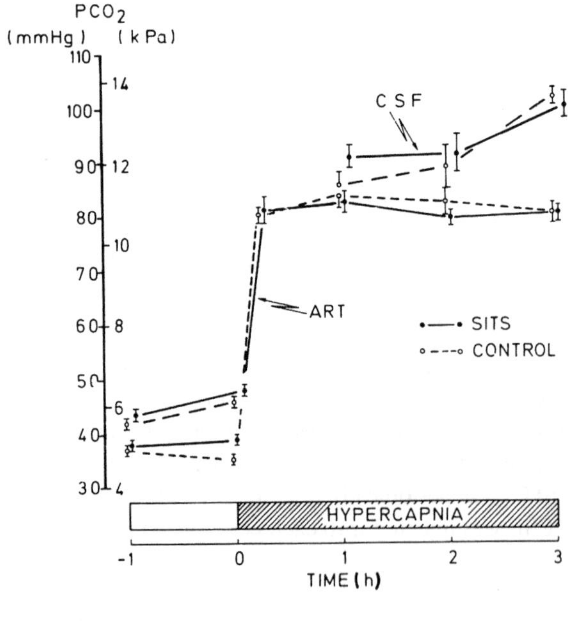

Fig. 1. Changes in arterial plasm: (*ART*) and cisternal cerebrospin fluid (*CSF*) p_{CO_2} vs time durin; acute hypercapnia (mean ± 1 SE group I (*control*) received intraventricular injections of mock C and group II, mock *CSF* containing *SITS*; changes in p_{CO_2} were similar in both groups

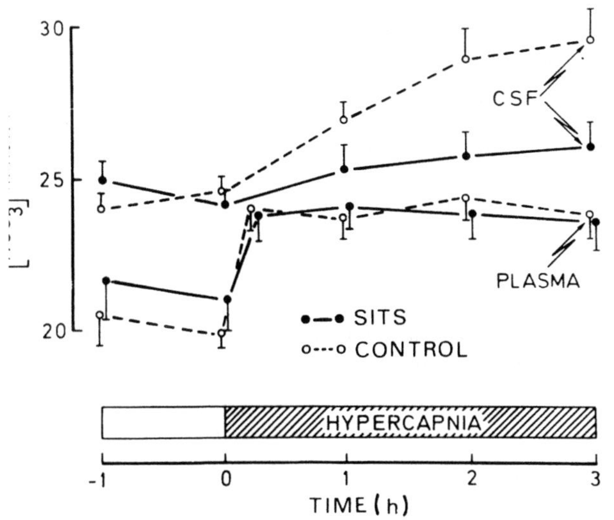

Fig. 2. Changes in arterial *plasma* and cisternal cerebrospinal fluid *(CSF)* [HCO_3^-] vs time during acute hypercapnia (mean ± 1 SE) in group I *(control)* and group II *(SITS)*; changes in *plasma* [HCO_3^-] were similar in both groups, but *SITS* diminished the rise in *CSF* [HCO_3^-] significantly

Table 1. Changes in arterial plasma and cisternal CSF acid-base variables vs time during acute respiratory acidosis superimposed on HCl acidosis; HCl was infused intravenously a few minutes before CO_2 elevation which was started just after 0 time. Group III received intraventricular injections of mock CSF and group IV mock CSF containing SITS. p_{CO_2} is in kPa and HCO_3^- and lactate concentrations in mmol/l (mean ± 1 SD), SITS inhibited the rise in CSF [HCO_3^-] (group IV); see text for further explanation

		Group III					Group IV				
	Time (h)	−1	0	1	2	3	−1	0	1	2	3
Arterial plasma	p_{CO_2}	5.2 ±0.2	5.1 ±0.4	10.5 ±0.6	10.3 ±0.5	10.1 ±0.4	5.1 ±0.3	4.9 ±0.3	10.3 ±0.3	10.3 ±0.2	10.4 ±0.4
	[HCO_3^-]	22.1 ±2.1	17.4 ±2.3	18.6 ±0.6	18.7 ±1.0	18.4 ±0.7	21.5 ±2.4	17.7 ±2.2	18.4 ±1.2	18.8 ±0.7	18.8 ±1.0
	pH	7.37 ±0.04	7.28 ±0.06	7.00 ±0.04	7.01 ±0.01	7.01 ±0.01	7.37 ±0.06	7.30 ±0.06	7.00 ±0.02	7.02 ±0.02	7.01 ±0.01
Cisternal CSF	p_{CO_2}	5.6 ±0.5	6.4 ±0.4	11.3 ±0.7	12.7 ±0.4	12.9 ±1.1	5.7 ±0.4	6.4 ±0.8	11.7 ±0.4	13.1 ±0.8	13.7 ±0.6
	[HCO_3^-]	25.6 ±2.0	25.7 ±1.5	27.8 ±1.2	29.0 ±0.9	28.8 ±1.5	24.8 ±1.5	24.0 ±0.9	24.2 ±1.1	23.7 ±1.1	23.5 ±1.2
	pH	7.42 ±0.03	7.37 ±0.04	7.15 ±0.04	7.13 ±0.01	7.11 ±0.04	7.40 ±0.03	7.33 ±0.05	7.08 ±0.03	7.02 ±0.02	6.99 ±0.03
	lactate	2.8 ±0.8	3.1 ±0.7	3.1 ±0.8	2.9 ±0.7	3.3 ±0.6	2.2 ±0.6	2.6 ±0.4	3.1 ±0.7	2.8 ±0.7	2.7 ±0.4

was prevented by intravenous HCl infusion. The CSF [HCO_3^-] in the SITS treated group (group IV) did not change significantly with time; this was in contrast to the gradual rise in CSF [HCO_3^-] noted in group III. Changes in cisternal CSF lactate concentrations were similar in both groups.

In summary, the results of the above two series with acute CO_2 elevation, with or without intravenous HCl infusion, are similar; the rise in CSF [HCO_3^-] in acute respiratory acidosis is diminished by intraventricular injection of SITS.

c) Isocapnic metabolic alkalosis; groups V und VI. Changes in plasma and cisternal CSF acid-base variables during isocapnic metabolic alkalosis are depicted in Tables 2 and 3. Changes in arterial plasma p_{CO_2} and [HCO_3^-] and cisternal CSF p_{CO_2} were

Table 2. Changes in arterial plasma acid-base variables in group V (control) and group VI (SITS) vs time in metabolic alkalosis; p_{CO_2} is in kPa and [HCO_3^-] in mmol/l (mean ± 1 SD). The changes were similar in both groups; see text for further explanation

Time	(h)	-1	0	3	6
Group V	p_{CO_2}	4.8 ±0.2	4.7 ±0.2	4.6 ±0.3	4.6 ±0.3
	[HCO_3^-]	20.3 ±1.6	19.8 ±1.8	33.7 ±1.5	34.2 ±1.2
	pH	7.37 ±0.03	7.37 ±0.04	7.60 ±0.03	7.61 ±0.03
Group VI	p_{CO_2}	4.8 ±0.2	4.8 ±0.2	4.6 ±0.3	4.7 ±0.2
	[HCO_3^-]	20.5 ±1.9	20.0 ±1.4	33.8 ±1.1	32.5 ±1.1
	pH	7.37 ±0.04	7.36 ±0.03	7.60 ±0.03	7.58 ±0.02

Table 3. Changes in cerebrospinal fluid acid-base variables in group V (control) and group VI (SITS) vs time in metabolic alkalosis; p_{CO_2} is in kPa and [HCO_3^-] in mmol/l (mean ± 1 SD); the rise in CSF [HCO_3^-] at 6 h is significantly increased in the presence of SITS; changes in other variables were similar; see text for further explanation

Time	(h)	-1	0	3	6
Group V	p_{CO_2}	6.4 ±0.4	6.7 ±0.5	6.6 ±0.4	7.1 ±0.4
	pH	7.37 ±0.03	7.32 ±0.04	7.38 ±0.05	7.36 ±0.04
	[HCO_3^-]	24.9 ±1.2	24.4 ±1.3	27.2 ±2.6	28.4 ±2.3
Group VI	p_{CO_2}	6.0 ±0.4	6.3 ±0.5	6.2 ±0.8	7.1 ±0.7
	pH	7.37 ±0.03	7.33 ±0.03	7.38 ±0.07	7.41 ±0.02
	[HCO_3^-]	24.6 ±1.0	23.5 ±1.2	26.8 ±1.9	32.3 ±2.6

similar in both groups, while the rise in cisternal CSF [HCO_3^-] was significantly higher in SITS treated animals; such differences in CSF [HCO_3^-], which were noted in the latter period of the experiment, were not due to differences in CSF p_{CO_2} or lactate concentrations which rose similarly in both groups. There were no significant changes in cisternal CSF [Na^+] with time in either group and decrements in CSF [Cl^-] accounted for the sum of the increments in CSF HCO_3^- and lactate concentrations. In SITS treated animals, however, the latter accounted only for 2/3 of the fall in CSF [Cl^-], indicating that other unmeasured anions had accumulated within CSF to achieve electroneutrality under such circumstances. The mean cisternal CSF SITS concentration was 0.53 ± 0.43 (1 SD) mmol/l at the end of the experiments with metabolic alkalosis.

The results of the present studies are compatible with the hypothesis that the regulation of CSF [HCO_3^-] in acute hypercapnia and in isocapnic metabolic alkalosis depends on mechanisms which are susceptible to inhibition by "SITS" and therefore probably involve the anion carrier system.

Within the last few years, a large amount of evidence has accumulated which indicates that in red blood cells a protein, the so-called band 3, is involved in transmembrane facilitated transport of Cl^- and HCO_3^-. Since this anion transport system has no intrinsic characteristics that might be useful in identifying its components, its recognition has been primarily due to application of anionic probes which combine with the membrane ligands, the fixed positive charges, and diminish the transmembrane fluxes of anions (Rothstein et al. 1976; Deuticke 1977).

Previous studies (Nattie and Romer 1978; Javaheri and Kazemi 1981) have shown that during acute (few hours) respiratory acidosis and iso-osmotic metabolic alkalosis, rises in cisternal CSF [HCO_3^-] are reciprocated by decrements in CSF [Cl^-]; the present study suggests that the Cl^--HCO_3^- exchange under such circumstances is carrier-mediated since it could be inhibited by SITS.

In acute respiratory acidosis, SITS diminished the rise in CSF [HCO_3^-]. The SITS inhibitable rise in CSF [HCO_3^-] in acute respiratory acidosis was also observed when the usual rise in plasma [HCO_3^-] due to acute respiratory acidosis was prevented by intravenous infusion of HCl. Under such circumstances SITS abolished the rise in CSF [HCO_3^-], which otherwise is invariably noted during acute respiratory acidosis.

These experiments, therefore, suggest that the compensatory rise in CSF [HCO_3^-] (influx of HCO_3^- into CSF) during acute CO_2 elevation is carrier mediated. Although we did not measure CSF [Cl^-] in this series of experiments, we contend that the changes in CSF [HCO_3^-] were due to reciprocal changes in CSF [Cl^-]; this reciprocal relationship has been shown to exist in acute respiratory acidosis as stated earlier.

Our conclusion that in acute respiratory acidosis SITS impairs the overall CSFH^+-homeostasis by diminishing the influx of HCO_3^-, is in accord with the observation of Ahmad and Loeschcke (this symp) that SITS diminishes the rise in cerebral extracellular fluid [HCO_3^-] as measured on the brain surface under such circumstances. It is noted, however, that the site(s) of the carrier mediated transport of Cl^- and HCO_3^- in acute respiratory acidosis remain(s) to be determined.

During metabolic alkalosis the rise in CSF [HCO_3^-] in SITS treated animals was significantly higher than that observed in the control group. In both groups cisternal CSF [Na^+] did not change significantly throughout the experiment and increments in CSF

[HCO_3^-] were associated with decrements in CSF [Cl^-]. We have previously reported similar observations in dogs with acute isocapnic metabolic alkalosis (Javaheri and Kazemi, 1981). The greater rise in CSF [HCO_3^-] and the greater fall in CSF [Cl^-] in the SITS treated group than their respective values in the control group suggest that a SITS inhibitable transport system might also be involved in regulation of CSF HCO_3^- and Cl^- concentration in metabolic alkalosis.

References

Cabantchik ZI, Rothstein A (1972) The nature of the membrane sites controlling anion permeability of human red blood cells as determined by studies with disulfonic stilbene derivatives. J Membr Biol 10:311–330

Cabantchik ZI, Rothstein A (1974) Membrane proteins related to anion permeability of human red blood cells. J Membr Biol 15:207–226

Deuticke B (1977) Properties and structural basis of simple diffusion pathways in the erythrocyte membrane. Rev Physiol Biochem Pharmocal 78:1–79

Javaheri S, Kazemi H (1981) Electrolyte composition of cerebrospinal fluid in acute acid-base disorders. Respir Physiol 45:141–151

Leusen I (1972) Regulation of cerebrospinal fluid composition with reference to breathing. Physiol Rev 52:1–56

Mitchell RA, Herbert DA, Carman CJ (1965) Acid-base constants and temperature coefficients for cerebrospinal fluid. J Appl Physiol 20:27–30

Nattie EE, Romer L (1978) CSF HCO_3^- regulation in iso-osmotic conditions: the role of brain P_{CO_2} and Plasma HCO_3. Respir Physiol 33:177–198

Rothstein A, Cabantchik ZI, Knauf D (1976) Mechanism of anion transport in red blood cells: role of membrane proteins. Fed Proc Fed Am Soc Exp Biol 35:3–10

Severinghaus JW (1966) Blood gas calculator. J Appl Physiol 21:1108–1116

Weyne J, Kazemi H, Leusen I (1979) Restoration of CSF [HCO_3^-] after its experimental lowering in normocapnic conditions. J Appl Physiol 47:369–376

Investigation of the Ional and Gaseous Exchange in the Brain: Measurement and Modelling

K. MÜCKENHOFF[1], A. LUTTMANN[2], H.H. LOESCHCKE[1], and H.R. AHMAD[1]

Introduction

Recent investigations of the mechanisms of ional exchange across the red cell membrane give evidence of different diffusional pathways for cations and anions. The mediation of an electroneutral bicarbonate-chloride transport by a transmembrane protein known as band 3 is confirmed (Cabantchik et al. 1978). The anions appear to cross the membrane by complexing with this protein, which is assumed to work as a diffusible carrier-anion complex (Knauf 1979). But more probably a portion of the band 3 protein performs a conformational change (Passow et al. 1980).

Wieth (1980) demonstrated, that the hydroxyl- and chloride-distributions between red blood cells are not affected by a potassium potential at the erythrocyte membrane induced by valinomycin.

The aim of the present studies on the red blood cells is to investigate the influence of the membrane potential upon the sodium-potassium exchange and upon the bicarbonate-chloride exchange and therefore on the dynamics of CO_2-binding in blood. In these experiments valinomycin has been given to the blood in order to produce a potassium diffusion potential. The treated erythrocytes may serve as a model for the glial cells, which are cells of a high membrane potential which is, according to the findings of Kuffler and Nicholls (1966), mainly a potassium diffusion potential. The main question to be answered is: Is there any interaction between cation- and anion exchange for instance through this potential?

It has recently been shown that, in the brain, bicarbonate exchanges against chloride not only across the blood-brain-barrier but also with brain cells (Ahmad et al. 1978). These results led to the conclusion that there exists a bicarbonate-chloride transport which is independent of the glial membrane potential. This would also mean that there is a cation exchange independent of the anion exchange. For experimental verification of this hypothesis in the present experiments the kinetics of the brain extracellular K^+-, Na^+- and H^+-activities were measured in anaesthetized and artificially ventilated cats.

1 Institut für Physiologie, Ruhr-Universität Bochum, Postfach 102148, D–4630 Bochum, Fed. Rep. of Germany
2 Institut für Arbeitsphysiologie an der Universität Dortmund, Ardeystr. 67, D–4600 Dortmund, Fed. Rep. of Germany

Central Neurone Environment
(ed. by M.E. Schläfke, H.P. Koepchen and W.R. See)

On the basis of the experimental results, a mathematical treatment of the ional and gaseous exchange processes between blood and the brain tissue is presented, taking into account that there is no uniform membrane concept as far as diffusion and transport of anions and cations across the interphases of the brain are concerned.

Ional Exchange in Blood and in the Brain

Material and Methods

Blood. The experiments were performed at 310 K with whole human blood. After addition of two different amounts of valinomycin (10^{-5} and 10^{-4} mol/l^{-1} blood) the time courses of plasma pH and of the potassium-, sodium-, chloride- and bicarbonate concentration as well as the blood cell-volume are determined. Samples of blood are drawn through a pH-electrode system and a pCO_2-electrode. The chloride concentration of true plasma is measured by a coulometric method. The extracellular potassium- and the intracellular sodium-concentrations are determined by flame photometry (Mückenhoff and Luttmann 1980). Blood cell volume was measured in a capillary tube after centrifuging blood for 6 min with 6000 *g*.

The time course of the H^+-ion activity after a respiratory disturbance was measured in a thin blood film covering the flat glass membrane of a pH-electrode (Luttmann et al. 1981). The CO_2 concentration in front of the blood film was changed stepwise from 4–8% CO_2 in air and vice versa.

Brain tissue. The experiments were performed in artificially ventilated cats anaesthetized with α-glucochloralose (40 mg/kg) and urethane (200 mg/kg). The H^+-ion, potassium- and sodium activities in the brain extracellular fluid were measured continuously following the method of Ahmad et al. (1978). Metabolic and respiratory acid-base-changes were induced either by intravenous injection of K_2SO_4 or addition of CO_2 to the inspired gas. pH was measured with a glass electrode. A conventional CO_2-electrode was used, consisting of a glass electrode and an Ag^+/AgCl reference, a silicon membrane and a bicarbonate buffer between membrane and electrode. In order to yield a response time of less than 1000 ms the signal of this electrode was linearized and frequency compensated (Luttmann et al. 1978). The extracellular potassium- and sodium activities were determined with ion sensitive surface electrodes which consisted of a silver plate inserted into a small PVC-cylinder and covered with sodium- and potassium-carrier impregnated PVC membranes (Mückenhoff et al. 1981).

Results and Discussion

Blood. In Fig. 1 the time course of the sodium-potassium exchange of blood in dependence upon the valinomycin concentration is shown. The cation concentration changes very slowly. At a higher valinomycin concentration, that means at a higher diffusion potential, the increase of the extracellular potassium- and of the intracellular sodium

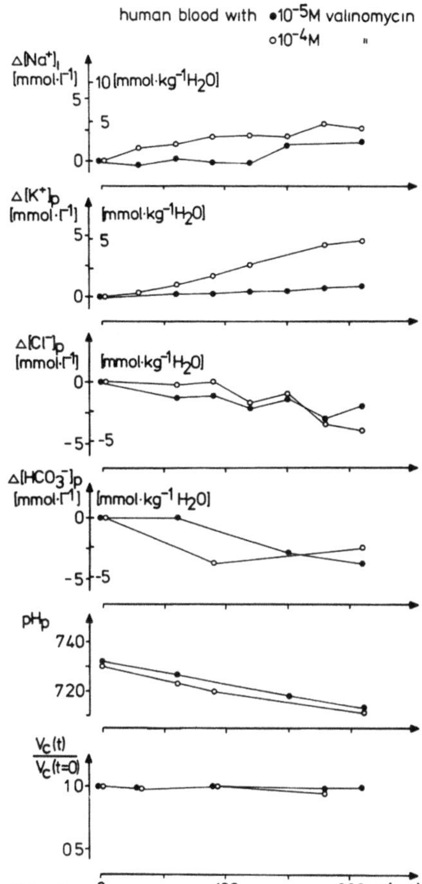

Fig. 1

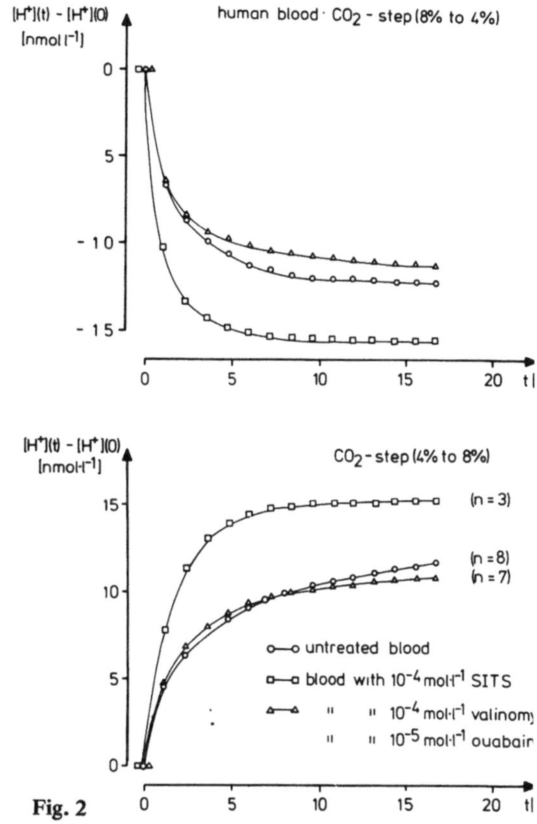

Fig. 2

Fig. 1. Time courses of the changes of intracellular $[Na^+]_i$, plasma $-[K^+]_p$, $-[Cl^-]_p$, $-[HCO_3^-]_p$, $-pH_p$ and of the relative cell volume after the addition of different amounts of valinomycin to whole human blood

Fig. 2. Time courses of the H^+-ion activity of blood plasma of untreated and differently treated human blood. CO_2 in front of a blood film was changed from 8% to 4% and vice versa

concentration is steeper. The measurement yields nearly a one to one exchange of sodium against potassium whereas the bicarbonate- and chloride concentrations remain nearly uninfluenced, especially in the beginning of the experiment. The blood cell volume is constant. Human erythrocytes with the same valinomycin concentration but additionally treated with ouabain in order to block the active sodium-potassium transport showed a slow outflux of chloride and potassium. This may be due to a higher membrane diffusion potential.

In Fig. 2 the time courses of the extracellular H^+-ion activity of blood after a stepwise CO_2-change is shown. A comparison yields no difference between untreated blood and blood with a valinomycin induced potassium diffusion potential. From these results we conclude, that there is no influence of the membrane potential on the fast bicarbonate-chloride exchange and therefore on the dynamics of CO_2-binding. The further question arises how far the time course of the binding process is influenced by an inhibited bicarbonate-chloride exchange. In Fig. 2 this is demonstrated with the change of the plasma H^+-ion activity due to a CO_2 change in SITS treated blood: a larger change with a faster time course than in blood with an intact bicarbonate-chloride takes place.

Brain. In Fig. 3 an original recording is shown representing the cation exchange at the blood brain barrier. In this experiment hydrogen- and potassium-ion activity are determined on the cortical surface of the brain of an artificially ventilated cat.

About 20 s after the start of an intravenous injection of K_2SO_4 the extracellular potassium concentration increases during the injection up to a maximum of $\Delta\,[K^+]_e$ = 1.2 mmol/l^{-1}. After the injection a return close to the control value is observed. With increasing K^+-activity the extracellular H^+-ion activity starts to rise and continues within the phase of potassium return.

An original recording of Fig. 4 shows the arterial pressure and the extracellular potassium- and H^+-ion activity on the surface of the parietal cortex measured during an approximately rectangular change of endtidal pCO_2. Due to CO_2 the H^+-ion activity is increased in the extracellular phase. At the same time there is an increase in the extracellular potassium activity. In another series of experiments under the same experimental conditions a decrease of the extracellular sodium activity was observed (Mücken-

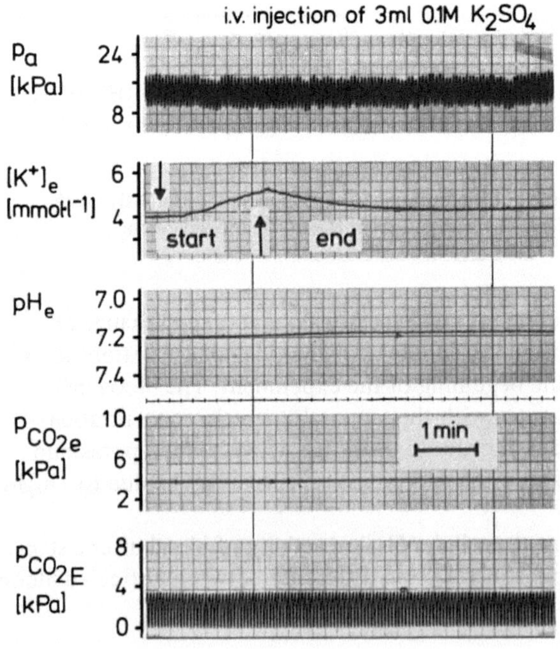

Fig. 3. Original recording during an intravenous injection of K_2SO_4 as indicated by the *arrows* in an artificially ventilated cat. *From top to bottom:* arterial blood pressure, extracellular $[K^+]$, pH and the CO_2-partial pressure on the surface of the parietal cortex; endexpiratory CO_2-partial pressure

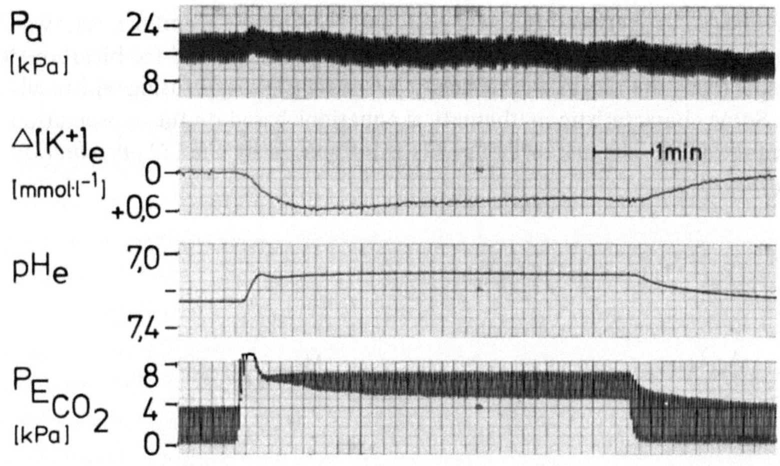

Fig. 4. Original recording of arterial pressure, extracellular Δ [K^+] and pH on the cortical surface of an artificially ventilated cat. The endexpiratory CO_2 is changed nearly rectangularly (*lower trace*)

hoff et al. 1981). It is concluded that there exists a passive sodium-potassium exchange at the blood brain barrier and at the glial mambrane; an active sodium-potassium transport system across the glial interphase effects the potassium homeostasis of the brain extracellular fluid. Therefore a disturbance in the extracellular potassium-activity is compensated quickly. With regard to the concept of the fixed charged model (Passow 1969) and to results of tracer experiments on human blood (Harris and Maizels 1951) our results indicate that an increase of the extracellular H^+-ion activity decreases the active sodium-potassium transport resulting in an increase of the extracellular potassium activity during respiratory acidosis.

Mathematical Analysis of the Ional and Gaseous Exchange in the Brain

Theory:

For the purpose of mathematical description the brain tissue is thought to consist of at least three compartments:

a) the *intracellular* fluid (ICF), represented mainly by the glial cells, in which CO_2 is produced, hydrated and dissociated into bicarbonate and H^+-ions. This reaction is catalyzed.
b) the *extracellular* fluid (ECF), an in vitro unbuffered fluid which doesn't contain protein.
c) the *plasma* compartment.

At both interphases, i.e. between ICF and ECF and between ECF and plasma, two independent ional exchange mechanisms for sodium and potassium and for bicarbonate and chloride respectively are assumed. Physically dissolved CO_2 is diffusing within all compartments. Some characteristic mathematical equations based on the conservation of mass may be presented: The intracellular CO_2 is influenced by the *CO_2-production-*, *binding-* and *diffusion:*

$$\frac{d[CO_2]_i}{dt} = \gamma_i (k_u[CO_2]_i - k_v[H_2CO_3]_i)$$

$$- \frac{A_{ei} L_{CO_2 ei}}{V_{H_2O_i}} (\alpha_i pCO_{2i} - \alpha_e pCO_{2e}) + \frac{\dot{M}_{CO_2 i}}{V_{H_2O_i}}$$

The CO_2-binding is formulated according to Roughton (1964) and Garg and Maren (1972).

Subscript i, e, p : intracellular, extracellular, plasma

Symbols a_{ei}, a_{ep} : physical parameter at the interphase between extracellular and intracellular fluid, extracellular fluid and plasma

- A_{ei} : area between extra-; and intracellular phase
- $[CO_2]$: concentration of physically dissolved $[CO_2] = \alpha \cdot pCO_2$
- α : solubility of CO_2
- k_u : velocity constant of hydration
- k_v : velocity constant of dehydration
- L_{CO_2} : permeability of CO_2
- $\dot{M}_{CO_2}$: CO_2-production
- V_{H_2O} : volume of water
- γ : factor of carbonic anhydrase activity (water: $\gamma = 1$)

Taking the equation of Garg and Maren (1972):

$$-\frac{d[CO_2]}{dt} = \frac{d[HCO_3^-]}{dt} = \frac{d[H^+]}{dt}$$

This represents the amount of H^+-ions produced during CO_2-hydration most of which is buffered by proteins and phosphate.

$$\frac{d[H^+]_i}{dt} = \frac{1}{\beta_i} [H^+]_i \frac{1}{\log e} \gamma_i (k_u[CO_2]_i - k_v[H_2CO_3]_i)$$

with

$$\beta_i = \frac{d[P^-]_i}{d(-\log[H^+]_i)} + \frac{d[HPO_4^{2-}]_i}{d(-\log[H^+]_i)}$$

$[P^-]$ = protein-anion concentration
$[HPO_4^{2-}]$ = dissociated phosphate concentration

The numerical value of β_i is used according to Middendorf and Loeschcke (1974).

For the description of ionic movements the Einstein-Smoluchowski equation is used:

$$\frac{d[C]}{dt} = -\frac{d}{dx}\left\{-D\frac{d[C]}{dx} + BX[C]\right\}$$

as derived by Smoluchowski (1923) for Brownian motion of particles in the presence of an external field of force. With mass conservation

$$\frac{d[C]}{dt} = -\frac{dJ}{dx}$$

we obtain J as the flux of particles

$$J = -D\frac{d[C]}{dt} + BX[C] \quad .$$

For the description of the *cationic* flux the modification of this equation according to Nernst-Planck is used

$$BX\,[C]_c = \frac{-z\,D_c}{RT}\,[C]_c\,\frac{d\varphi}{dx} \quad .$$

The fast *anion* transport is mediated by a diffusible carrier. This may be formally understood as a diffusion of the ionic particle under the influence of a force X which imposes a steady-state velocity v on the ion. According to Smoluchowski (1923) we use

$$BX\,[C]_a = v_a[C]_a$$

and insert it into the flux equation.

Assuming a linear potential profile φ (x) within the membrane (Goldman 1943) we derive the following expression by integration of the cationic flux equation:

$$J_c = -D_c\left([C]_{ci} - [C]_{ce}\,e^{-\frac{F}{R\cdot T}\Delta\varphi}\right)\frac{\frac{F}{R\cdot T}\frac{\Delta\varphi}{\Delta x}}{\left(1 - e^{-\frac{F}{R\cdot T}\Delta\varphi}\right)} \quad .$$

Integration of the anionic flux equation yields

$$J_a = \frac{D_a}{\Delta x}\left([C]_{ae} - [C]_{ai}\,e^{-\frac{v_a}{D_a}\cdot\Delta x}\right)\frac{\frac{v_a}{D_a}\cdot\Delta x}{\left(1 - e^{-\frac{v_a}{D_a}\cdot\Delta x}\right)} \quad .$$

The stationary state is characterized by a zero total ion current through the membranes

$$\sum_{\mu} (J_{a\mu} + J_{c\mu}) = 0 \ .$$

From the condition of electroneutrality of the system we deduce a cationic membrane diffusion potential:

$$\Delta\varphi = \frac{RT}{F} \ln \frac{D_K [K^+]_e + D_{Na} [Na^+]_e}{D_K [K^+]_i + D_{Na} [Na^+]_i}$$

which is a function of the concentration of sodium and potassium within the compartments and of the diffusion coefficients of the membrane for these particles, a so called Goldman potential. Furthermore we derive a distribution of bicarbonate and chloride. The distribution coefficient is d.

$$e^{-\frac{v_a}{D_a} \cdot \Delta x} = \frac{[Cl^-]_i}{[Cl^-]_e} = \frac{[HCO_3^-]_i}{[HCO_3^-]_e} = d$$

Subscripts	a, c	: anion, cation
	i, e	: intra-, extracellular phase
	μ	: chemical component
Symbols	B	: mechanical mobility
	[C]	: concentration
	D	: diffusion coefficient
	F	: Faraday's number
	J	: flux
	R	: gas constant
	T	: absolute temperature
	v	: convection velocity
	X	: force
	z	: charge number
	φ	: electrical potential

Assuming the conservation of bicarbonate the application of the derived expressions of CO_2-reaction and anionic diffusion leads to the following formulae

$$\frac{d[HCO_3^-]_i}{dt} = -\frac{d[CO_2]_i}{dt} + \frac{D}{\Delta x}\left(\frac{\left([HCO_3^-]_e - [HCO_3^-]_i\, d\right) \ln\left(\frac{1}{d}\right)}{1 - d}\right) \ .$$

For the intracellular chloride-concentration we obtain

$$\frac{d[Cl^-]_i}{dt} = + \frac{D}{\Delta x}\left(\frac{\left([Cl^-]_e - [Cl^-]_i\, d\right) \ln\left(\frac{1}{d}\right)}{1 - d}\right) \ .$$

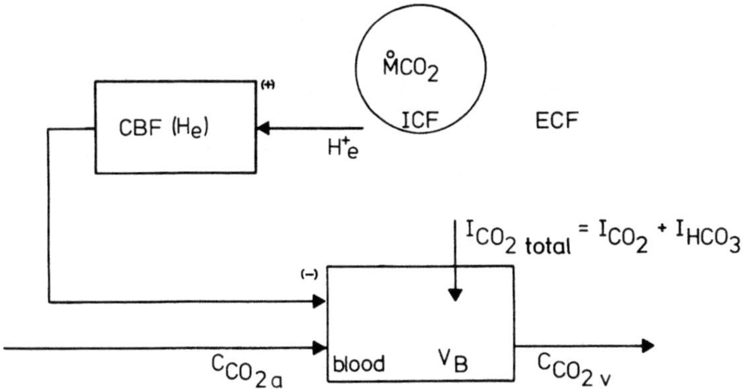

Fig. 5. Schematic diagram of the brain perfusion model. The CO_2 balance of the brain tissue is determined by the intracellular CO_2 production ($\dot{M}_{CO_2}$), total CO_2 flux into the blood ($I_{CO_2\,total}$), total CO_2 concentration in arterial and venous blood ($[CO_2]_{ta}$; $[CO_2]_{tv}$), the cerebral blood flow (CBF) and by the blood volume (V_B) of the brain tissue

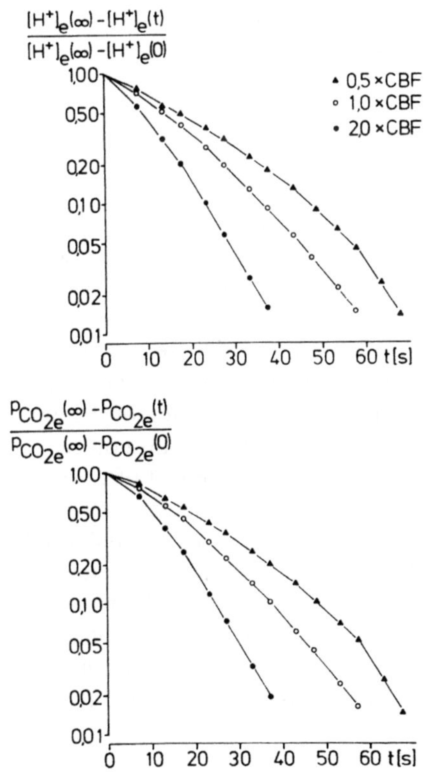

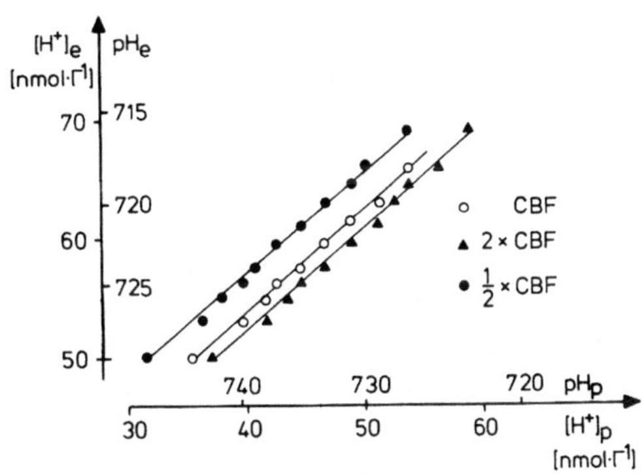

Fig. 7. Computed results of the dependence of the extracellular H^+-ion activity at varied cerebral blood flows. $[H^+]_p$ has been changed by variation of the CO_2-partial pressure in the arterial blood

◀ **Fig. 6.** Calculated influence of different cerebral blood flows on the dynamics of the extracellular H^+-activity (*top*) and the extracellular pCO_2 (*bottom*) after a stepwise increase of arterial pCO_2 at t = 0. Semilogarithmic plots

As shown in Fig. 5 the mathematical description of the exchange processes in the brain tissue is extended by the insertion of the blood flow (CBF), which has been formulated as a function of the extracellular H^+-ion activity (H^+_e). The formula of the H^+-ion dependence is based on the experimental results of Fencl et al. (1969). The CO_2-content and therefore the acid-base values of the perfusing blood are determined by its arterial composition as well as by the CO_2-influx from the tissue:

$$\frac{d[CO_2]_B}{dt} = \frac{CBF}{V_B}([CO_2]_{ta} - [CO_2]_{tv}) + \frac{I_{CO_2\ total}}{V_B}$$

Subscripts a, b, v : arterial, blood, venous
$[CO_2]_t$: total CO_2 concentration in blood
$I_{CO_2\ total}$: total CO_2-flux
CBF : cerebral blood flow
V_B : blood volume of the brain tissue

For the sake of computational simplification it was assumed that blood reaches its thermodynamic equilibrium during the passage of the brain capillaries and that it may be described by a mean CO_2-binding curve. For a formal description of the CO_2 dissociation curve of the blood the thermodynamic treatment and the numerical procedure according to Mückenhoff (1973) was used. The total model yields a stiff differential equation system. This was solved for various biological conditions by numerical integration using the method of Euler-Cauchy (Zurmühl 1965).

Results of Computation

Variation of the perfusion rate of the brain tissue affects the time course as well as the steady-state relation of the exchange between blood and the brain tissue. In Fig. 6 the computational results of a variation of the cerebral blood flow upon the H^+-ion activity and the CO_2-partial pressure of the brain extracellular fluid after a stepwise CO_2-change in the arterial blood is demonstrated in the time domain in a semilogarithmic plot. An increase of the CBF effects a lowered pCO_2. The time constants of these processes vary between 6 and 14 s. The steady-state solution predicts that with constant acid-base values of the blood a decrease of the perfusion rate produces acidification of the brain tissue as shown in Fig. 7. The second parameter influencing the time course of the exchange processes is the metabolic rate ($\mathring{M}CO_2$). Simulating an experiment of Loeschcke and Ahmad (1980) the arterial CO_2-partial pressure was assumed to increase stepwise. This leads to an increase of the extracellular H^+-ion activity and the extracellular pCO_2 as shown in the semilogarithmic plots of Fig. 8. The dynamic and the steady-state behaviour of this process is essentially influenced by the metabolic rate ($\mathring{M}CO_2$). The time constants vary between 10 and 15 s. Data for the metabolic rate of the brain were calculated after Thews (1960) as the arithmetic mean of the values for white and grey matter. Values for the cerebral blood flow were obtained from Mc Ilwain and Bachelard (1971).

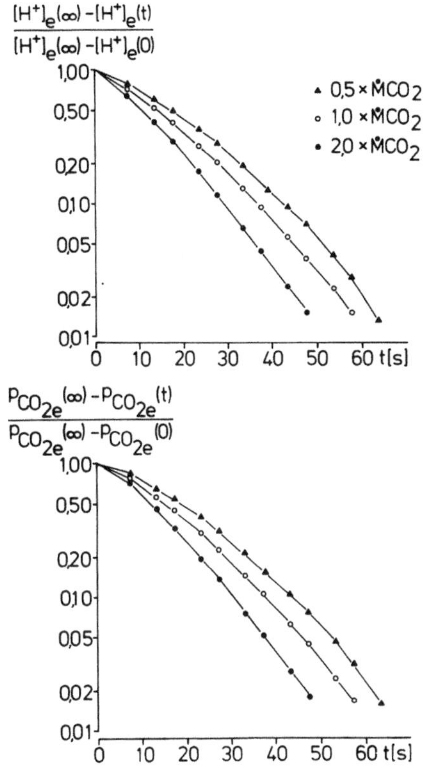

Fig. 8. Calculated influence of different cerebral CO_2-production on the dynamics of the extracellular H^+-activity (*top*) and the extracellular pCO_2 (*bottom*) after a stepwise increase of arterial pCO_2 at t = 0. Semilogarithmic plots

Discussion

The presented mathematical model has been formulated on the basis of experimental findings with three different passive exchange mechanisms for anions, for cations and for physically dissolved CO_2 in the brain tissue. It does not differentiate between glial cells and neurones, which behave differently as far as their membrane properties are concerned (Kuffler and Nicholls 1966).

From the computational results we conclude that in vivo brain perfusion (CBF) and metabolic rate ($\dot{M}CO_2$) influence time course and steady-state distributions of the exchanging substances. From a mathematical point of view the solution of the stiff differential equation system is very sensitive to a change of these parameters. So different values of CBF and $\mathring{M}CO_2$ may be one reason for the difference in time constants of pCO_{2e} and $[H^+]_e$ found in the experiments (T = 30 s, Loeschcke and Ahmad 1980) and by calculation (T = 10 s). Other reasons for this difference may be that in the calculation the CO_2 partial pressure is assumed to change stepwise which can hardly be achieved in the experiments. The measurements were performed on the surface and not within the brain tissue. This means that additional pathways were included, which may result in a delayed change of $[H^+]_e$ and pCO_{2e}. A discrepancy between model calculation and experiment can also be expected because it is assumed that blood is in thermodynamic equilibrium during its passage through the capillaries. Especially the CO_2 binding process in blood may take a longer time as the passage time in the brain capillaries. On the basis of the laws of diffusion Thews (1960) calculated a passage time in the range between 0.6–1.5 s.

Another important result is the prediction of a distribution of the diffusible anions between cells and extracellular fluid. This is very similar to distributions found in Donnan systems.

To summarize, the experiments presented here give evidence of different ional exchange mechanisms in blood, at the blood-brain barrier, and at the glial cell membrane. A mathematical model consisting of a differential equation system was formulated on the basis of these results, taking into account that a sodium-potassium diffusion potential does not influence the bicarbonate-chloride exchange. Its solution yields fast exchange processes for CO_2, H^+-, HCO_3^-, Cl^--ions in accord with the experimental findings.

Acknowledgements. Supported by the Deutsche Forschungsgemeinschaft (SFB 114 Bionach).

References

Ahmad HR, Loeschcke HH, Woidtke HH (1978) Three compartments model for the bicarbonate exchange of the brain extracellular fluid with blood and cells. In: Fitzgerald RS, Gautier H, Lahiri S (eds). Regulation of respiration during sleep and anaesthesia. Plenum Press, New York, p 195

Cabantchik Z, Knauf P, Rothstein A (1978) The anion transport system of the red blood cell: The role of membrane protein evaluated by the use of "probes". Biochim Biophys Acta (Biomembr Rev) 515:239–302

Fencl V, Vale JR, Broch JA (1969) Respiration and cerebral blood flow in metabolic acidosis and alkalosis in humans. J Appl Physiol 27:67–76

Garg LC, Maren TH (1972) The rates of hydration of carbon dioxide and dehydration of carbonic acid at 37°C. Biochim Biophys Acta 261:70–76

Goldman DE (1943) Potential, impedance and rectification in membranes. J Gen Physiol 22(2): 260–276

Harris EJ, Maizels M (1951) The permeability of human erythrocytes to sodium. J Physiol 113: 506–524

Ilwain Mc MC, Bachelard HS (1971) Biochemistry and the central nervous system. Churchill, Livingstone, Edinburgh

Knauf PH (1979) Erythrocyte anion exchange and the band 3 protein: Transport kinetics and molecular structure. Curr Top Membr Transp 12:249–363

Kuffler SW, Nicholls JG (1966) The physiology of neuroglial cells. Rev Physiol Biochem 57:1–90

Loeschcke HH, Ahmad HR (1980) Transient and steady state of bicarbonate chloride relationship of the brain extracellular fluid. In: Bauer C, Gros G, Bartels H (eds) Biophysics and physiology of carbon dioxide. Springer, Berlin Heidelberg New York, p 439

Luttman A, Mückenhoff K, Loeschcke HH (1978) Fast measurement of the CO_2 partial pressure in gases and fluids. Pflügers Arch 375:279–288

Luttmann A, Ahmad HR, Mückenhoff K, Loeschcke HH (1981) Influence of weak acids and bases on the pH and pCO_2 kinetics in blood. In: Piiper J, Scheid P (eds) Gas exchange function of normal and diseased lungs. Karger, Basel, p 193

Middendorf T, Loeschcke HH (1974) Approximative Berechnung der Pufferbase, einer Titrationsgeraden für Proteine und der CO_2-Dissoziationskurve des Gehirngewebes. Pflügers Arch 349: 1–8

Mückenhoff K (1973) Die Berechnung des Säuren-Basen-Status des menschlichen Vollblutes mit Hilfe eines numerischen Verfahrens. Diss Ruhr-Univ Bochum, Abt Biol
Mückenhoff K, Luttmann A (1980) Ein flammenfotometrisches Verfahren mit hoher Auflösung zur Bestimmung der Kationenkonzentration im Blut. Biomed Tech 25:346–348
Mückenhoff K, Ahmad HR, Luttmann A (1981) The kinetics of the potassium concentration in brain extracellular fluid during respiratory and metabolic disturbances. In: Lübbers DW, Acker H Buck RP, Eisenman G, Kessler M, Simon W (eds) Progress in enzyme and ion-selective electrodes. Springer, Berlin Heidelberg New York, p 116
Passow H (1969) Passive ion permeability of the erythrocyte membrane. Prog Biophys 19:423–467
Passow H, Kampmann L, Fasold H, Jennings M, Lepke S (1980) Mediation of anion transport across the red blood cell membrane by means of conformational changes of the band 3 protein. In: Lassen V, Ussing HH, Wieth JO (eds) Membrane transport in erythrocytes. Munksgaard, Copenhagen, p 345
Roughton FJW (1964) Transport of oxygen and carbon dioxide. In: Handbook of physiology, vol I: respiration. Washington DC, sect 3, vol I:31, p 767
Smoluchowski M (1923) Abhandlungen über die Brownsche Bewegung und verwandte Erscheinungen. Oswalds Klassiker der exakten Wissenschaften, No 207. Akad Verlagsges, Leipzig
Thews G (1960) Die Sauerstoffdiffusion im Gehirn. Pflügers Arch 271:197–226
Wieth JO (1980) Interaction between two types of pH-equilibration proton and hydroxal ion carriers. In: Lassen V, Ussing HH, Wieth JO (eds) Membrane transport in erythrocytes. Munksgaard, Copenhagen, p 512
Zurmühl R (1965) Praktische Mathematik. Springer, Berlin Heidelberg New York

Acid-Base Kinetics of Blood During CO_2- and O_2-Uptake

A. LUTTMANN[1] and K. MÜCKENHOFF[2]

Introduction

Studies on the acid base kinetics of blood were performed to determine the time blood needs to reach thermodynamic equilibrium after different kinds of acid base disturbances. In vivo such acid base changes occur during each circulation of the blood in the lung and tissue capillaries, when CO_2 and O_2 is exchanged and in working muscle when during heavy muscular exercise acidic substances diffuse into the blood. For reasons of respiration physiology it is important to know whether or not the blood acid base quantities have reached their final value when leaving the lung or passing the peripheral chemoreceptors or central chemosensitive structures, because the magnitude of ventilation and consequently the acid base status of the whole body is mainly determined by the actual acid base values at these structures.

Since the end of the last century the mechanism of gas transport by the blood was well investigated by a great number of scientists, and the knowledge of the chemical reactions and transport events that occur in blood after gas exchange was summarized by Roughton in 1964. In the late fifties a more kinetic aspect became predominant resulting in the question, whether or not the CO_2- or O_2-binding processes are completed when blood leaves the lung.

Theoretical considerations lead to the conclusion that the time course of the equilibration process after an acid base disturbance is mainly determined by the uncatalyzed CO_2 hydration in the plasma, supposing that no carbonic anhydrase is available to the plasma. This was first mentioned by Roughton (1935) and later by Hill et al. (1973) and Forster and Crandall (1975). Slow pH changes were measured first by Crandall et al. (1971) using a stopped flow method. Own determinations of the time course of several acid base parameters with a modified Hartridge Roughton continuous flow method (Luttmann 1977, Luttmann et al. 1980) showed a faster change of the plasma pH than was predicted by the theoretical considerations. Therefore, H^+ ion transfer between the plasma and cell phase is proposed which could be mediated by a transfer of weak acids or bases (Luttmann et al. 1980, 1981).

Recent experimental studies from Klocke (1978), Effros et al. (1978) and Crandall and O'Brasky (1978) indicate that carbonic anhydrase seems to be available to the plas-

1 Institut für Arbeitsphysiologie an der Universität Dortmund, Ardeystr. 67, D-4600 Dortmund, Federal Republic of Germany

2 Institut für Physiologie, Ruhr-Universität Bochum, Postfach 10 2148, D-4630 Bochum, Federal Republic of Germany

Central Neurone Environment
(ed. by M.E. Schläfke, H.P. Koepchen and W.R. See)

ma in the lung, reducing or abolishing slow pH changes in the plasma. Because estimates from these reports of the extent of CO_2 catalysis during lung transit vary widely, in a theoretical study by using a mathematical blood model (Luttmann 1980) the pH and pCO_2 time courses of the blood were calculated. In these calculations various catalyzing factors acting in the plasma during pulmonary capillary transit were assumed.

Material and Methods

All experiments were performed at 310 K with whole human or bovine blood. In some experiments various amounts of Na-lactate, Na-acetate or NH_4Cl were added in iso-osmotic solutions, at least 15 min before the experiment.

To produce a *respiratory disturbance* CO_2 gas was added using a modified Hartridge Roughton continuous flow method. In this method a stepwise increase in the total CO_2 content of the blood was obtained and the time course of pH_p, pCO_2, $[Cl^-]_p$, and the electrical resistance indicating the actual cell content were measured (for details see Luttmann 1977, Luttmann et al. 1980).

In other series of experiments the time course of the plasma pH was measured after a stepwise change of the pCO_2 or pO_2. These experiments were performed with a flat pH electrode (diameter 5 mm) combined with the reference electrode in the same shaft. The diaphragm was placed in the side of the electrode shaft. The H^+ ion-sensitive area was covered with 0.3 µl of blood, which was spread all over the membrane. The diaphragm at the reference electrode was connected with a small strip of moistened tissue paper. The gas composition in front of the blood film was changed stepwise. In one type of experiments the CO_2 fraction was altered from 0.04 to 0.08 in O_2 and vice versa, in another type the O_2 fraction was changed. Therefore a CO_2-O_2 or a CO_2-N_2 mixture, both having the same CO_2 partial pressure (6.2 kPa) could pass the blood film alternately. CO_2 partial pressure was measured in the gas leaving the pH electrode system by using a CO_2 electrode.

Theoretical Description

Figure 1 demonstrates the chemical reactions and exchange processes which happen during CO_2 uptake. On the basis of this diagram a mathematical model of the blood acid base kinetics, consisting of a system of differential equations and several boundary conditions was established (for details see Luttmann 1980). This model takes into account:

a) The hydration of CO_2 in the plasma and in the erythrocytes.
b) The carbamate formation and the dependency of the intracellular H^+ ion concentration upon the O_2-saturation.
c) The osmotically induced exchange of water.
d) An exchange of weak acids (or bases) down its concentration gradient.

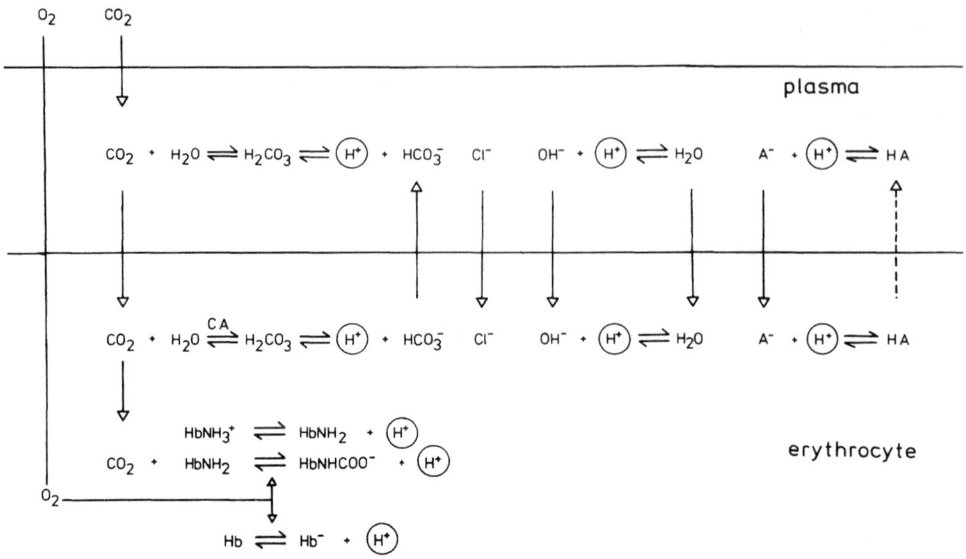

Fig. 1. Schematic diagram of chemical reactions and transport events occurring in blood during exchanges of CO_2 and O_2. For details see text

The description is obtained by making mass balances for each of the chemical species involved. The ional fluxes are described in terms of passive diffusion down electrochemical gradients. This may possibly not describe the exact exchange mechanism, because recent findings of Wieth (1980) demonstrate an electrosilent HCO_3^--Cl^- exchange across the red cell membrane. Nevertheless we used this description, because phenomenological permeability coefficients are available (Forster et al. 1980). Moreover the discrepancy will be small due to the small change in membrane potential. The model was used to simulate the chemical and transport events that occur in blood during and after gas exchange in the lung. Different extents of CO_2 catalysis in the lung capillaries were considered by assuming different catalyzing factors in the plasma during the lung passage time (T = 700 ms).

Results

Experimental Studies

Figure 2 demonstrates the time courses of pCO_2, pH and the changes of the electrical resistance and the chloride concentration in whole bovine blood obtained in the continuous flow experiments. After adding CO_2 the pCO_2 increases abruptly. Then the physically dissolved CO_2 is transformed into bound CO_2 resulting in a decrease of the pCO_2. During this process the H^+ ion concentration in the plasma increases whereas the chloride concentration decreases depending on the bicarbonate-chloride exchange. The electrical resistance, indicating the actual cell volume increases due to an osmotically induced water shift into the cells.

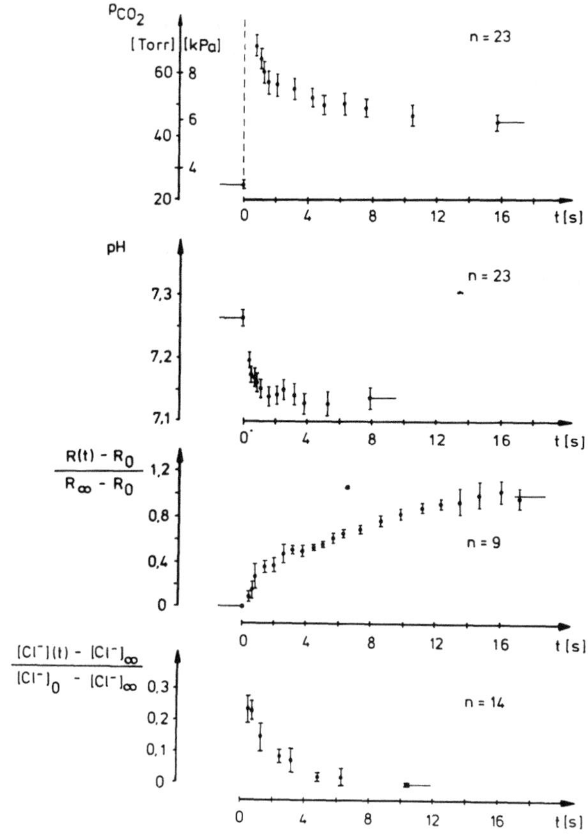

Fig. 2. Time courses of pCO_2, pH and the relative changes of the resistance and the chloride concentrations (mean values ± SEM) in whole bovine blood obtained in continuous flow experiments after a stepwise increase in total CO_2 content from 10 to 14 m mol • l^{-1} (Luttmann et al. 1980)

Comparison of the experimental data with model calculations, assuming an uncatalyzed CO_2 hydration in the plasma, shows a much faster increase in plasma H^+ ion concentration in the experiments than predicted from the calculations. The time for reaching 90% of the total change was measured to be smaller than 5 s, the calculated T_{90} however was 45 s. This was not caused by any plasma carbonic anhydrase activity from hemolysis as could be demonstrated in control experiments with carbonic anhydrase inhibitors (acetazolamide, benzolamide). To explain this discrepancy between the experimental findings and the theoretical considerations an H^+ ion transfer between the cell and the plasma phase was proposed. This H^+ ion transfer will occur, if a permeable weak acid or base is present and an H^+ disequilibrium between the cells and the plasma and therefore as a consequence a concentration gradient for the weak acid or base exists (Fig. 1). This will cause a flux of the permeable weak acid or base and consequently an H^+ ion transfer accompanied by a decrease in equilibration time. This hypothesis was proved experimentally in blood with an increased content of acetate, lactate or ammonia. For this purpose the pH time course in a thin film of human blood was measured after a stepwise change in the CO_2 fraction. Table 1 demonstrates the time for reaching 90% of the total change of the plasma H^+ ion concentration. An increased content of lactate, acetate or ammonia yields an acceleration of the equilibration process and a diminishing of T_{90}.

Table 1. Influence of lactate, acetate and ammonia on the time for reaching 90% of the total change of the plasma H^+ ion concentration (T_{90}) in a thin film of human blood after a stepwise change of the CO_2 fraction from 0.04 to 0.08 in O_2 and vice versa (mean values from 6 to 8 experiments)

	T_{90}[s]	
	CO_2 fraction changed from	
	0.04 to 0.08	0.08 to 0.04
Control		
[Lac] = 3,5 $mmol \cdot l^{-1}$	5.0	4.0
+ 7.5 $mmol \cdot l^{-1}$ Na lactate	1.9	1.6
+ 7.5 $mmol \cdot l^{-1}$ Na acetate	1.6	1.3
+ 7.5 $mmol \cdot l^{-1}$ NH_4Cl	1.2	1.1

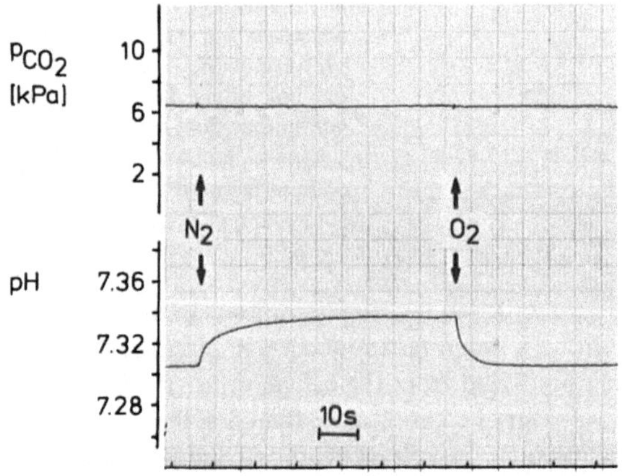

Fig. 3. Original recordings of the pCO_2 and pH time functions measured in a thin film of human blood after a stepwise change in pO_2. Gas in front of the blood film was changed from an O_2 - CO_2 mixture to a N_2 - O_2 mixture both having the same pCO_2

In another type of experiment an acid base disturbance was induced by changing the pO_2 (Christiansen – Douglas – Haldane effect). Figure 3 demonstrates an original recording of the pCO_2 and the pH time course. The gas in front of the blood film was changed from an O_2-CO_2 mixture to a N_2-CO_2 mixture, both having the same pCO_2 (see upper line of Fig. 3). During O_2 uptake H^+ ions are released from hemoglobin and the amount of oxylabile carbamate is diminished. The resulting pH change is shown in the lower line. The results of several experiments are summarized in Fig. 4. It clearly demonstrates the influence of an increased content of lactate on the time course of the plasma H^+ ion concentration and confirms the above mentioned H^+ ion transfer mechanism.

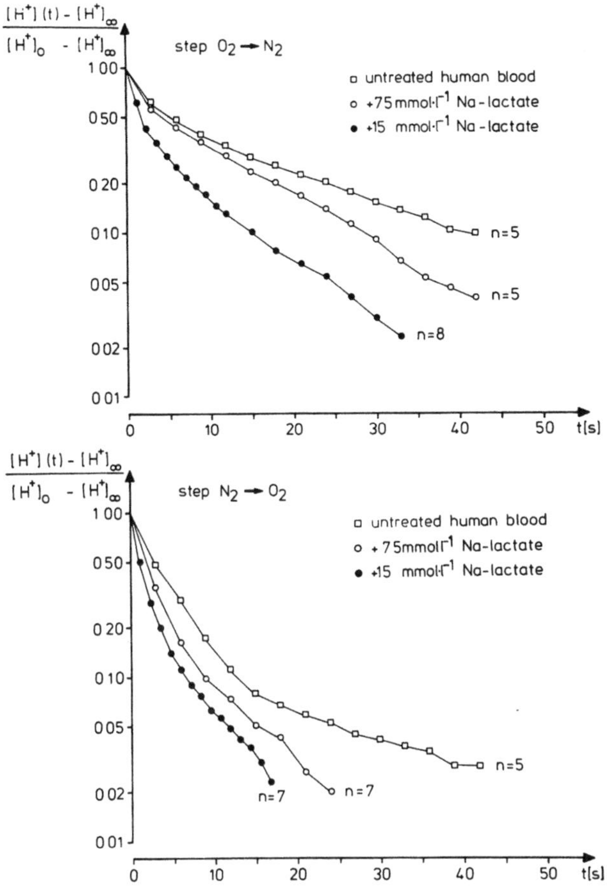

Fig. 4. Influence of different lactate contents on the pH time course in whole human blood after a stepwise change in pO_2 (blood film technique, mean values from 5 to 8 experiments)

Theoretical Calculations

The experimental studies were extended by theoretical calculations using the above mentioned blood model. Especially the effect of lung carbonic anhydrase on the acid base kinetics was analyzed. The calculations were carried out considering the following assumptions: CO_2 hydration is catalyzed in the cells (intracellular catalyzing factor = 500, an increase up to 5000 has no marked effect on the time functions). During lung passage time (T = 700 ms) CO_2 hydration is catalyzed in the plasma too (extracellular catalyzing factor CF between 1 and 100). After leaving the lung CF changes to 1. Blood gas values were supposed to start from venous level (pCO_2 = 6.27 kPa, pO_2 = 5.33 kPa) and to change to alveolar values (pCO_2 = 5.33 kPa, pO_2 = 13.3 kPa) instantaneously, when blood reaches the lung. Venous pH_p was assumed to be 7.375. Figure 5 shows the influence of different values of CF on the pH and pCO_2 time functions. If the CO_2

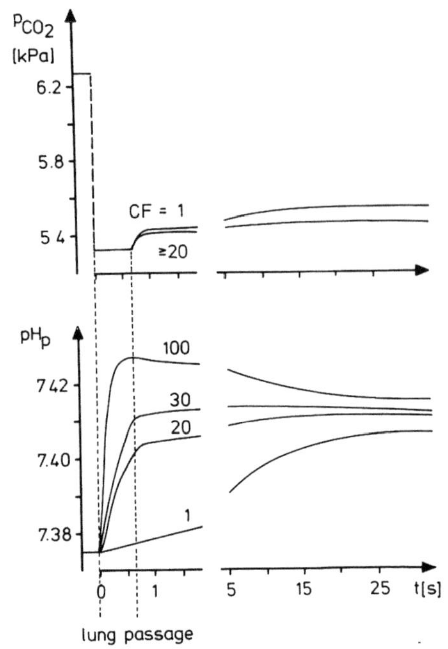

Fig. 5. Calculated pCO_2 and pH time courses in blood during and after lung transit. Blood gas values are assumed to change from venous (pCO_2 = 6.27 kPa, pO_2 = 5.33 kPa) to alveolar levels (pCO_2 = 5.33 kPa, pO_2 = 13.3 kPa) instantaneously when blood reaches the lung. During lung passage (T = 700 ms) different catalyzing factors CF are acting in the plasma. After leaving the lung CF is assumed to be 1. In the *left part* of the diagram the time function in the first seconds, in the right part the later state of reaction after blood has left the pulmonary capillary is shown

reaction is catalyzed by lung carbonic anhydrase, a fast increase of the pH_p is obtained during lung transit time. During this time a fast dehydration of bicarbonate in the plasma happens and due to the low buffer capacity of the plasma, a fast alkaline pH shift occurs. After blood has left the lung a slow uncatalyzed equilibration process happens accompanied by a small increase in pCO_2 and a slow pH change. The direction of this slow pH shift depends on the amount of lung carbonic anhydrase acting in the plasma during lung passage time.

Discussion

The conclusion reached from all series of experiments together with results obtained in earlier experiments with metabolic disturbances (Luttmann et al. 1981), is that in presence of weak acids or bases an H^+ ion transfer across the red cell membrane happens, which accelerates the equilibration process after any acid base disturbance. The influence of acetate and ammonia may be explained by the diffusion of the undissociated acetic acid or ammonia (Deuticke 1973, Klocke et al. 1972). For lactate, however, it has been shown from Dubinsky and Racker (1978) and from Deuticke et al. (1978) that in the red cell membrane a specific lactate transport system exists and the lactate anion transport is accompanied by a so-called H^+-symport or an OH^--antiport mechanism. From our experiments we cannot distinguish between the diffusion of undissociated lactic acid and a specific symport or antiport mechanism. In vivo this mechanism may play a role in severe muscular exercise in which an increased lactate content accelerates the equilibration after the uptake of the acids in the muscle and after CO_2 and O_2 exchange processes in the lung and body capillaries.

After a stepwise change in pO_2 the time course of the equilibration process depends on the direction of the pO_2 change (see Fig. 4). One reason for this difference may be the different velocity constant of the carbamate formation in oxyhemoglobin and desoxyhemoglobin as recently measured by Gros et al. (1981). For complete understanding further studies are necessary.

Summarizing the results of the experiments in vitro we can conclude that after a respiratory CO_2 change the plasma pH reaches 90% of the total change within 5 s, a stepwise increase in pO_2 yields a change of the H^+ ion concentration with a T_{90} of about 13 s. The transit time of the blood from the lung to the peripheral chemoreceptors or to the central nervous system is about 12 s (Grodins et al. 1967). So we can assume that under resting conditions blood has approximately reached its thermodynamic equilibrium when it passes these structures, even if no carbonic anhydrase is available in the lung. During exercise an increased lactate content may facilitate the equilibration process at the same time as the passage time is shortened.

Our model calculations demonstrate a small increase in arterial pCO_2 after blood has left the lung resulting in a positive difference between arterial and alveolar pCO_2 of about 0.1 kPa. This is in agreement with experimental findings of Scheid et al. (1972) obtained in diverse types of rebreathing experiments. The literature, however, is quite controversial (for reference see Piiper 1979).

Measurements of the pCO_2 time course in arterial blood were performed by Ponté and Purves (1980). They obtained a large decrease of pCO_2 with time up to 1.3 kPa when blood has left the lung. This cannot be explained with our model calculations, but the method applied seems to be critical, because they used a CO_2 consuming mass spectrometer sensor in a stopped flow experiment. Comparison of the calculated pH time course with experimental results shows that the lower curve (CF = 1) of Fig. 5 is in agreement with values observed in benzolamide treated animals by Crandall et al. (1977). The slow acidic pH change, calculated with higher carbonic anhydrase activity (CF > 30) shows agreement with observations by Rispens et al. (1980). In other experimental studies (Hill et al. 1977, Bidani and Crandall 1978) small alkaline pH changes were observed. From a theoretical point of view, both, an alkaline and an acidic shift can be explained depending on the amount of carbonic anhydrase in the lung. Here, further investigations on localisation and activity of lung carbonic anhydrase are needed to provide a complete description of CO_2 exchange.

Acknowledgements. Supported by the Deutsche Forschungsgemeinschaft (SFB 114 Bionach).

References

Bidani A, Crandall ED (1978) Slow postcapillary pH changes in blood in anesthetized animals. J Appl Physiol 45:674–680

Crandall ED, Klocke RA, Forster RE (1971) Hydroxyl ion movement across the human erythrocyte membrane. J Gen Physiol 57:669–683

Crandall ED, Bidani A, Forster RE (1977) Postcapillary changes in blood pH in vivo during carbonic anhydrase inhibition. J Appl Physiol 43:582–590

Crandall ED, O'Brasky JE (1978) Direct evidence for participation of rat lung carbonic anhydrase in CO_2 reactions. J Clin Invest 62:618–622

Deuticke B (1973) Transport of monovalent organic anions across the red cell membrane. In: Gerlach E, Moser K, Deutsch E, Wilmans W (eds) Erythrocytes, thrombocytes, leukocytes. Thieme, Stuttgart, p 81

Deuticke B, Rickert F, Beyer E (1978) Stereoselective SH-dependent transfer of lactate in mammalian erythrocytes. Biochim Biophys Acta 507:137–155

Dubinsky WP, Racker E (1978) The mechanism of lactate transport in human erythrocytes. J Membr Biol 44:25–36

Effros RM, Chang RSY, Silverman P (1978) Acceleration of plasma bicarbonate conversion to carbon dioxide by pulmonary carbonic anhydrase. Science 199:427–429

Forster RE, Crandall ED (1975) Time courses of exchanges between red cells and extracellular fluid during CO_2 uptake. J Appl Physiol 38:710–718

Forster RE, Obaid AL, Crandall ED, Itada N (1980) Cl^- and HCO_3^- movements across the red cell membrane. In: Bauer C, Gros G, Bartels H (eds) Biophysics and physiology of carbon dioxide. Springer, Berlin Heidelberg New York, p 285

Grodins FS, Buell J, Bart AJ (1967) Mathematical analysis and digital simulation of the respiratory control system. J Appl Physiol 22:260–276

Gros G, Wittmann B, Guggenberger L (1981) Carbamate kinetics of blood proteins. In: Piiper J, Scheid P (eds) Gas exchange function of normal and diseased lungs. Prog Respir Res, vol 16. Karger, Basel, p 205

Hill EP, Power GG, Longo LD (1973) Mathematical simulation of pulmonary O_2 and CO_2 exchange. Am J Physiol 224:904–917

Hill EP, Power GG, Gilbert RD (1977) Rate of pH changes in blood plasma in vitro and in vivo. J Appl Physiol 42:928–934

Klocke RA (1978) Catalysis of CO_2 reactions by lung carbonic anhydrase. J Appl Physiol 44: 882–888

Klocke RA, Anderson KK, Rotman HH, Forster RE (1972) Permeability of human erythrocytes to ammonia and weak acids. Am J Physiol 222:1004–1013

Luttmann A (1977) Steady state and transientes Verhalten von Säuren-Basen Parametern und Erythrozytenvolumen bei der Bindung von CO_2 im Blut (Entwicklung eines schnell ansprechenden CO_2-Meßsystems). Diss Ruhr-Univ Abt Biol, Bochum

Luttmann A (1980) Untersuchungen zum dynamischen Verhalten der Säuren-Basen-Parameter des Blutes nach respiratorischen und metabolischen Störungen. Habilitationssch Ruhr-Univ Abt Naturwiss Med, Bochum

Luttmann A, Mückenhoff K, Loeschcke HH (1980) Kinetics of acid-base parameters of blood after respiratory and metabolic disturbances. In: Bauer C, Gros G, Bartels H (eds) Biophysics and physiology of carbon dioxide. Springer, Berlin Heidelberg New York, p 321

Luttmann A, Ahmad HR, Mückenhoff K, Plaas-Link A (1981) Influence of weak acids and bases on the pH and pCO_2 kinetics in the blood. In: Piiper J, Scheid P (eds) Gas exchange function of normal and diseased lungs. Prog Respir Res, vol 16. Karger, Basel, p 193

Piiper J (1979) Blood-gas equilibration of CO_2 in pulmonary gas exchange of mammals and birds. Physiologist 22:54–59

Ponté J, Purves JM (1980) Changes in pH and pCO_2 with time in pulmonary post-capillary blood in cats. In: Bauer C, Gros G, Bartels H (eds) Biophysics and physiology of carbon dioxide. Springer, Berlin Heidelberg New York, p 315

Rispens P, Oeseburg B, Zock JP, Zijlstra WG (1980) Intraaortic decrease of blood pH. Pflügers Arch 386:97–99

Roughton FJW (1935) Recent work on carbon dioxide transport by the blood. Physiol Rev 15: 241–296

Roughton FJW (1964) Transport of oxygen and carbon dioxide. In: Handbook of physiology, vol I: Respiration. Am Physiol Soc, Washington, p 767

Scheid P, Teichmann J, Adaro F, Piiper J (1972) Gas-blood CO_2 equilibration in dog lungs during rebreathing. J Appl Physiol 33:582–588

Wieth JO (1980) Interaction between two types of pH-equilibrating proton and hydroxyl ion carriers. In: Lassen UV, Ussing HH, Wieth JO (eds) Membrane transport in erythrocytes. Munksgaard, Copenhagen, p 512

Localization of CO_2 Sensitive Units in the Rostral Medullary Chemosensitive Area of the Cat

L. ÇAKAR and M. TERZIOǦLU[1]

A number of investigations have been carried out recently with the purpose of localizing chemosensitive structures within areas (M: rostral, L: caudal, S: intermediate) on the ventrolateral surface of the medulla (Berndt et al. 1972 a, b, Mitchell et al. 1963 a, b, Pokorski 1976, Schlaefke and Loeschke 1967, Schlaefke et al. 1970, 1974, 1975, Trouth et al. 1973 a). Schlaefke et al. (1974, 1975, 1979) described about 180–200 nerve cells on each side of the intermediate area, rostromedial of the hypoglossal nerve exit (Schlaefke and Loeschcke 1967) located at 0–100 μm or 7–113 μm, most densely at 30–35 μm, below the surface, which they suggested as being involved in the central chemosensitive mechanism. Trouth et al. (1973 a), who induced an increase in ventilation by electrical stimulation of the ventral surface of the medulla, suggested the responsive substrate to be located medial to the hypoglossal roots and ventral to the inferior olive, which separates it from respiratory centers. Within this region characteristic cells were distributed at a distance of 8–450 μm, the greatest concentration being located at 150–200 μm, beneath the surface (Trouth et al. 1973 b).

In an interesting series of experiments, Berndt et al. (1972 a, b) described a mathematical model by which extracellular fluid pH (pH_e) could be estimated from acid-base parameters of blood and CSF. They postulated that CSF composition influences pH_e not farther than 400–500 μm below the surface and that structures reacting to variations in the $[H^+]$ of CFS must lie within this region. In the second series of experiments the same investigators tested the validity of this postulate by perfusing the ventral surface of the medulla of anesthetized, vagotomized and carotid chemo-deafferented cats with mock CSF of varying pH and recording the respiratory variables. On the basis of their results, the authors assumed the chemosensitive structures to be located at less than 1 mm below the ventral medullary surface.

Pokorski (1976) suggested that units responsive to changes in acid-base parameters lie in the first 1 mm layer of the ventral medulla, being most concentrated at 100–500 μm beneath the surface. Although direct and indirect evidence from these studies generally indicates a superficial localization of central chemosensitive structures within the circumscribed areas, other workers (Lipscomb and Boyarsky 1972, Pappenheimer et al. 1965) suggested that these units must lie at some distance below the medullary surface. According to Pappenheimer et al. (1965), the neural elements

1 Department of Physiology and Biophysics, Cerrahpasa Faculty of Medicine, University of Istanbul, Turkey

Central Neurone Environment
(ed. by M.E. Schläfke, H.P. Koepchen and W.R. See)

sensitive to changes in the $[H^+]$ of interstitial fluid are located 2/3 or 3/4 of the distance along the diffusion path between CSF and blood.

Lipscomb and Boyarsky (1972) recorded potentials from and at various depths beneath the ventrolateral medullary surface, while perfusing the region with solutions high in H^+-ion concentration. Since no change in potentials was observed on recording from the surface but only from deeper lying neurons within the bulbar respiratory center, they postulated that the perfusate reaches these primary respiratory neurons by diffusing through the arterioles and stimulates them directly.

From the above review, it is quite evident that the localization of chemosensitive structures within the delineated areas of the ventrolateral medulla is still in dispute. It is for this reason that we undertook the present study and attempted to localize the medullary chemosensitive units by means of potential recording.

Materials and Methods

Two series of experiments were conducted, differing in the mode of stimulation of the rostral chemosensitive area.

a) The first experimental series consisted of 28 cats divided ino 2 groups as follows: (1) 14 cats in the control, and (2) 14 in the peripheral chemoreceptor-denervated group.

Animals were anesthetized with intraperitoneal administration of 30 mg/kg of sodium pentobarbital. Fractional doses were injected when necessary. The trachea was cannulated and connected to an inspiratory-expiratory valve. The carotid sinus and vagus nerves were isolated for subsequent surgical procedures in the second group of animals. The ventral surface of the medulla, from the middle portion of the pons to the 12th cranial nerve roots, was then exposed by suboccipital craniotomy. The dura was cut and its edges reflected.

In the peripheral chemoreceptor-denervated animals, sinus nerves were sectioned and the vagal sheath was opened opposite the superior thyroid artery. The thin aortic nerve lying between the vagus and the sympathetic trunk was then cut bilaterally for the denervation of the aortic area.

Recording technique. For the recording of unit potentials, glass capillary microelectrodes filled with 2.5 M NaCl were used. The resistance of the electrodes was 1–5 megohms.

The electrode was stereotaxically inserted, by steps of 50 μm from the ventral surface, into the depth of the medulla. The cat was ventilated at each step for 5 min with air or a hypercapnic (6.5% CO_2 in air) gas mixture. The unit potentials and the respiratory movements were recorded on a Grass Model 7 Polygraph. The potentials were simultaneously monitored on a Tektronix 502 dual beam oscilloscope and, for records, photographed from the oscilloscope with a Grass kymograph camera.

b) The second experimental series consisted of 18 cats anesthetized with a mixture of chloralose (50 mg/kg) and urethane (250 mg/kg), administered intraperitoneally. The trachea, left femoral artery and vein were cannulated. The arterial blood pressure was recorded from the femoral artery by means of a transducer connected to the recorder. The chest was opened along the midline and the right internal thoracic artery was then cannulated. With the exception of the vertebral artery, all other branches of the right subclavian artery were ligated. The cat was then artificially ventilated by means of a Starling respiratory pump, connected to the tracheal cannula.

The peripheral chemoreceptors in this series were denervated by cutting both the vagi and the sinus nerves. The left phrenic nerve was isolated and strands from roots C_4--C_5 were prepared for recording of spontaneous respiratory activity.

The ventral surface of the medulla was exposed as in the first experimental series.

Recording technique. Ringer-Locke's solution, previously bubbled with 100% CO_2, was used as the test and that without CO_2 as the control solution. During the experiment, 1 ml of either of these solutions was administered *via* the vertebral artery. Before the injection of the test solution, the microelectrode was inserted into the depth of the medulla, by steps of 50 μm from the ventral surface. The potentials recorded at various depths during the injection of the test solution were compared with those obtained during the administration of the control solution.

Glass pipette electrodes, filled with Ringer-Locke's solution, were connected by chlorided platinum wires to a Grass microelectrode preamplifier. The isolated strands of the phrenic nerve were then placed on these electrodes. For the recording of the potentials from the chemosensitive area and from the phrenic nerve as well as of blood pressure, heart rate, ECG lead II, a jet-spray mingograph was used. Potentials were monitored on the oscilloscope screen as well as by means of an audio-monitor. A spike processor (digitimer 130) served to record the firing rate of the units.

The temperature of the cat was kept at circa 38°C by means of a thermistor-controlled heating blanket.

In order to localize the points or regions of potential recording within the chemosensitive area, a needle was inserted along the path of the microelectrode at the end of each experiment. The isolated brain stem was fixed in 10% formaldehyde, frozen and cross-sectioned at 50 μm intervals. The sections were then stained with methylene blue. From the needle trace in the histological sections and the stereotaxic coordinates, the region of penetration of the microelectrode and the point or the region of potential recording could thus be estimated.

Results

a) *The response of medullary* CO_2 *chemosensitive units to the breathing of hypercapnic gas mixtures: The control group.* While the electrode tip was on the surface of the rostral medullary chemosensitive area, no change in potentials was observed on switching from air to CO_2 breathing. On the other hand, respiratory frequency increased from a mean of 19.3 ± 7.7/min in the air phase to 24.7 ± 9.9/min in hypercapnia. From these values the mean percentage increase of this parameter in the hypercapnic phase was found to be 27.3 ± 12.9%.

When the electrode was inserted by steps of 50 μm from the ventral surface into the depth of the medulla, no increase in firing rate of the units on breathing of hypercapnic gas mixtures was observed until depths of 400 μm were attained. From thereon, an increase in discharge frequency in response to hypercapnia occurred as the electrode was inserted into deeper levels. The highest firing rates in response to hypercapnia were recorded at depths of 600–1000 μm. Although there was a diminution in the frequency of potentials and hence in sensitivity to CO_2 below 1000 μm, the response persisted until depths of 1400 μm below the surface were attained.

Peripheral chemoreceptor-denervated group. When the peripheral chemoreceptor-denervated animals were allowed to breathe 6.5% CO_2 in air, respiratory frequency was increased from a mean of 16.9 ± 6.2/min in the air phase to 18.2 ± 8.2/min in hypercapnia. From the absolute values of this parameter in air and the hypercapnic phases, the mean percentage increase in hypercapnia was calculated to be 11.6 ± 5.3%. This lower magnitude of increase in respiratory frequency of the

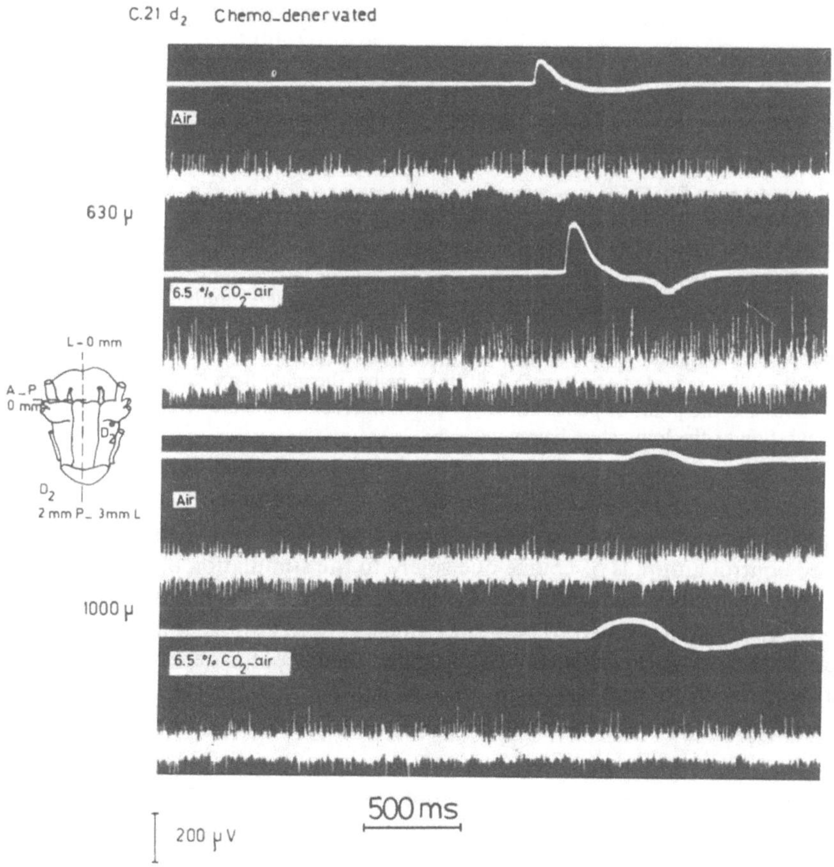

Fig. 1. Spontaneous respiratory movements (*upper trace*) of the anesthetized, peripheral chemoreceptor-denervated cat and potentials (*lower trace*) recorded at the indicated depths below the surface of its rostral chemosensitive area, while the animal was breathing air or the hypercapnic gas mixture. Stereotaxic coordinates and the point of penetration of the microelectrode D_2, (●) are given in the schematic representation of the brain stem on the left

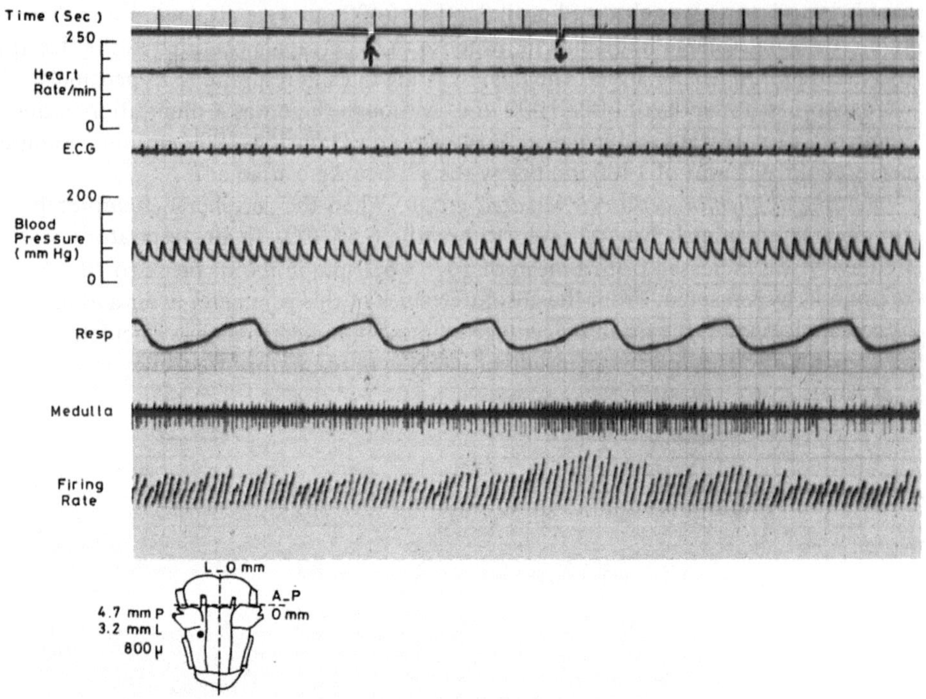

Fig. 2. The effect of rapid injection via the vertebral artery of CO_2-equilibrated Ringer-Locke's solution on the indicated parameters of the artificially ventilated cat. Injection started at ↑ and stopped at ↓ on the time scale. The *last two tracings* give the actual potentials and the firing rates of the units, recorded at a depth of 800 μm below the medullary surface. The stereotaxic coordinates and the point of penetration (●) of the microelectrode are given in the schematic representation of the brain stem below

peripheral chemoreceptor-denervated animals, when compared with that of the similarly treated control group, is of course due solely to the stimulation of the chemosensitive areas and to the lack of input from the chemoreceptors. The results of potential recording, however, seemed to be quite similar in both groups of animals. Thus, the response to CO_2 breathing in the peripheral chemoreceptor-denervated animals was first observed at a depth of 470 μm below surface, was highest between 600–1000 μm and persisted at a lower magnitude until depths of 1342 μm (Fig. 1).

b) The response of medullary chemosensitive units to the administration of CO_2 rich solutions via the vertebral artery. The results obtained in the second experimental series were quite comparable to those of the first. A rapid injection of the test solution containing CO_2 into the vertebral artery brought about an increase in potentials, recorded at depths of 500–1300 μm below the medullary surface (Figs. 2 and 3). The increased firing rate was first observed 2–3 s after starting the injection, continued at about the same rate throughout the period of delivery and rapidly diminished to the pre-stimulation rate a few seconds after the end of injection (Fig. 2). At the same time, an increase in the phrenic nerve impulse traffic was observed, indicating augmented

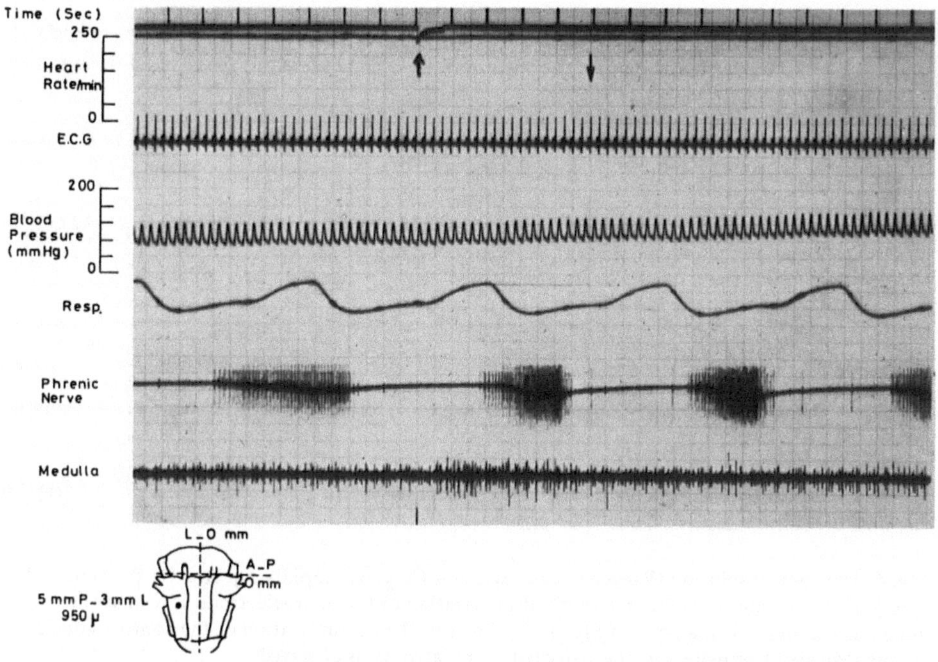

Fig. 3. The effect of rapid injection via the vertebral artery of CO_2-equilibrated Ringer-Locke's solution on the indicated parameters of the artificially ventilated cat. Injection started at ↑ and terminated at ↓. Note the increase in phrenic nerve impulse traffic and the augmented discharge of the chemosensitive units, shown in the lower two tracings. The point and the depth of penetration (●) of the microelectrode as well as the stereotaxic coordinates are given in the schematic representation of the brain stem below

activity of the respiratory centers (Fig. 3). The injection of the control solution via the same route, however, had no effect on the activity of the medullary CO_2 chemosensitive structures, located at depths of 500–1300 μm below the surface (Fig. 4).

The blood pressure was generally stable at an average of 160 ± 30 mm Hg. The heart rate was never affected by injection of the test solution and was generally quite stable at an average of 230 ± 15/min (Fig. 3).

The results of both series of experiments indicate that within the rostral medullary chemosensitive region the area responsive to CO_2 seems to lie 3–7 mm caudal to the ponto-medullary border and to extend bilaterally 2–5 mm from the median line. It should be stated, however, that not all structures within the region were responsive to CO_2, since in only one-third of a total of 280 penetrations did we encounter chemosensitive units.

Discussion

The results of the present study indicate that units responsive to CO_2 are located at depths of 400–1400 μm, most densely at 600–1000 μm, beneath the surface of the

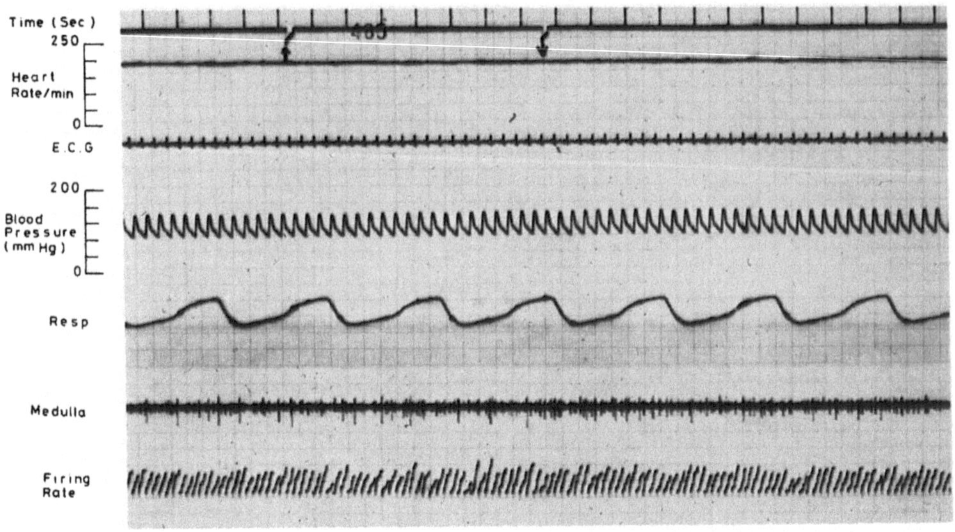

Fig. 4. The control solution (Ringer-Locke's without CO_2) was rapidly injected via the vertebral artery between ↑ and ↓. Point and depth of penetration of the microelectrode as well as the stereotaxic coordinates same as in Fig. 4. No effect of the administration of the control solution on the indicated parameters of the artificially ventilated cat is observed

rostal chemosensitive area in the cat. That these units are stimulated by CO_2 in blood and in the interstitial fluid perfusing the area is suggested from the results of both experimental series. Thus, the inhalation of hypercapnic gas mixtures by both the control and the peripheral chemoreceptor-denervated cats induced increases of comparable magnitude in the firing rates of the units, located at corresponding levels below the medullary surface. This finding indicates that, under our experimental conditions, CO_2 sensitivity of the region is independent of impulses mediated via the peripheral chemoreceptors. Further, the immediate response of the units to the injections of CO_2 rich solutions into the vertebral artery of the peripheral chemoreceptor-denervated, artificially ventilated cat offers additional evidence for their being directly stimulated by CO_2 in the perfusing fluid.

Our results differ from those of most investigators (Berndt et al. 1972 a, b, Mitchell et al. 1963 a, b, Pokorski 1976, Schlaefke and Loeschcke 1967, Schlaefke et al. 1970, 1974, 1975, 1979, Trouth et al. 1973 a, b) concerning the localization of chemosensitive structures within the circumscribed areas of the brain stem. In some of these studies, the responsive structures were localized very superficially (Schlaeke et al. 1974, 1975, 1979) while in others (Berndt et al. 1972 b, Pokorski 1976, Trouth et al. 1973 a), the lower boundary of the chemosensitive area was found to extend deeper in the brain stem to 450 μm (Trouth et al. 1973) or to less than 1 mm below the surface (Berndt et al. 1972 b). Pokorski (1976), who recorded nonphasic unit potential from area M in spontaneously breathing cats during metabolic acidosis and alkalosis, concluded that the responsive units lay within the first 1 mm layer of the ventral medulla, being more concentrated between 100–500 μm beneath the surface. In spite of these findings, Pokorski (1976) seemed to support the concept of Mitchell et al. (1963 a, b) and Loeschcke (1965) concerning the superficial localization of the chemosensitive structures.

On the other hand, other investigators (Lipscomb and Boyarsky 1972, Pappenheimer et al. 1965) offered convincing evidence for a deeper localization of chemosensitive structures within the brain stem. Although our results agree with those of Lipscomb and Boyarsky (1972) in that the responsive units are located deep in the medulla, we do not believe that they are bulbar respiratory neurons. Anatomically, our site of potential recording was the rostral part of the ventrolateral medulla, and the area responsive to CO_2 was found to lie 3–7 mm caudal to the ponto-medullary border and to extend 2–5 mm bilaterally from the median line. This area, which corresponds to area M and the overlapping anterior portion of area S included the nucleus preolivarius and the nucleus trapezoid body. It should be added that only 1/3 of the units within the delineated area were responsive to CO_2.

Acknowledgement. Part of this study was carried out in the Department of Physiology, The Middlesex Hospital Medical School, while one of the authors (L. Çakar) was on a year's leave of absence from Istanbul University. Our grateful thanks are therefore due to Professor Eric Neil for his interest, advice and critical comments during the progress of this work.

References

Berndt J, Berger W, Mückenhof K (1972a) Untersuchungen zum zentralen chemosensiblen Mechanismus der Atmung, I. Die Bestimmung des extracellulären pH im Gehirngewebe. Pflügers Arch 332:127–145

Berndt J, Berger W, Berger K, Schmidt M (1972b) Untersuchungen zum zentralen chemosensiblen Mechanismus der Atmung, II. Die Steuerung der Atmung durch das extracelluläre pH im Gewebe der Medulla oblongata. Pflügers Arch 332:146–170

Lipscomb WT, Boyarsky LL (1972) Neurophysiological investigations of medullary chemosensitive areas of respiration. Respir Physiol 16:362–376

Loeschcke HH (1965) A concept of the role of intracranial chemosensitivity in respiratory control. In: McC Brooks C, Kao FF, Lloyd BB (eds) Cerebrospinal fluid and the regulation of ventilation. Blackwell, Oxford, p 183

Loeschcke HH, Mitchell RA, Katsaros B, Perkins JF, Konig A (1963) Interaction of intracranial chemosensitivity with peripheral afferents to the respiratory centers. Ann NY Acad Sci 109: 651–660

Mitchell RA, Loeschcke HH, Massion WH, Severinghaus JW (1963a) Respiratory response mediated through superficial chemosensitive areas on the medulla. J Appl Physiol 18:523–533

Mitchell RA, Loeschcke HH, Severinghaus JW, Richardson BW, Massion WH (1963b) Regions of respiratory chemosensitivity on the surface of the medulla. Ann NY Acad Sci 109:661–681

Pappenheimer JR, Fencl V, Heisey SR, Held D (1965) Role of cerebral fluids in the control of respiration as studied in unanesthetized goats. Am J Physiol 208:436–450

Pokorski M (1976) Neurophysiological studies on central chemosensor in medullary ventrolateral areas. Am J Physiol 230:1288–1295

Schlaefke ME, Loeschcke HH (1967) Lokalisation eines an der Regulation von Atmung und Kreislauf beteiligten Gebietes an der ventralen Oberfläche der Medulla oblongata durch Kälteblockade. Pflügers Arch gesamte Physiol Menschen Tiere 297:201–220

Schlaefke ME, See WR, Loeschcke HH (1970) Ventilatory response to alterations of H^+-ion concentration in small areas of the ventral medullary surface. Respir Physiol 10:198–212

Schlaefke ME, Kille JF, Folgering H, Herker A, See WR (1974) Breathing without central chemosensitivity. In: Umbach W, Koepchen HP (eds) Central rhythmic and regulation. Hippokrates, Stuttgart, p 97
Schlaefke ME, Pokorski M, See WR, Prill RK, Loeschcke HH (1975) Chemosensitive neurons on the ventral medullary surface. Bull Physiopathol Resp 11:277–284
Schlaefke ME, Kill JF, Loeschcke HH (1979) Elimination of central chemosensitivity by coagulation of a bilateral area on the ventral medullary surface in awake cats. Pflügers Arch 378: 231–241
Trouth CO, Loeschcke HH, Berndt J (1973a) A superficial substrate on the ventral surface of the medulla oblongata influencing respiration. Pflügers Arch gesamte Physiol Menschen Tiere 339:135–152
Trouth CO, Loeschcke HH, Berndt J (1973b) Histological structures in the chemosensitive regions on the ventral surface of the cat's medulla oblongata. Pflügers Arch 339:171–183

Microstimulation Within the Intermediate Area and It's Influence on Respiratory Neurons

K. GOTO[1], W.R. SEE[2], and M.E. SCHLAEFKE[2]

Introduction

There are only a few papers dealing with the functional connection between the superficial chemosensitive structures of the medulla and respiratory and reticular neurons (Peskow and Piatin 1976, Davies and Loeschcke 1977, Davies 1980). It is known that local electrical stimulation on the ventral medullary surface causes both respiratory and cardiovascular effects (Loeschcke et al. 1970, Trouth et al 1973). Therefore, in order to gain information on central pathways from the chemosensitive substrates to respiratory neurons, we investigated the effect of microelectrical stimulation in the intermediate area on respiratory neurons in the nucleus retroambigualis (NRA). This nucleus is relatively easy to reach from the ventral surface and contains cells which receive information from the respiratory rhythm generator but are not chemosensitive themselves (Mitchell and Herbert 1974).

Methods

The experiments were performed on artificially ventilated cats anesthetized with chloralose-urethane (50/200 mg kg^{-1}) and paralyzed with pancuronium-bromide. Tungsten-electrodes (tip diameter ca. 1 μ) were inserted into the NRA from the ventral medullary surface at the level of the obex. Extracellular action potentials showing respiratory rhythm were confirmed to originate in the soma-dendritic part of the neuron following standard procedures. Respiratory unit activities were recorded together with phrenic nerve activity, endtidal pCO_2, and femoral arterial blood pressure on a direct writing polygraph, unit (after processing through a counter) and phrenic nerve activity (after rectification and integration) also on magnetic tape. At the end of each experiment, an electrolytic lesion was performed to identify the site of recording. For the analysis of respiratory unit activity an electronic computer was used. Period (P), duration (D), number of spikes (N) and mean discharge frequency (N/D) each per respiratory unitary

1 Department of Pharmacology II, Jikei University, Tokyo, Japan; present adress: Department of Pediatrics, Jikei University, Tokyo, Japan
2 AG Physiologie der Regulation, Ruhr-Universität, D-4630 Bochum, Fed. Rep. of Germany

Central Neurone Environment
(ed. by M.E. Schläfke, H.P. Koepchen and W.R. See)

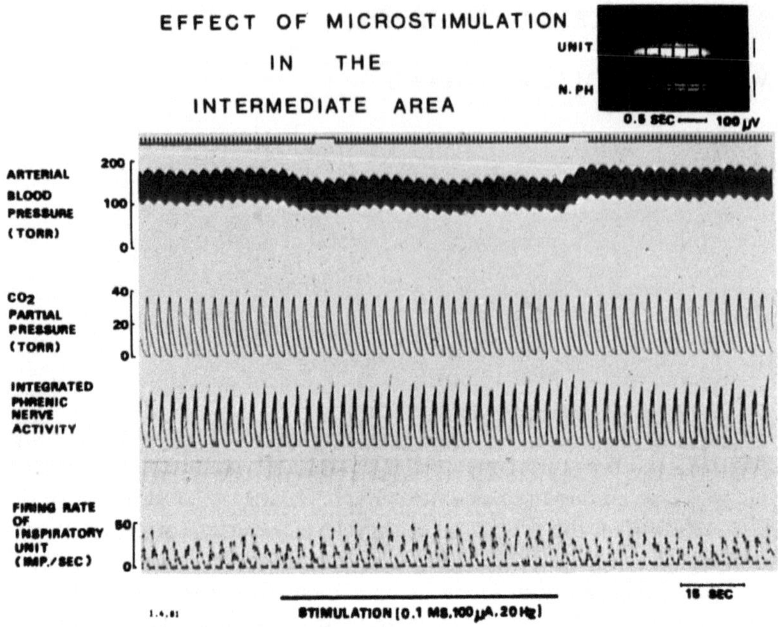

Fig. 1. The effect of micro-electrical stimulation (20 Hz, 0.3 ms, 110 μA) on the surface of the intermediate area for 1 min. From *top*: arterial blood pressure (torr), endtidal CO_2 partial pressure (torr), integrated phrenic nerve activity (arb. units) and firing rate of an inspiratory activity (impulses/s). Stimulation in the intermediate area (*lowest line*) caused a decrease of arterial blood pressure and an incremant of integrated phrenic nerve and inspiratory unit activity in the nucleus retroambigualis. Peripheral chemoafferents intact. *Insert* in the upper right hand corner: oscilloscope tracing of inspiratory unit activity (UNIT) and phrenic nerve activity (N.PH)

burst and the amplitude of integrated phrenic nerve activity were measured in 10 consecutive respiratory cycles before and during microstimulation (square wave: 100–120 μA, 0.1–0.3 ms, 5–100 Hz) in the surface of the intermediate area (6 mm caudal from the foramen caecum, 3–4 mm lateral from the midline). In order to evaluate the effect of microstimulation on respiratory unit and phrenic nerve activity, stimulation vs. control ratios of mean values of these 5 parameters were calculated.

Results

Microstimulation in the intermediate area for 1 min caused a decrease of arterial blood pressure and an increase of integrated phrenic nerve activity (Fig. 1). In only one case an increment of blood pressure was observed during stimulation.

Respiratory Unit Activity

Respiratory unit activities (inspiratory: 4, expiratory: 1, expiratory-inspiratory phase-spanning: 1, expiratory frequency modulated: 1) were recorded from the NRA ipsilatera

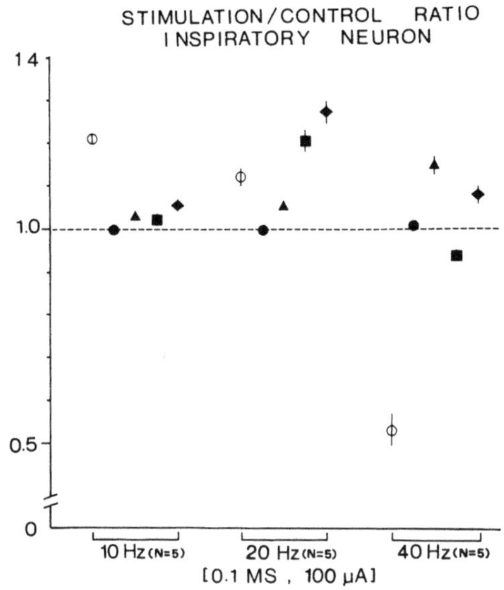

Fig. 2. Mean values (± SE) of stimulation vs control ratios (*ordinate*) for four parameters i.e. ●: period, ▲: duration, ■: mean discharge frequency, and ◆: number of spikes each per inspiratory unitary discharge and ○: integrated phrenic nerve activity. *1.0* means control level. Stimulating pulse duration and current were kept constant whereas stimulating frequency was changed from 10 to 100 Hz. Peripheral chemoafferents intact

to the stimulating site. Two inspiratory neurons were recorded contralaterally. There was no tendency in any of the recorded respiratory neurons to change it's phase relation to phrenic nerve discharge during stimulation.

Inspiratory Neurons

Figure 2 represents the mean values (± SE) of stimulation vs. control ratios of 4 parameters in inspiratory unit activity and integrated phrenic nerve activity. The inspiratory neuron shows an increase in (D) and (N) at stimulating frequencies up to 40 Hz whereas (P) remains close to control values. Stimulation with 100 Hz leads to a complete inhibition of the unitary discharge. All of the inspiratory neurons in the NRA showed the same response characteristic during stimulation. After bilateral carotid sinus nerve section, the mean values of (N) and (N/D) were diminished to 50%–60% of the values with sinus nerves intact. Microstimulation caused an increase in (D) and (N) of inspiratory neurons in the same manner as before sinus nerve section. The latencies of the inspiratory unit response were between 7–10 ms, while phrenic nerve activities had latencies of between 11–15 ms (Fig. 3). Local application of bupivacainhydrochlorid, a local anaesthetic, to the ventral medullary surface caused a complete cessation of both phrenic nerve and inspiratory unit activity.

Expiratory Neuron

In an expiratory unit both (D) and (N) were decreased during stimulation (110 μA, 0.2 ms, 5–20 Hz).

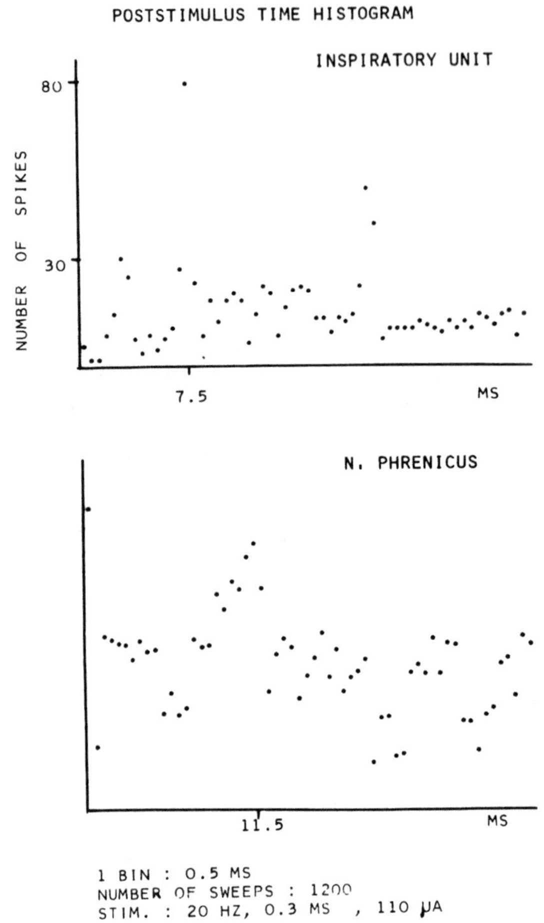

Fig. 3. Poststimulus time histogram of an inspiratory unit and phrenic nerve activity during microstimulation (20 Hz, 0.3 ms, 110 μA) in the intermediate area. The latency which shows the highest peak in the histogram of the inspiratory neuron is 7.5 ms, that of phrenic nerve activity, 11.5 ms. 1 bin: 0.5 ms, number of sweeps: 1200

Expiratory Frequency Modulated Neuron

Stimulation caused a decrement of mean discharge frequency both in the inspiratory and in the expiratory phase at stimulation frequencies of up to 10 Hz (110 μA, 0.3 ms). At frequencies higher than 20 Hz both increased. The extent of the increment of mean discharge frequency in the inspiratory phase was larger than in the expiratory phase.

Expiratory-Inspiratory Phase-Spanning Neuron

Stimulation (110 μA, 0.3 ms, 5–100 Hz) increased both (D) and (N) in this unit. The latency of this unit response was between 6–7 ms. After bilateral carotid sinus nerve section the firing pattern of this neuron changed into an inspiratory-expiratory phase-spanning discharge type. Moreover, stimulation now caused a decrement of (D) and (N). After bilateral vagus nerve section this neuron ceased firing.

Discussion

These first preliminary results show that microelectrical stimulation in the intermediate area causes an increment of inspiratory phase related neuron group activity in the nucleus retroambigualis before and after sectioning peripheral chemoafferents. On the other hand, stimulation decreased expiratory phase related neuron group activity in this nucleus. According to Schwanghart et al. (1974), the application of low concentrated novocaine to the ventral brain stem surface caused a decrement of (N), (D) and (N/D) in inspiration unit activity and an increment of (D) in expiratory unitary discharge. The application of higher dosages of novocaine provoked complete cessation of inspiratory unit activity. The majority of expiratory neurons became tonic during novocaine block. There is uncertainty about the recording site of the respiratory neurons, however, their results coincide well with our results of electrostimulation in the intermediate area and also with the application of a local anaesthetic.

Bilateral carotid sinus nerve section altered the firing pattern of an expiratory-inspiratory phase-spanning to an inspiratory-expiratory phase-spanning discharge pattern. After further bilateral vagotomy this neuron ceased firing. Lipski et al. (1975) reported that there is a projection of the sinus nerve to neurons in the vicinity of the nucleus ambiguus.

From our experimental results, it is suggested that respiratory neurons in the nucleus retroambigualis may be involved in a central pathway related to the respiratory function of the intermediate area. However, it seems unlikely that this nucleus receives direct input from the central chemosensor which is in accord with the suggestion of Mitchell and Herbert (1974) derived by other methods. It is further implied that afferent input of sinus nerves and vagi may influence the firing pattern in some of the phase-spanning neurons.

Acknowledgements. We wish to express our gratitude Mr. U. Schutter and Mrs. E. Goto for their help in the calculations. We also thank Mr. J. F. Kille for his histological work.

This research was supported by the Deutsche Forschungsgemeinschaft (Sonderforschungsbereich 114 Bionach).

References

Davies RO (1980) Evidence that neural paths from the caudal and cranial chemoreceptor zones of the ventral medulla converge in the intermediate zone. Proc Int Un Physiol Sci 14:371

Davies RO, Loeschcke HH (1977) Neural activity evoked by electrical stimulation on the chemosensitive areas on the ventral medullary surface. Proc Int Un Physiol Sci 13:164

Lipski J, McAllen RM, Spyer KM (1975) The sinus nerve and baroreceptor input to the medulla of the cat. J Physiol 251:61–78

Loeschcke HH, Lattre De J, Schlaefke ME, Trouth CO (1970) Effects on respiraton and circulation of electrically stimulating the ventral surface of the medulla oblongata. Respir Physiol 10:184–197

Mitchell RA, Herbert DA (1974) The effect of carbon-dioxide on the membrane potential of medullary respiratory neurons. Brain Res 75:345–349

Peskow BJ, Piatin WF (1976) Reactions of neurons of the respiratory center to local cooling of the ventral surface of the medulla oblongata. Sachenov Physiol J USSR 62/7

Schwanghart F, Schroeter R, Klüssendorf D, Koepchen HP (1974) The influence of novocaine block of superficial brain stem structures on discharge pattern of specific respiratory and unspecific reticular neurons. In: Umbach W, Koepchen HP (eds) Central rhythmic and regulation. Hippokrates, Stuttgart, pp 104–110

Trouth CO, Loeschcke HH, Berndt J (1973) A superficial substrate on the ventral surface of the medulla oblongata influencing respiraton. Pflügers Arch 339:135–152

Further Studies on the Interaction of the Central Chemosensitive Drive and the Respiratory Drive in Hyperthermia

W.R. SEE[1], H. FOLGERING[2] and M.E. SCHLAEFKE[1]

Introduction

Previous experiments have shown that in the normothermic animal apnea is induced if central and peripheral chemoreceptor afferents to the respiratory rhythm generator are eliminated (Schlaefke and Loeschcke 1967, Schlaefke et al. 1969, 1974). Under these conditions neither nonchemical respiratory afferents nor other respiratory stimuli as e.g. arousal reactions or muscular work can produce rhythmic respiratory motor output (Schlaefke et al. 1969, See 1973, Spode and Schlaefke 1975, Hukuhara et al. 1976). The only exception observed was the hyperthermic animal (Schlaefke and Loeschcke 1967). In a series of experiments in anaesthetized, spontaneously breathing cats we could show that electrical stimulation of the anterior hypothalamus produced panting. If under these circumstances chemical afferents were blocked, panting not only persisted, but was greatly enhanced. The ensuing ventilation however could no longer guarantee stable acid-base conditions, a severe respiratory acidosis was the consequence (See 1976 a, b). The findings were at least partly in accordance with those from Albers (1961a,b), Kappey et al. (1962), Pleschka et al. (1965) and Pleschka (1969) who found in the panting dog a shift of the CO_2 threshold to low and even negative values. Von Euler et al. (1970) and Clark and von Euler (1972) assumed for the hyperthermic cat a 'frequency setting mechanism' to be the main governor of respiration. From our experiments we concluded that under heat stress hypothalamic structures gain a considerable influence on ventilation, using the respiratory apparatus for the dissipation of heat. Their action however is counterbalanced by those regulatory mechanisms of ventilation responsible for the acid-base balance.

To establish varying controlled ventilatory and acid-base conditions independent of experimental manipulations, in the present series of experiments we tried to produce thermal panting by whole body temperature elevation in anaesthetized, paralyzed and artificially ventilated cats after section of carotid sinus nerves and vagi and elimination of central chemosensitivity by reversible cold block of the intermediate area (S) on the ventral medullary surface.

1 AG Physiologie der Regulation, Ruhr-Universität Bochum, D-4630 Bochum, Fed. Rep. of Germany
2 Dept. of Physiology, University of Nijmegen, Nijmegen, The Netherlands

Central Neurone Environment
(ed. by M.E. Schläfke, H.P. Koepchen and W.R. See)

Methods

We used 10 female cats with a body weight between 1.7–2.85 kg (mean 2.42 kg ± 0.37 S.D.). The animals were anaesthetized with ketamine-hydrochloride (10 mg kg^{-1} i.m.) and Chloralose and Urethane (up to 40 mg kg^{-1} Chloralose and up to 250 mg kg^{-1} Urethane i.v.) and paralyzed with Pancuronium-bromide (0.15 mg kg^{-1} h^{-1} i.v.). A tracheal tube and catheters in one femoral artery and vein were inserted. A phrenic nerve was dissected, cut and immersed in liquid paraffin at body temperature for subsequent recording. The ventral medullary surface was exposed using the method described in detail elsewhere (Schlaefke and Loeschcke 1967). Bilateral cold block of the intermediate area was achieved with a thermode. The animal's body temperature was regulated by an automatic controlling system and brought to the desired temperature with a heating pad.

Previous experiments had shown that panting was difficult to induce in the Chloralose-Urethane anaesthetized cat by whole body themperature elevation. It turned out that it could more readily be established if prior to heating 4-Aminopyridine, a possible centrally acting transmitter releaser (Thesleff 1980), was injected intravenously. At normal body temperature this leads to an increase of both respiratory frequency and phrenic amplitude. The drug, if applied locally to the chemosensitive structures on the ventral medullary surface has the same effects and also increases the activity of chemosensitive units in the intermediate area (See et al. 1978). We injected 0.52 mg kg^{-1} ± 0.35 SD) (range 0.17–1.5 mg kg^{-1}) intravenously.

The following variables were measured and recorded on a polygraph: arterial pressure with a strain gauge, endexpiratory CO_2 with an infrared analyzer, integrated phrenic nerve activity (after rectification) (arb. units), rectal temperature, thermode tip temperature, and tracheal oxygen content.

The experiments were all performed under hyperoxic conditions. Prior to heating control cold blocks of the intermediate area were performed before and after cutting sinus nerves and vagi in order to verify the correct position of the thermode. After heating and establishing panting several cold blocks were executed at various endexpiratory CO_2 levels. Whenever possible the body temperature was then allowed to decrease to normal levels and another control cold block was performed.

Results

The results essentially confirm those obtained in spontaneously breathing cats. At body temperatures between 41–42°C (314–315 K) phrenic burst frequencies had risen from 12.2 min^{-1} (± 1.9 SE) at normal temperatures to 72 min^{-1} (± 6.3 SE) while phrenic amplitude had decreased from 35.6 (± 1.8 SE) to 30.4 (± 3.8 SE), indicative of a panting-like breathing pattern. The induction of cold block of the intermediate area under these conditions had typically the effect as shown in Fig. 1 in an original recording. Respiratory frequency at 42.9°C (315.9 K) in this animal was 84 min^{-1} [12 min^{-1} at 38.6°C (311.6 K)]. Cold block led to a further decrease of phrenic amplitude and a marked further increase of respiratory frequency to 216 min^{-1}. After releaving the cold

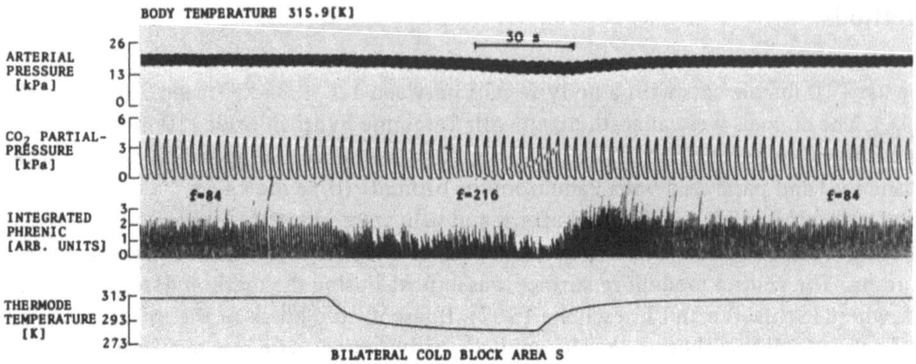

Fig. 1. Original recording from an anaesthetized, artificially ventilated, panting cat at high body temperature before, during, and after bilateral cold block of the intermediate area. Sinus nerves and vagi are cut. The traces show from top to bottom: arterial pressure (*kPa*), tracheal CO_2 partial pressure (*kPa*), integrated phrenic nerve activity (arbitrary units), and thermode tip temperature (*K*). Note the considerable increase of respiratory frequency and decrease of phrenic amplitude during cold block

block the respiratory pattern returned to pre-cooling levels within 1–2 min. Arterial pressure dropped slightly during cold block, which is a general observation when cooling the intermediate area.

The influence on panting of elimination of central chemosensitivity in different ranges of endtidal CO_2 is shown in Fig. 2 for phrenic amplitude and Fig. 3 for respiratory frequency. Values are given as mean ± SE for the respective parameters at high body temperature (range 40.9–43.5°C (313.9–316.5 K) before and during cold block of the intermediate area. In a CO_2 range between 25–40 Torr (3.3–5.3 kPa) phrenic amplitude decreased from 30.4 ± 3.8 to 5.5 ± 1.3 during cold block (Fig. 2 middle panel) whereas respiratory frequency increased from 72 min^{-1} ± 6.3 to 298 min^{-1} ± 15, in a few cases to more than 400 min^{-1} (Fig. 3, middle panel). Elevating endtidal CO_2 [range 40–60 Torr (5.3–8 kPa)] led to a reduction of panting expressed in an increase of phrenic amplitude to 40.2 ± 10.2 and a decrease of respiratory frequency to 55 min^{-1} ± 8.2. During cold block however, panting was restored with values similar to those observed under cold block at normal CO_2. Phrenic amplitude was 4.8 ± 2.1 and respiratory frequency 308 min^{-1} ± 31.2 (Figs. 2 and 3, right panel). When reducing endtidal CO_2, thus diminishing the CO_2 induced influence on panting, the direction of changes observed in phrenic amplitude and respiratory frequency was comparable to that seen with cold block in normal CO_2 ranges. Phrenic amplitude was diminished to 23.2 ± 3.3 and respiratory frequency rose to 209 min^{-1} ± 28.8. However, under this condition cold block provoked a further inhancement of panting with a phrenic amplitude of 10.1 ± 2.1 and respiratory frequency of 287 min^{-1} ± 16, indicative of a supressing action on panting of structures underlying the intermediate area at very low CO_2 values (Figs. 2 and 3, left panel). The similarity of phrenic amplitudes and respiratory frequencies during cold block in different CO_2 ranges makes evident that no influence on panting through CO_2 can be exerted without chemoreceptive structures.

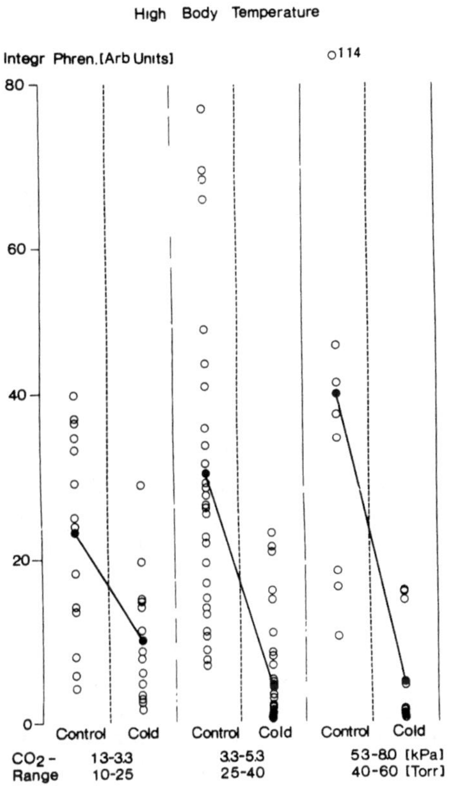

Fig. 2

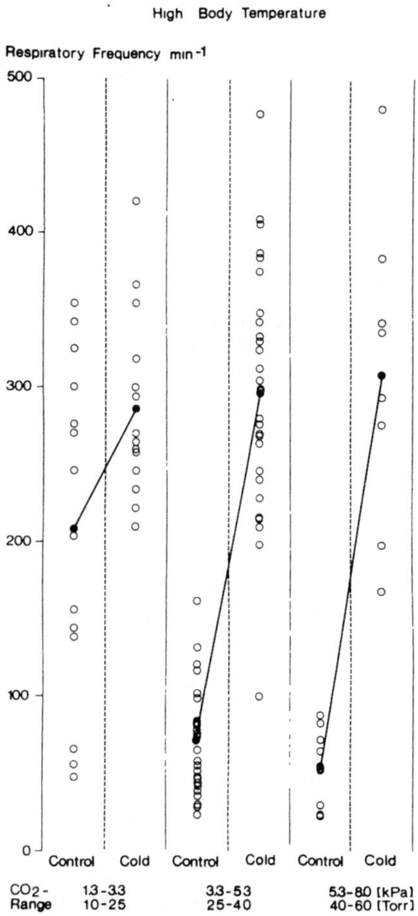

Fig. 3

Fig. 2. Graphic representation of integrated phrenic nerve amplitude at high body temperature under different experimental conditions in three different endexpiratory CO_2 ranges. *Open circles* values of single experimental runs. *Filled circles connected* mean values of all runs under the respective condition. The three different CO_2 ranges from left to right: 1.3–3.3 kPa, 3.3–5.3 kPa, 5.3–8.0 kPa. Within each range control runs are shown (*control*) cold block of the intermediate area (*cold*)

Fig. 3. Same as Fig. 2 for respiratory frequency min^{-1} as derived from phrenic bursts

In observing the pattern of breathing in pre-panting stages, especially in the transient phase of cold blocks we detected in most cases two superimposed respiratory rhythms as shown in Fig. 4. In this example one can see in the phrenic recording upon releasing the cold block that on top of the restored slower rhythm the high frequency respiratory rhythm of about 280 min^{-1} remains visible for about 30 s.

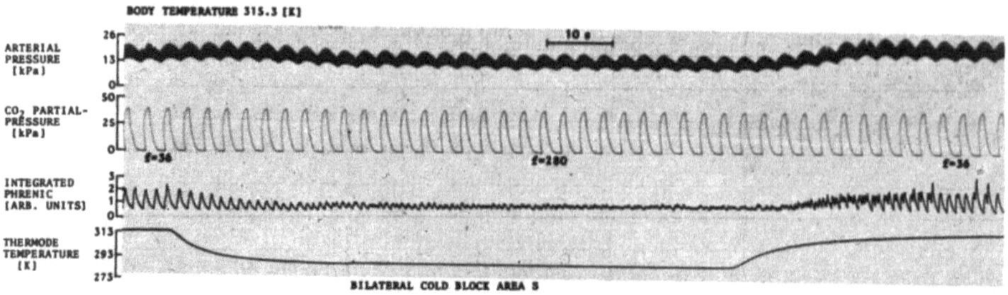

Fig. 4. Original recording as in Fig. 1. Note the increase of respiratory frequency during cold block and the superposition of two distinct respiratory rhythms of different amplitude and frequency in the transient phase of releasing the cold block

Discussion

The results of this study confirm those derived from spontaneously breathing cats in which panting was induced by local hypothalamic heating (See 1976b) or electrical stimulation (See 1976 a, b) in showing that (a) heat stress alone can produce rhythmic respiratory center output and that (b) under heat stress there exists a concurring respiratory drive from central chemosensitivity serving the requirements of acid-base balance. The latter was demonstrated by the enhancement of panting during cold block of the intermediate area.

Quantitative differences to other studies as, e.g. the very high panting frequencies during cold block in our preparation (comp. Fig. 3) can most likely be attributed to anaesthesia (Lim and Grodins 1955, Albers 1961a, Pleschka and Wang 1975, See 1976 a, b) and/or the method used to induce heat stress (whole body temperature elevation vs local hypothalamic heating) (Lim and Grodins 1955, Jessen and Ludwig 1971, See 1976 a, b). It is also possible that recording ventilation with a spirometer is not sufficiently sensitive to detect very small and frequent respiratory movements and that phrenic nerve recording prooves to be the adequate technique for determining panting.

The effect of changing endexpiratory CO_2 on panting is in accord with results of numerous other authors in a variety of species. Increasing CO_2 led to a reduction of panting (Albers 1961b, Kappey et al. 1962, Cohen 1964, Pleschka 1969, von Euler et al. 1970) whereas lowering CO_2 had an opposite effect, i.e., enhancement of panting (Whalen 1955, Albers 1961 a, b, Pleschka et al. 1965, Pleschka 1969, von Euler et al. 1970, Hales et al. 1970, Ruiz 1973, Hales 1974). These effects however were abolished by blocking central chemosensitivity, in fact, panting without chemical afferents seemed to be completely independent of CO_2. Neither in normal nor in low or high endexpiratory CO_2 ranges did the ventilatory parameters during cold block differ markedly. This indicates that a direct effect of e.g. hypocapnia during panting on the respiratory rhythm generator or hypothalamic structures enhancing panting as proposed by von Euler et al. (1970) is unlikely or no longer effective. Our results confirm findings of See (1976b) who concluded that the well known depressant effect of an increased CO_2 on panting is dependent on the intactness of chemosensitive structures. In the spontaneously breathing cat panting during cold block was insufficient to guarantee alveolar ventilation with the consequence of a severe respiratory acidosis, which however did not depress panting as could have been expected.

Even at CO_2 values well below the CO_2 threshold for manifest respiratory output activity under normal conditions, there was still an augmenting effect on the already increased panting when cooling the intermediate area. Pleschka et al. (1965) suggested for the dog under heat stress a reduction of the set point for CO_2. Relative to such a reduced set point there might exist a ventilatory drive from central chemosensitive structures even at very low CO_2 values in the panting animal. This would counteract the temperature dependent drive, mainly governing respiratory frequency. There is evidence that even under normal conditions (Bainton et al. 1978, Lahiri et al. 1978 and Cherniack et al. 1979) the operational range of nervous structures underlying the intermediate area extends well below the CO_2 threshold. This reasoning is in line with the proposal that in analogy to the peripheral chemoreceptors (Katsaros 1965, Schlaefke et al. 1979) a basic drive of central chemosensitive structures exists independent of CO_2 (See 1976 b, Schlaefke et al. 1979). In the light of these various findings it seems unlikely that by blocking the structures underlying the intermediate area an efferent pathway from the hypothalamus relaying to phrenic motoneurones was eliminated. This possibility appears remote also with respect to the fact that when blocking central chemosensitivity we did not observe a respiratory arrest but an enhancement of panting, the latter most likely being driven from the hypothalamus (See 1976 a, b).

We have at te moment no explanation to offer as to how the two different respiratory rhythms seen in many cases, expecially in transient phases of either temperature or cold block, come about. It looks as if the respiratory rhythm generator is at the same time confronted with two distinct and different inputs having an opposite effect on either tidal volume and respiratory frequency. The resulting respiratory output activity then resembles an addition of these inputs (i.e. frequencies and tidal volumes). Neither can we offer an explanation as to where the integration or formation of the two rhythms visible in the common output occurs.

In conclusion the present study has shown:

1. heat stress in contrast to other known respiratory afferents, can produce rhythmic respiratory center output in the absence of chemoreceptive afferents,
2. in all endexpiratory CO_2 ranges tested, extending from 10 to 60 Torr (1.3–8.0 kPa) the ventral medullary structure mediating the central chemosensitive drive has an inhibitory action on panting, i.e. serves the generation of tidal volume and thus alveolar ventilation at the same time reducing respiratory frequency,
3. CO_2 induced changes of panting in the peripherally chemodenervated cat are dependent upon the intactness of neural structures underlying the intermediate area even at very low CO_2 levels.

Acknowledgements. Supported by the Deutsche Forschungsgemeinschaft (SFB 114, Bionach).

References

Albers C (1961a) Der Mechanismus des Wärmehechelns beim Hund I. Die Ventilation und die arteriellen Blutgase während des Wärmehechelns. Pflügers Arch 274:125–147

Albers C (1961b) Der Mechanismus des Wärmehechelns beim Hund III. Die CO_2-Empfindlichkeit des Atemzentrums während des Wärmehechelns. Pflügers Arch 274:166–183

Bainton CR, Kirkwood PA, Sears TA (1978) On the transmission of the stimulating effects of carbon dioxide to the muscles of respiration. J Physiol (London) 280:249–272

Cherniack NS, Euler von C, Homma I, Kao FF (1979) Graded changes in central chemoreceptor input by local temperature changes on the ventral surface of medulla. J Physiol (London) 297:191–211

Clark FJ, Euler von C (1972) On regulation of depth and rate of breathing. J Physiol (London) 222:267–295

Cohen MI (1964) Respiratory periodicity in the paralyzed, vagotomized cat: hypocapnic polypnea. Am J Physiol 206 (4): 845–854

Euler von C, Herrero F, Wexler I (1970) Control mechanisms determining rate and depth of respiratory movements. Respir Physiol 10:93–108

Hales JRS (1974) Physiological responses to heat. In: Robertshaw D (ed) MTP Int Rev Sci Physiol Ser 1, vol 7, p 107

Hales JRS, Bligh J, Maskrey M (1970) Cerebrospinal fluid acid-base balance during respiratory alkalosis in the panting animal. Am J Physiol 219 (2):469–473

Heisey SR, Adams T, Hofman W, Riegle G (1971) Thermally induced respiratory responses in the unanaesthetized goat. Respir Physiol 11:145–151

Hukuhara T jr., See WR, Fritsche P, Schlaefke ME (1976) Respiratory drive and the ascending reticular activating system. Pflügers Arch 365:R39

Jessen C, Ludwig O (1971) Spinal cord and hypothalamus as core sensors of temperature of the conscious dog. II. Addition of signals. Pflügers Arch 324:202–216

Kappey F, Albers C, Schmidt R (1962) Die ventilatorische CO_2-Reaktion des Hundes während der Wärmetachypnoe. Pflügers Arch 275:312–326

Katsaros B (1965) Die Rolle der Chemoreceptoren des Carotisgebietes der narkotisierten Katze für die Antwort der Atmung auf isolierte Änderung der Wasserstoffionenkonzentration und des CO_2-Druckes des Blutes. Pflügers Arch 282:157–178

Lahiri S, Mokashi A, Delaney RG, Fishman AP (1978) Arterial PO_2 and PCO_2 stimulus threshold for cacrotid chemoreceptors and breathing. Respir Physiol 34:359–375

Lim PK, Grodins FS (1955) Control of thermal panting. Am J Physiol 180:445–449

Pleschka K (1969) Der Einfluß der Temperatur auf die elektrische Aktivität des N. phrenicus. Pflügers Arch 308:357–381

Pleschka K, Wang SC (1975) The activity of respiratory neurons before and during panting in the cat. Pflügers Arch 353:303–315

Pleschka K, Albers C, Heerd E (1965) Der Einfluß der Temperatur auf die CO_2-Schwelle des Atemsystems. Pflügers Arch 284:142–158

Ruiz AV (1973) Ventilatory response of the panting dog to hypoxia. Pflügers Arch 340:89–99

Schlaefke ME, Loeschcke HH (1967) Lokalisation eines an der Regulation von Atmung und Kreislauf beteiligten Gebietes an der ventralen Oberfläche der Medulla oblongata durch Kälteblockade. Pflügers Arch 287:201–220

Schlaefke ME, See WR, Massion WH, Loeschcke HH (1969) Die Rolle spezifischer und unspezifischer Afferenzen für den Antrieb der Atmung, untersucht durch Reizung und Blockade von Afferenzen an der decerebrierten Katze. Pflügers Arch 312:189–205

Schlaefke ME, Kille J, Folgering H, Herker A, See WR (1974) Breathing without central chemosensitivity. In: Umbach W, Koepchen HP (eds) Central rhythmic and regulation. Hippokrates, Stuttgart, p 97

Schlaefke ME, See WR, Herker-See A, Loeschcke HH (1979) Respiratory response to hypoxia and hypercapnia after elimination of central chemosensitivity. Pflügers Arch 381:241–248

See WR (1973) Die Rolle chemischer und nichtchemischer Afferenzen des N. vagus als Atmungsantriebe. Dipl Abt Biol, Ruhr Univ, Bochum

See WR (1976a) Respiratory drive in hyperthermia, interaction with central chemosensitivity. In: Loeschcke HH (ed) Acid-base homeostasis of the brain extracellular fluid and the respiratory control system. Thieme, Stuttgart, p 122

See WR (1976b) Über den Beitrag des Hypothalamus zur Atmung nach Ausschaltung der chemischen Atemantriebe. Diss Abt Biol, Ruhr Univ, Bochum

See WR, Folgering H, Schlaefke ME (1978) Central respiratory and cardiovascular effects of the ACh-releaser 4-Aminopyridine (4-AP). Pflügers Arch 377:R20

Spode R, Schlaefke ME (1975) Influence of muscular exercise on respiration after central and peripheral chemodenervation. Pflügers Arch 359:R49

Thesleff S (1980) Aminopyridines and synaptic transmission. Neuroscience 5:1413–1419

Whalen WC (1955) Some effects of hyperventilation on respiratory patterns. Am J Physiol 183: 445–450

Training of Central Chemosensitivity in Infants with Sleep Apnea

M.E. SCHLAEFKE and F. BURGHARDT[1]

Introduction

Central chemosensitivity is understood as an H^+-sensitive structure within the ventral medullary surface layer surrounding the intermediate area (area S) rostrally and caudally (Schlaefke 1981). Elimination of the intermediate area alone is sufficient to abolish the central H^+-sensitivity completely. Within this area neurones increase their spontaneous tonic discharge frequency with a decrease of pH, measured on the ventral medullary surface, and may also respond to afferent electrical stimulation of femoral or tibial nerves (Schlaefke 1976 a, b, Schlaefke et al. 1979 a). Data from in vitro measurements and pharmacological studies led to the assumption that central chemosensitivity is represented by cholinergic synapses and that input from muscle afferents or other sources increases their activity (Dev and Loeschcke 1979, Fukuda and Loeschcke 1979, Spode 1980).

The resting respiratory drive in awake cats is strongly diminished by bilateral lesions of the intermediate area. In such cats the respiratory system still responds to hypoxia and a number of unspecific stimuli. The symptoms are characterized by respiratory acidosis, periodic breathing, and long apneic phases, and thus show an analogy to the Ondine's Curse Syndrome (OCS), (Schlaefke et al. 1979 b).

In an 8 month old child having suffered from these symptoms, the arcuate nucleus on the ventral side of the medulla oblongata was missing bilaterally (Folgering et al. 1979). This led us to investigate several brains from victims of the Sudden Infant Death Syndron (SIDS) with the result that we found parts of the superficially located nucleus rostromedial of the hypoglossal root exit missing either unilaterally or bilaterally. On the base of these observations we presumed that central chemosensitivity in the human may be represented by a structure at an analogous location as in the cat, rostromedial of the hyp glossal nerve, and that in certain cases of SIDS or OCS this structure may be involved.

Using the concept of the central chemosensitive mechanism one should be able, in cas of it's insufficient function, to improve the activity by stimulation of certain afferent pat ways. This we tried in two babies of 5 weeks and 2 years, and in one girl of 15 years of ag (Schlaefke et al. 1981 a, b).

---

1 Arbeitsgruppe Physiologie der Regulation, Ruhr-Universität, D-4630 Bochum, Fed. Rep. of Germany

Central Neurone Environment
(ed. by M.E. Schläfke, H.P. Koepchen and W.R. See)

Subjects and Method

The girl whom we saw for the first time at an age of 5 weeks did not breathe during sleep and had phases of insufficient breathing when awake, from birth.

The other girl whom we met at an age of 2 years did not breathe at all from birth, but after several months had started breathing when awake. During sleep, blood gas values indicated a further increase of the existing respiratory acidosis, and the long apneic phases required artificial respiration. At an age of two years the girl was usually artificially ventilated in the first 2 h and in the last 2 h of night sleep or longer if indicated by blood gas parameters or clinical signs.

The oldest girl was introduced to us when she was 12 years old. She was healthy up to her tenth year, when she started suffering from an unknown central disease which was later diagnosed as the consequence of a gradually developing tetraplegia, caused by an anomalia of the first and second vertebra. After successful surgical treatment she recovered, however, the respiratory system remained insufficient. She was able to breathe when awake, tolerating a severe respiratory acidosis without any dyspneic sensations, but did not breathe when tired or during sleep. One year after the surgery she was sufficiently recovered to be dismissed from the hospital, but returned once a week for artificial ventilation, which made possible about 10 h of sleep. She was unable to become "refreshed" by her sleep if not artificially ventilated and was unable to reach phases of REM sleep. All three patients were provided with a tracheostoma.

In these children the respiratory response to CO_2 was analyzed during sleep. During the measurement they inhaled a gas mixture of 40% O_2 in N_2. Endtidal CO_2 partial pressure was measured with an infrared-analyzer and ventilation with a spirometer.

The children were stimulated during sleep. This was done by blowing an air jet on the face within the region of the infra-orbital nerve exit. The respiratory effect was strengthened by the combination of a first air jet with a 1.5 s insufflation of humidified CO_2 into the tracheal tube interspaced by 0.5 s. The paired stimul were applied every 10 s. In every tenth pair, the air jet was omitted. During stimulation the children inhaled 22–25% humidified oxygen.

Results

In the 5 week old girl (I) two respiratory response curves for CO_2 were recorded before the stimulation was started. All data were obtained during sleep (Fig. 1). The curves (2.5.81 and 24.3.82) indicate the dependence of the CO_2 partial pressure on the ventilatory volume. After recording the second curve (2.5.81) stimulation was started; we do not know, however, whether it was performed systematically. The mother was told to train the baby five times per day for 5–10 min. We performed longer training periods (half an hour twice a day) on 2 days per week in May and June. In June we recorded for the first time a positive CO_2-response (Fig. 1). Figure 2 shows how we tried to disconnect the baby from the respirator during sleep. The recording on top (endtidal PCO_2) shows the artificial ventilatory frequency and shallow rapid spontaneous movements following every artificially induced tidal volume. Application of an air jet on the

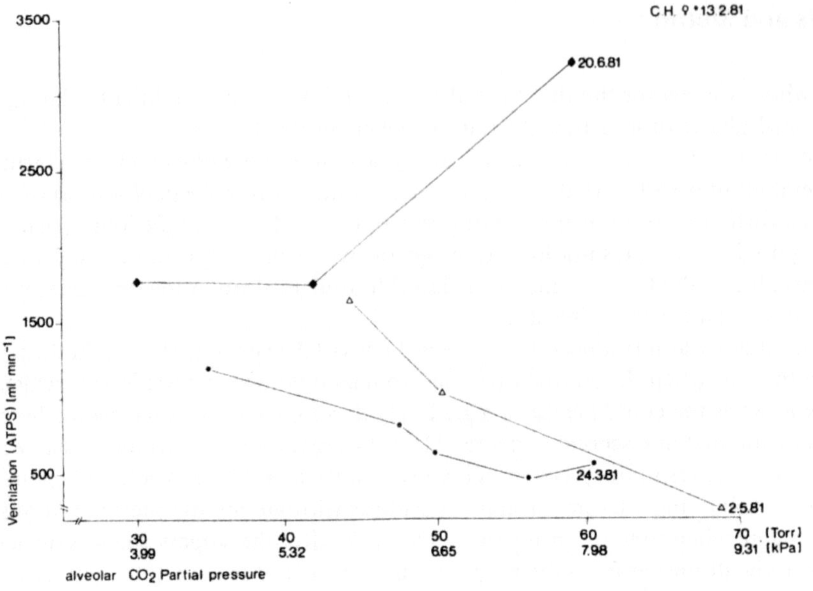

Fig. 1. Ventilatory response of patient I with increasing alveolar PCO_2 during inhalation of 40% O_2 in N_2 at different ages while sleeping. Training with combined stimuli started sporadically in the beginning of May and was extended in the middle of May. In June the baby was able to breathe spontaneously during sleep for 4 h without interruption

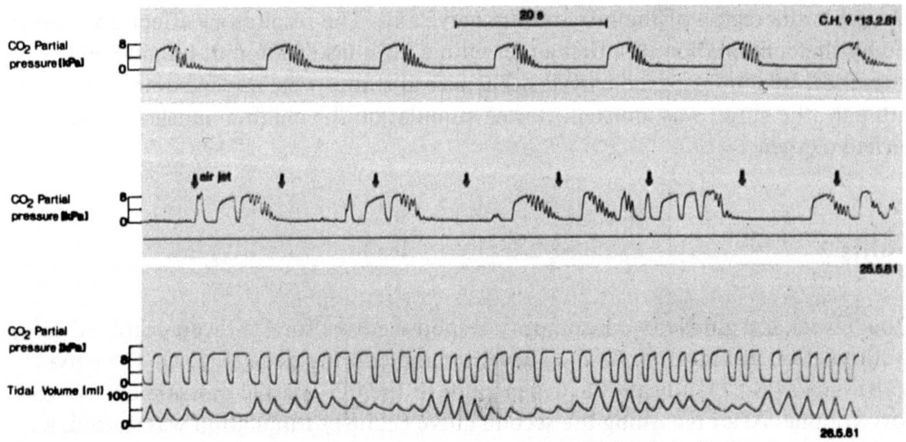

Fig. 2. Procedure to disconnect patient I from the respirator during sleep when not breathing spontaneously. *Top* endtidal PCO_2 during artificial ventilation; *middle* respiratory effect of short air jet stimuli (*arrows*) on to the region of the infraorbital nerve exit during artificial ventilation (same frequency as top recording). *Bottom* spontaneous breathing during sleep without stimulation (recorded after extended training sessions)

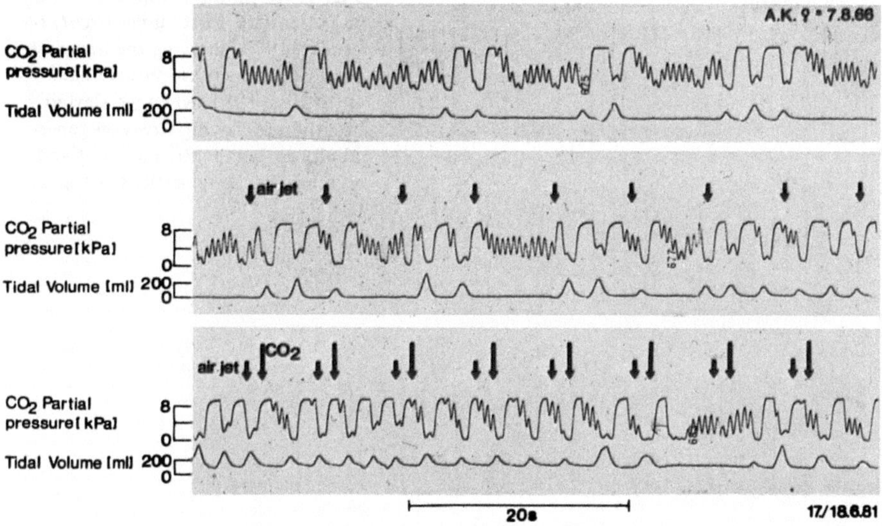

Fig. 3. Patient III. Spontaneous breathing (endtidal PCO_2, tidal volume) during sleep (*top*), the effect of the air jet stimuli (*middle*), and combined stimulation (*small arrows* air jet; *large arrows* CO_2)

nose-mouth angle caused spontaneous respiratory movements within the expiratory pauses given by the respirator (middle). Similar effects on spontaneous breathing were obtained in the 2 year old girl (II) and also in the 15 year olf girl (III), (Fig. 3). The respiratory effects were intensified or stabilized after various times (e.g. 20 min) of the combined stimulation (air jet/CO_2), (Fig. 3: bottom). The bottom recording of Fig. 2 was taken after an intensive training from patient I and presents the capability to breathe during sleep, although still periodically. After 14 days and nights of training in August the baby was able to breathe continuously during sleep. However, by interruption of the training and after an attack of unknown origin the baby "unlearned" the ability to keep her blood gas values within a tolerable range during sleep. Three months later the training was started again with the additional measure to allow the baby to breathe spontaneously as long as her alveolar PCO_2 did not exceed 51 mmHg during sleep. She is now again able to breathe spontaneously for 3 h without interruption and without stimulation during sleep. Her blood gas values under these conditions are PCO_2:42 mmHg and PO_2:100 mmHg. Today we must leave it open whether another 14 days of intensive training in the described way may enable the child again to breathe spontaneously through the whole night. At present however, the baby is able to do so as long as the combined stimulation is applied during hypoventilatory phases and phases of rapid shallow breathing.

From the oldest girl (III) we learned that she considered herself undisturbed by the stimulation and subjectively felt refreshed after sleeping with combined stimulation. She started to sleep more deeply and showed no interruption of her night sleep as she had done before when not artificially ventilated. Her respiratory response to CO_2 had been improved, and she did not ask for the respirator for the 12 days in which she was

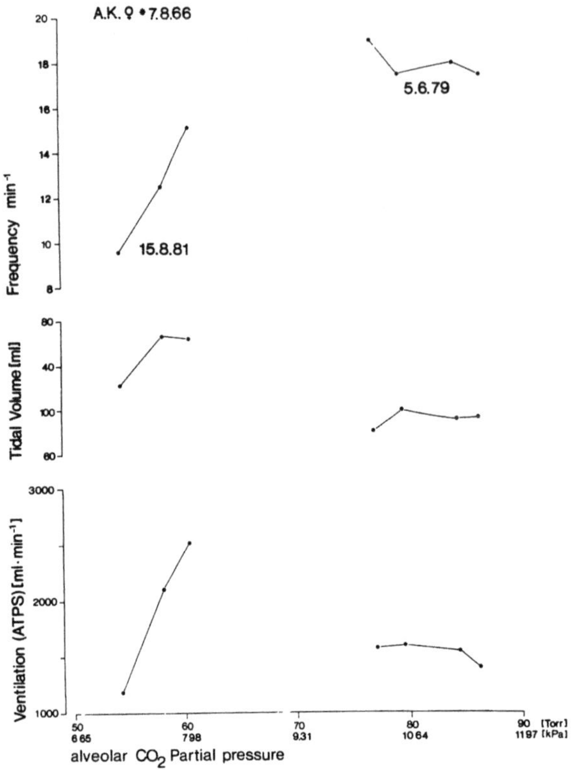

Fig. 4. Respiratory response to CO_2 in patient III. First curves (*right*) recorded 1979, showing the high resting PCO_2 which had not changed until June 1981, when we started the therapy, laying stress on general physiotherapeutic measures and gymnastics. Same patient as Fig. 3

trained (Fig. 4), whereas before, she had required artificial ventilation every third or fourth night.

The application of the combined stimulation was supported by an intensive respiratory physiotherapy including gymnastics, massages, vibration, percussion, and cold applications. The strengthening, of the muscle, hereby induced may have its own positive influence on the respiratory capability to respond to CO_2 as observed temporarily in patient III, 3 months after intensive physiotherapy only. Now, the baby, not having had the combined stimulus programme, is unable to regulate her acid-base parameters by ventilation during sleep and is dependent upon the respirator. The combined training will be performed within the nearest future and so will possibly allow the discrimination of the effects of physiotherapy and the training with combined stimulation.

First approaches in animal experimental studies now seem to support the hypothesis that the insufficient function of central chemosensitivity can be improved by afferent stimulation in the described way.

Discussion

Insufficient respiration during sleep could be improved by using a combined stimulation method consisting of an unspecific (air jet) and a specific (CO_2) stimulus, which allowed training of the respiratory muscles during sleep within a tolerable alveolar PCO_2 range. Since the hitherto flat respiratory response to CO_2 became positive with the application of this method a training mechanism could apply to central chemosensitivity also, in a sense that stimulation in the described manner has a facilitatory effect on or induces the homeostatic function. Whether or not the CO_2-sensitivity of the peripheral chemoreceptors does participate in such a facilitatory effect must remain open. In our subjects their sensitivity to hypoxia was principally intact. From the cat model we know that intact peripheral chemoreceptors guarantee ventilation after elimination of central chemosensitivity, however, their contribution to the respiratory response to CO_2 is minimal under these conditions.

Which mechanism could account for the facilitatory effect? Based on data of former studies on central chemosensitivity we postulate that the adequate modulating function of H^+-ions on the transmitter regulation within the central chemosensitive synapses needs a sufficient amount of basic activity (Dermietzel 1976, Fukuda and Loeschcke 1979, Fukuda et al.1978; Schlaefke 1981). The convergence of various afferent pathways within the chemosensitive network, forming synapses with neurones responsible for the processing of respiratory chemosensitivity (e.g. within the paragigantocellular nucleus) could provide favourable conditions for a certain plasticity of the central chemosensitive mechanism (Bloom 1974, Eccles 1964, Schlaefke 1973, 1981, Schlaefke et al 1969, 1974, 1979a,b). In newborns with immature synapses within this area, the air jet stimulation may elevate the activity of functioning synapses and induce activity of hitherto silent synapses. Under the condition that during sleep the synaptic activation is below the threshold necessary for H^+-ions to be effective, the training of synaptic activity during sleep could be the decisive step for the stabilization of respiration. The facilitation may also comprise other systems. For instance the sleep control system could adapt its own threshold, enabling the baby to stay asleep inspite of breathing.

In the case of complete loss or nonexistence of the central chemosensitive neurones as a possible cause for SIDS or OCS, or as in our animal model in which area S is coagulated bilaterally, we have to consider that CO_2-sensitivity must be taken over by another anatomical substrate . In that case reticular or other neurones influencing or belonging to the respiratory system, and primarily unable to respond to CO_2, would replace the central chemosensitive mechanism (Marino and Lamb 1975, Fukuda and Loeschcke 1977). First data from our own laboratory indicate that such a possibility exists.

In our study on humans and in the animal experimental approach, the training effect on central chemosensitivity was more prominent and stable if we applied the air jet not alone but in combination with the CO_2 stimulus. This makes it unlikely that a more general activation through the air jet accounts for the observed positive development of the CO_2-response. The CO_2 stimulus may initiate respiratory drive at the movement it finds the site of it's action at threshold level. Further, due to the breath elicited by the air jet, conditions for the production of synchronized respiratory activity by CO_2 are favourable (Cohen 1979).

The stimulation pattern resembles the concept of classical conditioning. In this context we have to ask for the correct sequence of the conditioned and the unconditioned stimulus, since classical conditioning fails in the case of conditioned stimulus following the unconditioned one (Pavlov 1960). Whether CO_2 is the one or the other seems equivocal. In the patients with no chemosensitivity at all, CO_2 should act as the conditioned stimulus. Since CO_2 did inhibit ventilation in the beginning, however, it seemed of no use to us to apply it as the first stimulus because it would not reach the site of action in the non-breathing patient. Inducing breathing by the air jet stimulus however, made it possible for CO_2 to be inhaled. In the case of a rudimentary CO_2-sensitivity, CO_2 could also play the role of the unconditioned stimulus. Whether or not the model of classical conditioning or backward conditioning is applicable to the procedure by which central chemosensitivity "learned" its function has to await further studies in animal models (Spetch et al. 1981).

Being aware of the as yet small number of experimental data the question of how to overcome a lack of central chemosensitive drive in connection with OCS and SIDS seems to us nevertheless to be worth a discussion in the light of the reported observations.

Acknowledgements. We wish to thank Drs. G. Metz and B. Nebel, Abteilung für Anaesthesie und Intensivmedizin, Kreiskrankenhaus Emmendingen/Freiburg, for continuous encouragement in the clinical work and for their personal engagement in the treatment of patient III. Supported by the Deutsche Forschungsgemeinschaft (SFB 114 Bionach).

References

Bloom FE (1974) Dynamics of synaptic modulation. Introduction. In: The neurosciences vol III. Cambridge (Mass) London, p 904

Cohen MI (1979) Neurogenesis of respiratory rhythm in the mammal. Physiol Rev 59:1105–1173

Dermietzel R (1976) Central chemosensitivity, morphological studies. In: Loeschcke HH (ed) Acid-base homeostasis of the brain extracellular fluid and the respiratory control system. Thieme, Stuttgart, p 52

Dev NB, Loeschcke HH (1979) A cholinergic mechanism involved in the respiratory chemosensitivity. Pflügers Arch 379:29–36

Eccles JC (1964) The Physiology of synapses. Springer, Berlin Heidelberg New York

Folgering H, Kuyper F, Kille JF (1979) Primary alveolar hypoventilation (Ondine's curse syndrome) in an infant without external arcuate nucleus. Case Report. Bull Physio-pathol Respir 12/4: 659–666

Fukuda Y, Loeschcke HH (1977) Effect of H^+ on spontaneous neuronal activity in the surface layer of the rat medulla oblongata in vitro. Pflügers Arch 371:125–134

Fukuda Y, Loeschcke HH (1979) A cholinergic mechanism involved in the neuronal excitation by H^+ in the respiratory chemosensitive structures of the ventral medulla oblongata of rats in vitro. Pflügers Arch 379:125–135

Fukuda Y, Honda Y, Schlaefke ME, Loeschcke HH (1978) Effect of H^+ on the membrane potential of silent cells in the ventral and dorsal surface layers of the rat medulla in vitro. Pflügers Arch 376:229–235

Marino PL, Lamb TW (1975) Effects of CO_2 and extracellular H^+ iontophoresis on single cell activity in the cat brainstem. J Appl Physiol 38/4:688–695

Pavlov IP (1960) Conditioned reflexes: An investigation of the physiological activity of the cerebral cortex. In: Anrep GV (ed and transl). Dover, New York. (first publ 1927)

Schlaefke ME (1973) 'Specific' and 'nonspecific' stimuli in the drive of respiration. Acta Neurobiol Exp 33:149–154

Schlaefke ME (1976a) Central chemosensitivity: Neurophysiology and contribution to regulation. In: Loeschcke HH (ed) Acid-base homeostasis of the brain extracellular fluid and the respiratory control system. Thieme, Stuttgart, p 66

Schlaefke ME (1976b) Central chemosensitivity.Verh Ges Lungen Atmungsforsch 6:14–20

Schlaefke ME (1981) Central chemosensitivity: A respiratory drive. Rev Physiol Biochem Pharmacol 90:171–244

Schlaefke ME, See WR, Massion WH, Loeschcke HH (1969) Die Rolle 'spezifischer' und 'unspezifischer' Afferenzen für den Antrieb der Atmung, untersucht durch Reizung und Blockade von Afferenzen an der decerebrierten Katze. Pflügers Arch 312:189–205

Schlaefke ME, Kille JF, Folgering H, Herker A, See WR (1974) Breathing without central chemosensitivity. In: Umbach W, Koepchen HP (eds) Central rhythmic and regulation. Hippokrates, Stuttgart, p 97

Schlaefke ME, See WR, Kille JF (1979a) Origin and afferent modification of respiratory drive from ventral medullary areas. Wenner-Gren Cent Int Symp Ser 32:25–34

Schlaefke ME, Kille JF, Loeschcke HH (1979b) Elimination of central chemosensitivity by coagulation of a bilateral area on the ventral medullary surface in awake cats. Pflügers Arch 379: 231–241

Schlaefke ME, Hukuhara T Jr, See WR (1981a) Loss of central chemosensitivity, experimental studies on a clinical problem. In: Hutas I, Debreczeni LA (eds) Advances in physiol Sci, vol 10. Respiration. Pergamon Press, New York, pp 609–616

Schlaefke ME, Burghardt F, Mückenhoff K, Heimann J (1981b) Technisch unterstütztes Atemtraining bei Säuglingen mit mangelnder CO_2-Empfindlichkeit der Atmung. Biomed Techn, vol 26 (Ergänzungsbd 1981)

Spetch ML, Wilkie DM, Pinel JPJ (1981) Backward conditioning: A reevaluation of the empirical evidence. Psychol Bull 89:163–175

Spode R (1980) Ausschaltung der zentralen und peripheren chemosensiblen Atemantriebe bei der anaesthesierten Katze während Muskelarbeit. Med Diss, Ruhr Univ, Bochum

Adrenergic, Cholinergic, and Peripheral Chemoreceptor Effects on Thermal Panting in Cats

H. FOLGERING[1]

In several species ventilation not only serves acid-base homeostasis but also thermal homeostasis. Thermal panting is a quick and efficient way to dissipate heat. The central nervous system is specially protected against acid-base disturbances, which are sensed immediately by the central chemoreceptors, and against overheating because the carotid blood is cooled in the rete mirabile before entering the brain. Thermal panting almost invariably leads to respiratory alkalosis. On the other hand, the alkalosis is a prerequisite for panting. Adding CO_2 in the inspiratory air diminishes panting. Apparently during heat stress, the acid-base homeostasis is abandoned in favor of thermostatic mechanisms (Richards 1970, Pleschka 1969) . The functional pathways that mediate these conflicting interests of thermoregulation and acid-base control are the inputs from the hypothalamus and from the central and peripheral cheoreceptors, respectively, into the respiratory neuronal organization. Schlaefke (1973) has shown that when all chemo-afferents to the "respiratory centers" are eliminated, ventilation stops. Only afferent activity from the hypothalamus is able to maintain a rhythmic output from the respiratory neuronal organization. Thermal and electrical stimulation of the hypothalamus induces panting in cats which are completely chemodenervated (See, 1976). We have shown that cooling of the intermediate chemosensitive area at the ventral medullary surface in panting animals considerably enhances this panting, i.e., the respiratory frequency increases and the tidal volume decreases. After chemodenervation we found no effect of the P_{A,CO_2} level on panting (See et al., this symp).

The purpose of this study was to investigate what kind of neurotransmitter (cholinergic or noradrenergic) might be involved in the process of panting, and what is the role of the peripheral chemoreceptors.

Methods

Fourteen cats of either sex, weighing 2.4–3.6 kg, were used in these experiments. After premedication with ketamine HCl (25 mg) they were anesthetized with chloralose-urethane (50/200 mg/kg), paralyzed with pancuronium bromide, tracheostomized and artificially ventilated. Catheters were placed in the femoral artery and vein. One of the

1 Physiology, University of Nijmegen, P.O.Box 9101, 6500 HB Nijmegen, The Netherlands

Central Neurone Environment
(ed. by M.E. Schläfke, H.P. Koepchen and W.R. See)

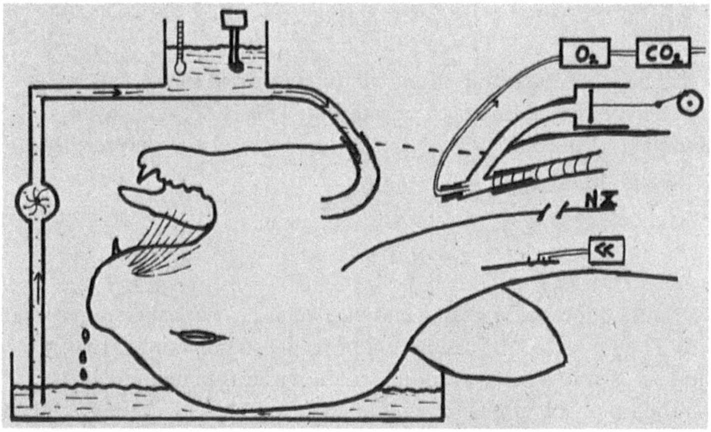

Fig. 1. The experimental arrangement. The cat is in supine position. The rostral part of the esophagus is cannulated and flushed with warm saline. The saline flows freely from the mouth and nose. The vagal nerves are cut. Electric activity of the phrenic nerve is recorded. The trachea is cannulated for artifical ventilation

phrenic nerves was dissected and prepared for recording purposes; the phrenic nerve activity was quantified according to the method described by Smolders et al. (1975). "Neural minute ventilation" (V_N) was calculated by multiplying the quantified burst activity or neural tidal volume by the burst frequency. The criteria for panting were a decrescendo pattern of the phrenic nerve discharge (Hilaire and Manteau, 1974) and a burst frequency of approximately 80/min^{-1}.

Because of the pharmacological aspects of this study, it was not feasible to use the transmitter releaser 4-Aminopyridine to induce panting as we did in See's study (See, this symp). Panting was induced by whole body heating to 39–40°C and by flushing the oro- and nasopharynx with warm saline (43°C) via a cannula inserted in the esophagus rostrally of the tracheostomy. The warm saline flowed freely out of the mouth and nose (Fig. 1). In this way the carotid rete mirabile was heated and consequently the brain rostrally from the mesencephalon. The vagal nerves were cut in order to prevent respirator effects on the phrenic nerve discharge via Breuer and Hering reflexes. The carotid sinus nerves were dissected and a loose ligature was placed around them for later sectioning.

In the panting cats, CO_2 and O_2 response curves were taken with intact and severed carotid sinus nerves.

The effect of peripheral chemoreceptor stimulation on panting was studied by intravenous application of doxapram HCl (1–2 mg). The effect of elimination of the peripheral chemoreceptors was studied by cutting the carotid sinus nerves.

The following drugs were given intravenously during panting before and after severing the sinus nerves: acetylcholine 1-10 μg, atropine 0.5 mg, noradrenaline 1–10 μg, phentolamine 5–10 μg, isoprenaline 10–100 μg and metoprolol 2–5 mg.

For statistics the unpaired t-test was used because it was not possible to do all tests on all animals. Values of p of 0.05 or less were considered to be significant.

Results

The effects of various maneuvers on panting often will be described in the following sections in terms of enhancement or depression of panting. This is defined as an increase respecitvely decrease of phrenic nerve burst frequency and a concomittant decrease respectively increase in neural tidal volume.

CO_2-response Curves

Increasing P_{A,CO_2} invariably depressed panting and vice versa. A typical CO_2-response curve is shown in Fig. 2. The net effect of changes in burst frequency and amplitude did not always result in such a smooth V_N response as shown in this figure but the tendencies were always the same. Peripheral chemodenervation caused an upward shift in the CO_2-response curves.

O_2-response Curves

Hypoxia depressed panting in all cats. After peripheral chemodenervation there was no qualitative or quantitative effect of hypoxia on the pattern of panting (Fig. 3).

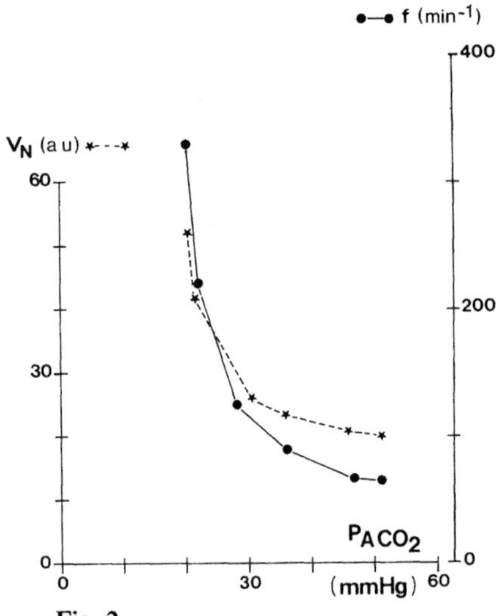

Fig. 2

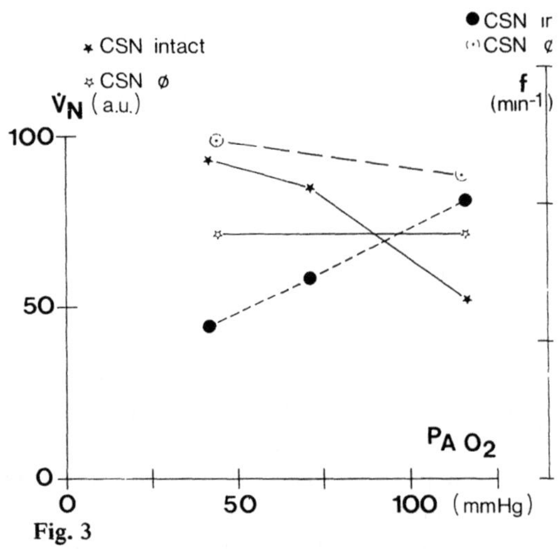

Fig. 3

Fig. 2. CO_2-response curve of a panting cat. *f* burst frequency of the phrenic nerve activity; V_N neu equivalent of minute ventilation, amplitude of the quantified burst activity multiplied by burst freq cy; *a.u.* arbitrary units. Panting is enhanced by lowering the P_{A,CO_2}; this is a positive feedback. *Asterisks* V_N, *circles* f

Fig. 3. O_2-response curves in a panting cat with intact and servered Carotid Sinus Nerves. Hypoxia depresses panting in the intact animals. Symbols as in Fig. 2. *Filled symbols* carotid sinus nerve intac *open symbols* carotid sinus nerves cut (CSN Ø)

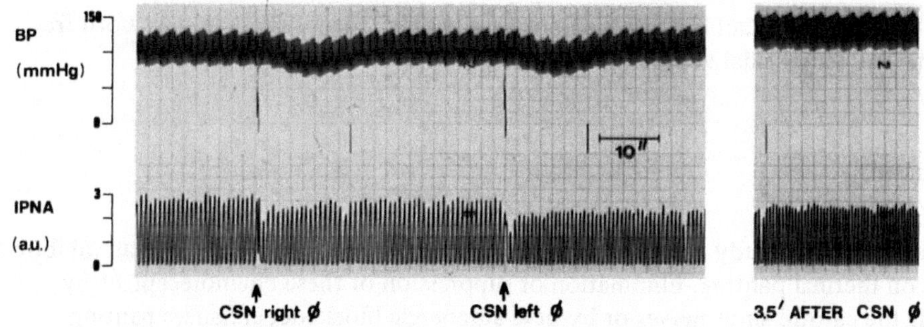

Fig. 4. Recording of the effects of severing the Carotid Sinus Nerves. *BP* systemic blood pressure; *IPNA* Integrated Phrenic Nerve Activity. Cutting the sinus nerves (CSN ∅) enhances panting

Stimulation and Elimination of Peripheral Chemoreceptors

Intravenous injections of Doxapram HCl depressed panting when the carotid sinus nerves were intact (Table I). After cutting the sinus nerves the changes in frequency and neural tidal volume were significantly less.

The severing of the carotid sinus nerve enhanced panting consistently the burst frequency increased by 95 ± 36% and neural tidal volume decreased by 29.6 ± 16.3% (Fig. 4).

Drug Effects (are shown in Table 1)

Acetylcholine depressed panting in intact and in peripherally chemodenervated cats. Atropine, noradrenaline and phentolamine never had any significant effect on panting. Isoprenaline depressed panting in intact and chemodenervated cats. Metoprolol en-

Table 1. Relative changes in panting parameters between control situation immediately before administration of drugs and at the moment of maximal effect after the administration

Drug	Δ Frequency (%) CSN intact	CSN cut	Sign (p)	$\Delta V_{T,N}$ (%) CSN intact	CSN cut	Sign (p)
Doxapram HCl	- 34.8 ± 18.8	- 5.0 ± 3.6	0.05	+ 77.5 ± 54.8	+ 11.5 ± 4.7	0.07
Atropine	- 1.3 ± 10.6	- 1.5 ± 2.5	ns	- 18.0 ± 12.7	+ 4.5 ± 0.5	ns
Acetylcholine	- 28.3 ± 10.4	- 6.8 ± 4.9	0.01	+ 35.0 ± 9.7	+ 17.2 ± 18.6	0.05
Noradrenaline	- 4.4 ± 5.8	- 6.4 ± 3.6	ns	- 7.2 ± 8.2	+ 1.2 ± 7.9	ns
Phentolamine	+ 25.3 ± 40.8	- 9.6 ± 10.2	ns	- 17.6 ± 32.8	+ 10.5 ± 16.3	ns
Isoprenaline	- 39.1 ± 17.4	- 19.0 ± 11.7	0.05	+ 118.0 ± 98.0	+ 20.8 ± 31.4	0.07
Metoprolol	+ 72.9 ± 52.0	- 17.8 ± 19.3	0.01	- 19.3 ± 28.0	- 14.8 ± 6.5	ns

"Δ frequency" indicates relative changes in phrenic nerve burst frequency.
"$\Delta V_{T,N}$" is relative change in Integrated Phrenic Nerve Activity being the equivalent of tidal volume. Doses of the drugs are given in the methods section. "Sign" indicates the level of significance (p value) of the difference between the effects with intact and with severed carotid sinus nerves (unpaired *t*-test)

hanced panting in intact cats; in the peripherally chemodenervated animals both frequency and neural tidal volume were depressed.

Discussion

The results of this study show that the peripheral chemoreceptor input has an inhibitory effect on thermal panting. Elimination or suppression of these chemoreceptors by cutting the carotid sinus nerves or by beta-adrenergic blockade, enhances panting. Stimulation by hypoxia, hypercapnia, doxapram HCl, acetylcholine, isoprenaline diminishes panting. These results confirm earlier findings of Ruiz (1973) in dogs, where hypoxia also depressed panting, whereas peripheral chemoreceptor stimulation by doxapram did not (Ruiz 1975). The cholinergic effects on panting are partially mediated by the peripheral chemoreceptors. However, after peripheral chemodenervation there still remains an effect of acetylcholine which must be central. This central effect may be mediated by the central chemoreceptors which are most probably cholinergic (Dev and Loeschcke 1979, or by a direct effect on the hypothalamus. Cholinergic mechanisms in the thermoregulatory centers in the hypothalamus have been described by Feldberg et al. (1966). The fact that atropine has no effect, though it easily penetrates the blood brain barrier, might suggest nicotinergic receptor effects.

The alpha-adrenergic drugs noradrenaline and phentolamine have no effect on panting. The absence of any effect after peripheral chemodenervation should be interpreted carefully. These drugs do not very easily penetrate the blood-brain barrier. Further experiments are necessary.

Beta-adrenergic drugs isoprenaline and metoprolol affect the activity of the peripheral chemoreceptors (Folgering et al. 1980). The effects in the cats with intact sinus nerves should be interpreted in this perspective. The beta-adrenergic effects in peripherally chemodenervated cats are expressions of central effects either on the "respiratory centers", on hypothalamic structures or on connections between the two. In an earlier study we have shown that beta-adrenergic drugs affect the respiratory center activity (Folgering 1980).

In conclusion, thermal panting is attenuated by peripheral chemoreceptor input and by cholinergic and beta-adrenergic mimetic drugs. The drug effects are established by peripheral chemoreceptor as well as central mechanisms. Panting is not affected by alpha-adrenergic drugs.

References

Dev NB, Loeschcke HH (1979) A cholinergic mechanism involved in the respiratory chemosensitivity of the medulla oblongata in the cat. Pflügers Arch 379:29–36

Feldberg W, Hellon RF, Meyers RD (1966) Effects on temperature of monoamines injected into the cerebral ventricles of anesthetized dogs. J Physiol (London) 186:416–423

Folgering H (1980) Central beta-adrenergic effects on the control of ventilation in cats. Respiration 39:131–138

Folgering H, Ponte J, Purves M (1980) Beta-adrenergic receptors and carotid body chemoreception. J Physiol (London) 303:28P

Hilaire H, Monteau R (1974) Activité des motoneurones phréniques au cours de la polypnée thermique ou hypocapnique, J Physiol (Paris) 68:193–203

Pleschka K (1969) Der Einfluß der Temperatur auf die elektrische Aktivität des Nervus Phrenicus. Pflügers Arch 308:357–381

Richards SA (1970) The biology and comparitive physiology of thermal panting. Biol Rev 45: 223–264

Ruiz AV (1973) Ventilatory response of the panting dog to hypoxia. Pflügers Arch 340:89–99

Ruiz AV (1975) Effects of cyanide and doxapram during panting. Pflügers Arch 361:79–81

Schlaefke ME (1973) Specific and non-specific stimuli in the drive of respiration. Acta Neurobiol Exp 33:149–154

See WR (1976) Über den Beitrag des Hypothalamus zur Atmung nach Ausschaltung der chemischen Atemantriebe. Doct thesis, Ruhr-Univ, Bochum

Hypoxic Modulation of Central Chemosensitivity

H. KIWULL-SCHÖNE and P. KIWULL[1]

Introduction

Since the observations of Loeschcke et al. (1958) there is growing evidence that ventilation is closely related to the pH in the extracellular fluid (ECF) at the ventral surface of the medulla oblongata. Ahmad et al. (1976) introduced a non-invasive technique for measuring the pH of the ECF directly. They have shown that either respiratory or metabolic acid-base changes in the blood compartment are sufficiently quick and reliably reflected by pH_{ECF}. Furthermore, there seems to be a close linear correlation between the changes in tidal volume and changes in pH_{ECF} following a step rise in P_ACO_2. This is valid not only for the steady state (Kiwull-Schöne and Kiwull 1981) but also for the dynamic responses of both variables (Ahmad and Loeschcke 1982), as long as the peripheral chemoreflexes do not interfere (Kiwull-Schöne and Kiwull 1980). More recently, Shams et al. (1981) have shown that within certain limits the same statement is valid for metabolic acidosis occurring in the blood as a result of intravenous infusions of mineral acids.

Hypoxia induced lactic acid production, due to anaerobic glycolysis, may be regarded as a naturally occuring metabolic acidosis. Two questions arose: First would this kind of metabolic acidosis be a stimulus to central chemosensitivity, and second, if so, to which extent would metabolic acidosis be compensated by increased pulymonary ventilation and hence decreased P_{CO_2}. The present study was performed in peripherally chemodenervated animals in order to eliminate the possible influence of the carotid and aortic chemoreceptors.

Methods

The experiments were carried out in anaesthetized rabbits (n = 9) and cats (n = 14). After an initial intravenous dose, a nearly constant level of general anaesthesia was maintained by continuous infusion of sodium pentobarbital (for details see

---

1 Institut für Physiologie, Ruhr-Universität Bochum, D-4630 Bochum, Fed. Rep. of Germany

Central Neurone Environment
(ed. by M.E. Schläfke, H.P. Koepchen and W.R. See)

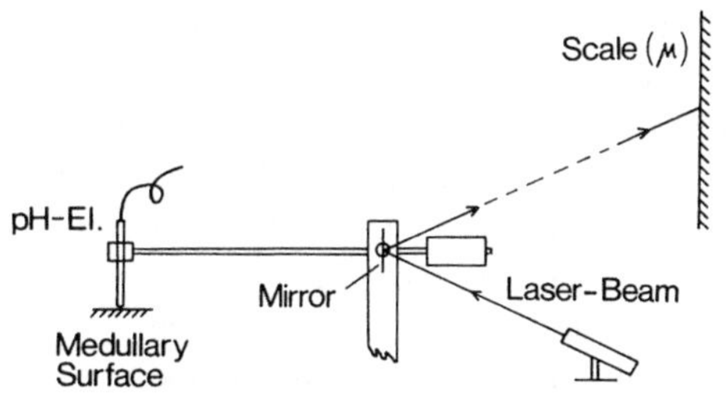

Fig. 1. Schematic representation of the method indicating the pH electrode position on the medullary surface with high accuracy. This is achieved by a Laser beam reflected by a mirror on the axis of the electrode balance over a long (5 m) distance to a scale

Kiwull-Schöne and Kiwull 1979, 1981). Blood coagulation was prevented by Heparin-Riker. Ventilatory variables ($\dot{V}$, V_T, f) were determined by pneumotachography in an open circuit, end-tidal CO_2 pressure (P_ACO_2) by infrared absorption, and arterial O_2 pressure (PaO_2) by polarography. The PaO_2 was monitored continuously in the rabbits (Kiwull-Schöne and Kiwull 1979), whereas in the cats it was determined from blood samples taken under ventilatory steady state conditions.

The arterial pH (pHa), CO_2 pressure ($PaCO_2$), and the actual and standard bicarbonate concentrations (HCO_{3a}^-, HCO_{3st}^-) were measured by the Astrup technique. Under hypoxic conditions the pH_a values were corrected according to the Hb concentration and O_2-Hb saturation (Kiwull-Schöne and Kiwull 1981). Lactic acid concentrations were analysed enzymatically (Boehringer Biochemica Test Combination) from blood and from CSF samples.

The medullary surface pH (pH_m or pH_{ECF}) in the cats was measured by a balanced pH-electrode, following the technique of Ahmad et al. (1976). However, we introduced special refinements that should help to ascertain that the measured pH value in fact represents that of the extracellular fluid (ECF). First, microcirculation must remain unaffected. Second, the electrode must be placed very close to the brain surface and remain in the same position throughout the experiment. A virtual minimal weight of the electrode was easily achieved by adjusting the balance to a horizontal position. The near touch to the brain surface was achieved by using a micromanipulator under visual microscopic control. In order to get continuous information about the position of the electrode, we developed an additional optical control system as shown in Fig. 1. It consists of a LASER-beam reflected by a mirror at the axis of the balance. The beam was focussed over such a long distance that it was possible to detect deviations of the electrode (e.g., by CSF production) within the 10 μ range.

Several variables were averaged for the entire rabbit and cat populations. These data are expressed in the text as means with standard error (± S.E.M.). A significant difference refers to P_d ($\leqslant 0.05$, if calculated by a paired t-test).

Results

Ventilatory Reaction to Arterial Metabolic Acidosis

If in carotid-chemodenervated rabbits, the arterial PO_2 was decreased stepwise (each step lasting 10 min) from hyperoxic to severe hypoxic values, a strong acidosis developed in the arterial blood. The pH_a values decreased from 7.398 ± 0.016 at P_aO_2 = 80 kPa to 7.225 ± 0.029 at P_aO_2 = 3.3 kPa. The hypoxic value of P_aCO_2 did not significantly differ from 4.43 ± 0.16 kPa during hyperoxia. The concomitant increase of the lactate concentration from 3.9 ± 0.9 to 9.1 ± 0.7 mmol · l^{-1} and decrease of the standard HCO_3^- concentration from 21.2 ± 0.9 to 14.8 ± 1.0 mmol · l^{-1} points to the fact that the observed acidosis was entirely of metabolic origin.

In spite of the profound acid-base disturbances, ventilation did not significantly change from the hyperoxic value of 344.9 ± 39.5 ml · min^{-1} · kg^{-1} to the hypoxic value of 341.9 ± 41.8 ml · min^{-1} · kg^{-1}. Tidal volume was even significantly depressed. Thus, there is a complete lack of ventilatory reaction to metabolic acidosis in the arterial blood.

Ventilatory Reaction to Medullary Metabolic Acidosis

The typical course of the experiments in peripherally chemodenervated cats is shown in Fig. 2. The cat was breathing alternatively hyperoxic and hypoxic gas mixtures each for about half an hour. At each steady state level of arterial oxygen pressure (hyperoxia 25 kPa, hypoxia between 8.7 and 9.5 kPa), the endtidal P_{CO_2} was increased to one hypercapnic value of about 2 kPa above normocapnia. The records were obtained during the 7th minute of each CO_2 condition. Comparing the first and the last steady state record, both during normocapnic hyperoxia, there was a considerable fall in the medullary surface pH. The acidotic shift of about 0.08 pH units definitely did not cause any changes in either tidal volume or end-tidal P_{CO_2}.

The evaluation of another experiment yields similar results, as shown in Fig. 3. During alternatively changing hyperoxic and hypoxic periods, the normocapnic steady state values of ventilation, endtidal P_{CO_2}, arterial pH and medullary

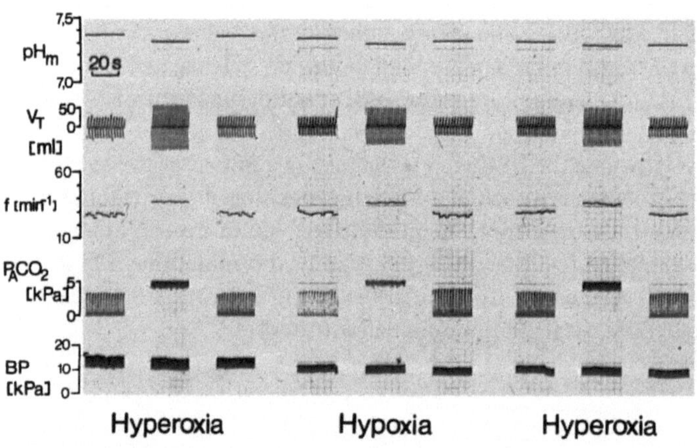

Fig. 2. Steady state recordings of medullary surface pH, tidal volume, respiratory frequency, end-tidal P_{ACO_2} and arterial blood pressure in a spontaneously breathing, peripherally chemodenervated cat (2.3 kg). During each period of hyperoxia (P_aO_2 = 8.4–9.1 kPa) a normocapnic, hypercapnic and again normocapnic condition is shown. Note the continual acidotic shift of the normocapnic pH_{ECF} values without considerable changes in V_T

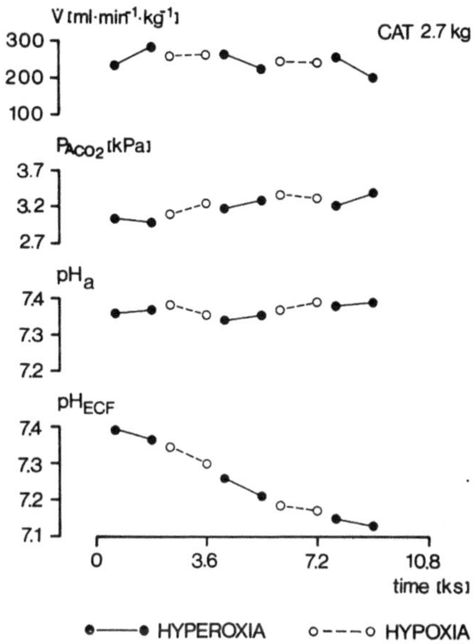

Fig. 3. Steady state values of ventilation, endtidal P_{CO_2}, arterial pH and pH in the ECF of the brain in a peripherally chemodervated cat, plotted against the time from the beginning of the experiment. Each pair of values represents the normocapnic condition before and after hypercapnia, either during hyperoxia (P_aO_2 = 23.2–24.1 kPa) or hypoxia (P_aO_2 = 9.6–11.5 kPa)

surface pH, are plotted against time from the start of the experiment. If the first and the last hyperoxic runs are compared, $\dot{V}$ fell slightly by about 17%, correspondingly, the P_ACO_2 increased by 0.36 kPa. However, since the arterial P_{CO_2} (P_aCO_2) decreased rather than increased, a shift of pH_a towards alkalinity could be observed (+ 0.028 units), indicative of a diminished arterio-alveolar P_{CO_2} difference. In this case, no signs of metabolic acidosis could be detected in the arterial blood, since the standard HCO_3^- concentration even slightly increased. There was however a continual fall in pH_{ECF} throughout the experiment, maximal during the first hypoxic and during the second hyperoxic period. Interestingly, this occurred again without any driving effect upon ventilation.

Comparison of Ventilatory Reactions to Respiratory and Metabolic Medullary Acidosis

In several respects the examples shown in Figs. 2 and 3 are typical for the whole population (Fig. 4). The data were achieved under comparable normocapnic hyperoxic blood gas conditions and averaged in 14 cats. The pH_{ECF} values fell from 7.336 ± 0.017 at the start, to 7.076 ± 0.043 at the end of the experiments. Concomitantly, no significant change in ventilation was observed: it even decreased (from 331.0 ± 20.1 to 305.8 ± 26.6 ml · min^{-1} · kg^{-1}) rather than increased.

The most striking acidotic shift in the medullary pH occurred in the course of the hypoxic periods, as can be seen in Fig. 4. Within the same time interval of about 30 min, the average fall of pH_{ECF} was 0.050 units during hypoxia but only 0.015 units during hyperoxia. Since during hyperoxia the P_aCO_2 after the hypercapnic run was slightly higher than before (+ 0.23 kPa), only one third of the acidotic shift can be considered to be of metabolic origin. On the other hand, P_aCO_2 before and after hypercapnia remained practically the same during the hypoxic periods, so that

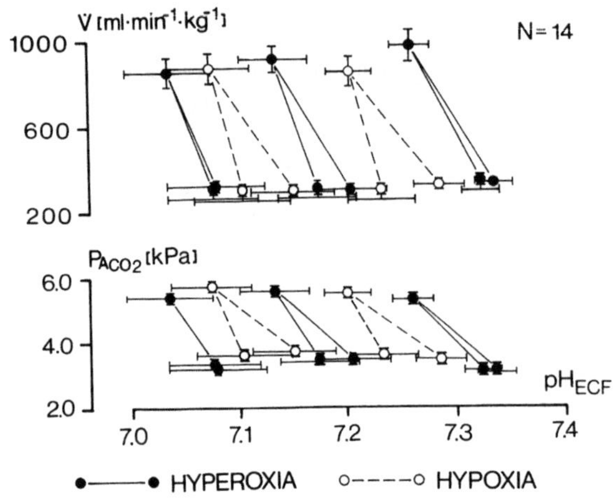

Fig. 4. Steady state responses of ventilation and entidal P_{CO_2} to application or removal of hypercapnia as functions of the medullary surface pH. Mean values (± S.E.M.) of spontaneously breathing, peripherally chemodenervated cats, alternatively during hyperoxia (P_aO_2 = 24.0 ± kPa) and hypoxia (P_aO_2 = 9.8 ± 0.2 kPa). Note the acidotic shift (time course from right to left) of the normocapnic pH_{ECF} values, particularly during the hypoxic periods

the shift in pH_{ECF} may be attributed totally to local metabolic acidosis under this condition.

Other than in the rabbit experiments mentioned above, pH_a was fairly stable throughout. The same was true for the concentrations of standard HCO_3^- and lactate in the arterial blood.

Although no ventilatory reaction to metabolic acidosis at the ventral surface of the brain-stem could be observed, the CO_2-responsiveness of ventilation remained unvariably high (Fig. 4). The steady state ventilatory responses to changes in arterial P_{CO_2} ranged from 447 to 430 ml · min^{-1} · kg^{-1} · kPa^{-1} at the start and at the end of the experiment, respectively. The average rise in V per unit CO_2-induced decrease in pH_{ECF} was 11859 ml · min^{-1} · kg^{-1} · pH^{-1}. A decrease of pH_{ECF} by 0.260 units would thus be expected to augment ventilation nine times above the resting value. However, a similar drop in pH_{ECF} of metabolic origin was unable to change ventilation at all.

Discussion

The most prominent finding of this presentation is the lack of ventilatory responsiveness to metabolic acidosis at the brainstem surface. This is rather unexpected in the light of the current "reaction theory" of central respiratory chemosensitivity as reviewed by Loeschcke (1980). His theory was supported by a variety of experimental data over the last twenty years (e.g., Loeschcke et al. 1958, Schlaefke and See 1980), which invariably led to the conclusion that central chemosensitivity and hence ventilation are stimulated by an increase of the H^+ activity close to the ventral medullary surface.

Although there remains some controversy concerning the depth of the chemosensitive structures below the brainstem surface, several authors agree that ventilation is correlated neither with blood pH nor with cerebrospinal fluid pH, but is a unique function of the extracellular-fluid pH within the brainstem (Pappenheimer et al.

1965, Berndt 1974, Berkenbosch et al. 1978). Recently, Ahmad et al. 1976, succeeded in measuring the latter variable directly. The extracellular fluid pH was shown to decrease not only by inhalation of CO_2 but also as a consequence of intravenously infused fixed acids in artificially ventilated cats (Ahmad et al. 1976) and dogs (Javaheri et al. 1981).

With these findings in mind, the question arises, why peripherally chemodenervated rabbits (see results) do not hyperventilate in response to lowered blood pH caused by hypoxia. Hypoxic depression of central respiratory neurons seems to be unlikely, since a comparable hypercapnia induced pH_a-response of ventilation was shown not to be attenuated by hypoxia (Kiwull-Schöne and Kiwull 1981). Furthermore, during the hyperoxic recovery after the hypoxic period, the metabolic acidosis in the arterial blood persisted, but ventilation had only insignificantly increased by about 20% above the initial control value. A comparable drop in pH_a, if induced by CO_2, would have augmented ventilation by 100–200% above the resting level.

According to general experience, metabolic acidosis as a whole seems to be a less powerful stimulus to ventilation than respiratory acidosis. This does not necessarily contradict the assumption that ventilation is a unique function of the brain-stem extracellular pH under both acid-base conditions, as long as the closed-loop situation of the respiratory control system is considered (Loeschcke 1973). However, the question remains, whether in the above mentioned experiments the obviously large deviations in blood pH were not sufficiently reflected by pH_{ECF}.

The aim of further experiments (see results) was to test this possibility by measuring the pH value at the ventral medullary surface directly. The medullary surface pH is regarded as being closely correlated to or even identical with the pH of the adjacent freely exchanging extracellular fluid (Ahmad et al. 1976, Ahmad and Loeschcke 1982). By using this method, we found that in peripherally chemodenervated cats a considerable local metabolic acidosis developed at the ventral medullary surface during the hypoxic periods. However, this was not followed by the expected hyperventilation, i.e. ventilation did not behave as a unique function of pH_{ECF}.

The lack of ventilatory reaction to metabolic acidosis cannot be ascribed to a functional failure in either central chemosensitivity or of the respiratory centers, since hypercapnia was invariably followed by hyperventilation with nearly unchanged sensitivity (Fig. 4). Furthermore, it is unlikely that the local metabolic acidosis was caused by disturbed micro-circulation, since upmost care was taken to minimize the electrodes virtual weight (see methods). Local metabolic acidosis developed predominatly during hypoxia, but only if the peripheral chemoreflexes were additionally eliminated (Kiwull-Schöne and Kiwull 1980). This fact also argues against a systematic error caused by the experimental set-up.

The possibility remains that under certain circumstances the acid-base conditions surrounding the chemosensitive structures were not reflected at the ventral medullary surface or vice versa. It has been reported, however, by Cragg et al. (1977), using a pH-microelectrode technique, that, in chemodenervated hypoxic cats, not only the brainstem surface but also deeper structures became acidotic. In both, their and our experiments, the concomitant acid-base situation in the blood remained nearly undisturbed. Cragg et al. (1977) discuss the possibility that in vasodeafferented animals a stronger dependence between cerebral blood flow and blood pressure may exist.

In this context, the degree of hypoxia may determine the degree of hypotension (Kiwull-Schöne and Kiwull 1979). Under these conditions, the environment of the chemosensitive structures at various depths below the medullary surface should become more or less acidotic so that the question still remains if or how the chemosensors can distinguish between different sources of H^+-ions. Shams et al. (1981) recently also found different strengths of ventilatory reactions to either metabolic or respiratory acidosis at the brainstem surface. However, this phenomenon could only be observed with prolonged intravenous acid infusion. The main difference between the infusion experiments and ours with low oxygen inhalation may be that during hypoxia the pH within the cells primarily and immediately decreases due to anaerobic glycolysis.

One possibility to distinguish between different sources of extracellular H^+ would be to take the intracellular pH as reference value. If it is assumed that the chemosensory structures are affected by the intracellular-extracellular pH gradient, a simulataneous decrease of pH in both compartments would remain ineffective. The earlier findings of Loeschcke and Katsaros (1959) point in that direction. The authors could elicite an increase in tidal volume even though the extracellular pH had been kept experimentally constant. This was possible because at the same time, the intracellular compartment was made selectively alkaline by application of ammonium chloride.

Summary

In peripherally chemodenervated animals, ventilation reacts in a different manner to comparable changes in medullary surface pH, if induced either by respiratory or by metabolic acid-base disturbances. There is a lack of ventilatory reaction to medullary acidosis induced by hypoxia, indicating that ventilation is not a unique function of pH_{ECF} under this condition. Due to anaerobic glycolysis, a concurrent drop in intracellular and extracellular pH is suspected to prevent the central chemosensitive elements from sensing the pH-change in their environment.

Acknowledgements. We would like to thank Mrs. P. Wollenberg for her expert technical assistance, Ms. M.-Th. Franckson for the drawings, Mrs. M. Duparc for typing and Mr. I.F.R. Ukat for reading and discussing the manuscript.

References

Ahmad HR, Loeschcke HH (1982) Transient and steady state responses of pulmonary ventilation to the medullary extracellular pH. Pflügers Arch (in press)

Ahmad HR, Berndt J, Loeschcke HH (1976) Bicarbonate exchange between blood, brain extracellular fluid and brain cells at maintained P_{CO2}. In: Loeschcke HH (ed) Acid base homeostasis of the brain extracellular fluid and the respiratory control system. Thieme, Stuttgart, p 19

Berkenbosch A, Goede de Y, Olievier CN, Quanier Ph P (1978) Influence of the CSF bicarbonate concentration on the ventilatory response to CO_2 in relation to the location of the central chemoreceptors. Respir Physiol 35:215–236

Berndt J (1974) The one-receptor-theory of central chemosensitivity of respiration. In: Koepchen HP, Umbach W (eds) Central rhythmic and regulation. Hippokrates, Stuttgart, p 93

Cragg P, Patterson L, Purves MJ (1977) The pH of brain extracellular fluid in the cat. J Physiol (London) 272:137–166

Javaheri S, Herrera L, Kazemi H (1979) Ventilatory drive in acute metabolic acidosis. J Appl Physiol 45:913–918

Javaheri S, Clendening A, Papadakis N, Brody JS (1981) Changes in brain surface pH during acute isocapnic metabolic acidosis and alkalosis. J Appl Physiol 51:276–281

Kiwull-Schöne H, Kiwull P (1979) The role of the vagus nerves in the ventilatory response to lowered P_aO_2 with intact and eliminated carotid chemoreflexes. Pflügers Arch 381:1–9

Kiwull-Schöne H, Kiwull P (1980) Central and peripheral components in the dynamic CO_2 response of the medullary surface pH and ventilation during hypoxia. In: Betz E, Grote J, Heuser D, Wüllenweber R (eds) Pathophysiology and pharmacotherapy of cerebrovascular disorders. Witzstrock, Baden-Baden, p 129

Kiwull-Schöne H, Kiwull P (1981) The influence of hypoxia on the central CO_2 (H^+) sensitivity of ventilation. In: Belmonte C, Pallot DJ, Acker H, Fidone S (eds) Arterial chemoreceptors. Univ Press, Leicester, p 502

Loeschcke HH (1973) The respiratory control system: Analysis of steady state solutions for metabolic and respiratory acidosis-alkalosis and increased metabolisms. Pflügers Arch 341: 23–42

Loeschcke HH (1980) Chemical alterations of cerebrospinal fluid acting on respiratory and cirulatory control systems. In: Wood JH (ed) Neurobiology of cerebrospinal fluid. Plenum Press, New York, p 29

Loeschcke HH, Katsaros B (1959) Die Wirkung von in den Liquor cerebrospinalis eingebrachtem Ammoniumchlorid auf Atmung und Vasomotorik. Pflügers Arch 270:147–160

Loeschcke HH, Koepchen HP, Gertz KH (1958) Über den Einfluß von Wasserstoffionenkonzentration und CO_2-Druck im Liquor cerebrospinalis auf die Atmung. Pflügers Arch 266:565–585

Pappenheimer JR, Fencl V, Heisey SR, Held D (1965) Role of cerebral fluids in the control of respiration as studied in unanaesthetized goats. Am J Physiol 208:436–450

Schlaefke M, See WR (1980) Ventral surface stimulus response in relation to ventilation and cardiovascular effects. In: Koepchen HP, Hilton SM, Trzebski A (eds) Central interaction between respiratory and cardiovascular control systems. Springer, Berlin Heidelberg New York, p 56

Shams H, Ahmad HR, Loeschcke HH (1981) The dependence of ventilation on H^+ activity in the extracellular fluid of the brain in respiratory or metabolic acidosis. Pflügers Arch 389:R 53

The Response Characteristic of Peripheral Chemoreceptors and Their Physiological Relevance

A. PLAAS-LINK and H.H. LOESCHCKE[1]

In 1960, Yamamoto suggested that arterial pCO_2 oscillations play a role in the control of respiration. Within breath fluctuations of blood gas tensions of pO_2 (Purves 1966), pCO_2 (Plaas-Link et al. 1976) and pH (Band et al. 1969) have been measured. It remains unclear how the peripheral chemoreceptors react to the small and quick changes of pCO_2. Several authors observed different receptor characteristics according to their technique of measurement. In the present study we tried to approximate the experimental conditions to natural conditions in meeting the following requirements:

1. The solution to perfuse the receptor should be blood containing the natural red cell fraction. (Fitzgerald et al. 1968, Ponté and Purves 1974).
2. The amplitudes of the changes in the solutions should be small, because they will influence the response characteristic of the receptors. (Ponté and Purves 1974).
3. The sinus nerves should be intact, because efferent control of the chemoreceptor affects the chemoreceptor response. (Majcherczyk and Willshaw 1973, Goodman 1974).
4. The experiments should be done under normoxia, because hypoxia or hyperoxia will affect the neural discharge of the receptor. (Black and Torrance 1967, Gray 1968, Fitzgerald and Parks 1971, Gehrich and Moore 1973).

Some authors attributed the modulation of chemoreceptor discharge in dependence of respiratory frequency to concomitant oscillations in pCO_2. (Gehrich and Moore 1973, Eldridge 1972). It is generally accepted that the glomus receptor will not respond to quick pO_2 oscillations. (Black and Torrance 1967, McCloskey 1968, Fitzgerald et al. 1969, Ponté and Purves 1974, Folgering et al. 1978). The response of the peripheral chemoreceptor to pH changes is slower and smaller than to changes of pCO_2 (Ponté and Purves 1974).

Methods

The experiments were carried out in cats anesthetized with 30 mg kg^{-1} α-gluco-Chloralose and 160 mg kg^{-1} Urethan. Additionally 100 IE/1.6 kg^{-1} Heparin were infused intravenously at the beginning and 100 IE per hour during the experiments.

1 Institut für Physiologie, Ruhr Universität Bochum, D-4630 Bochum, Fed. Rep. of Germany

Central Neurone Environment
(ed. by M.E. Schläfke, H.P. Koepchen and W.R. See)

For artificial ventilation we gave 0.001–0.003 ml Flaxedil i.v. The cats breathed spontaneously during sinus nerve recordings. Tidal volume (ml) was calculated by integration of the pneumotachogram signal. Blood pressure (kPa) was measured in the femoral or in the common carotid artery. The arterial pCO_2 (kPa) and pO_2 (kPa) were measured with a masspectrometer (Fa. Leybold, Heraeus). The oscillations of pCO_2 or pH to the carotid bodies of the cat were produced with alternating mixtures of (1) hypercapnic blood and (2) unchanged blood coming directly from the aortic artery. The pCO_2 of the first fraction was increased with a gas exchanger. The size of the oscillations, their frequency and the slope could be varied. The blood pressure variations of the perfusing blood were slow and small.

The sinus nerves were complete and intact and desheathed for recording. The integrated sinus nerve activity (ISNA) is presented in arbitrary units.

Results

The first figure illustrates the amplitude of oscillations in pCO_2 (Fig. 1 a) and in pH (Fig. 1 b) in relation to the respiratory frequency in an artificially ventilated cat. The size of pH oscillations decreases with an increase of respiratory frequency. These amplitudes are measured under resting conditions while during exercise the amplitudes rise by 5 to 8 times, depending on the CO_2 production. The peak to peak amplitude divided by a half of its period length was considered as the mean slope. The slope of natural oscillations under resting conditions was found to be 0.005 to 0.01 pH units per second. The corresponding slope of pCO_2 oscillations was 0.08 to 0.17 kPa per second. The slope depends on functional residual capacity, the red cell fraction and the mean pCO_2 (Plaas-Link et al. 1976). These results were obtained in 8 cats and 250 mean values of periods of 5 oscillations.

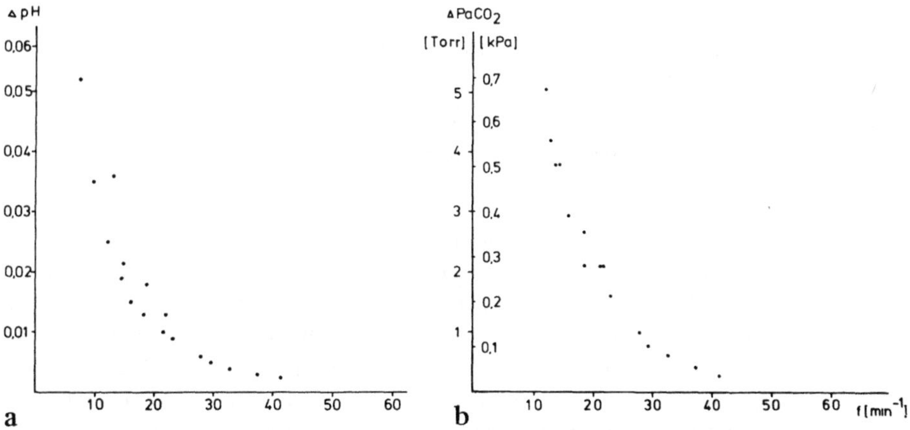

Fig. 1 a, b. Peak to peak arterial oscillation amplitudes of pH (**a**) and pCO_2 (**b**) depending on respiratory frequency. The mean pCO_2 was kept constant at 4.8 kPa pCO_2

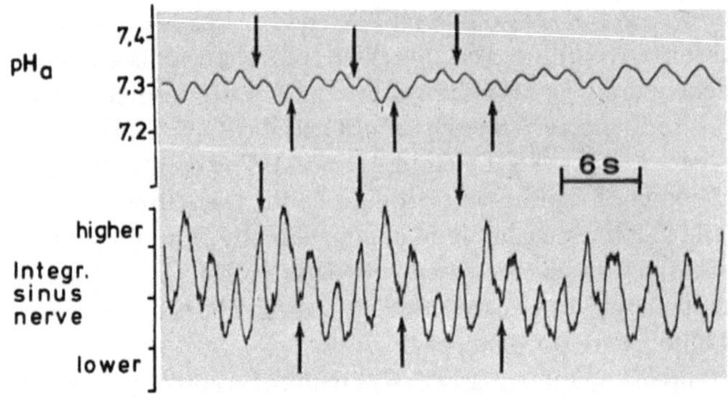

Fig. 2. The slope dependent receptor response to variations in pH and pCO_2. The oscillations with increasing pH are marked with ***arrows pointing upward*** and they cause the similarly marked ISNA's. The oscillations with falling pH are marked with ***arrows pointing downward*** and cause higher ISNA's, although the pH is not more acid

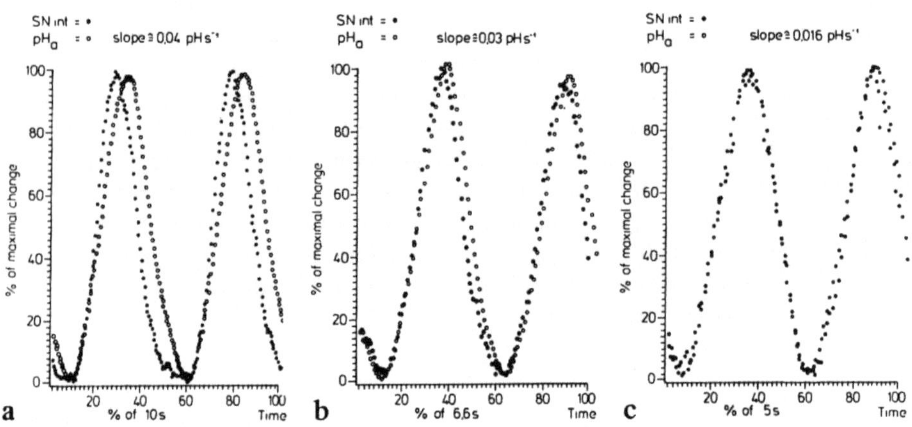

Fig. 3 a–c. The influence of the slope of pH-oscillations on the response characteristic of the glomus caroticum. The averaged signals of 30 pH oscillations *(circles)* and the corresponding 30 values of the integrated nerve activity *(dots)* are represented on the ordinate. The abscissa shows the time for two period lengths. With a slope of 0.04 pH units s^{-1} (Fig. 3a) a differential response is obvious. With lower slopes the differential response characteristic disappears

The pH oscillations in Fig. 2 vary in their absolute value. The arrows pointing upward mark pH oscillations with increasing pH and the arrows pointing downward mark oscillations with decreasing pH. The pH of the oscillations with arrows pointing upwards is as low or even lower than the pH value of the oscillations marked with arrows po ing downward at the moment indicated. If the level of pH alone is determing the ISNA, tl oscillations marked with arrows pointing upward should cause equal or greater integrated sinus nerve activity (ISNA) than the oscillations marked with arrows pointing downward. However the oscillations marked with arrows pointing downward cause higher ISNA's. Figure 2 suggests that the response characteristic is different for falling and increasing pH (in fact pCO_2) signals. This indicates adaptation of the response of the glomus receptors:

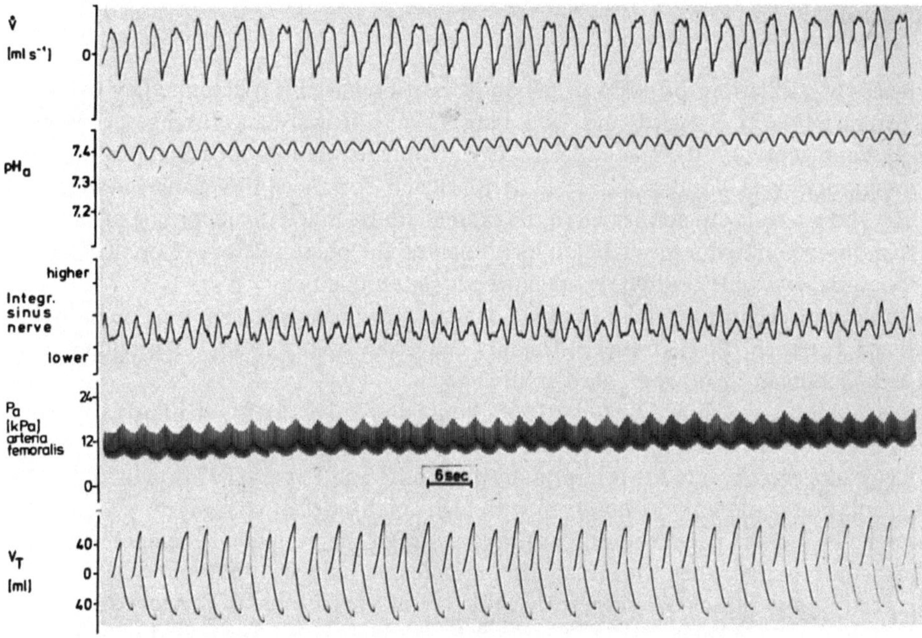

Fig. 4. The importance of the phase relation oscillations on the effect of ventilation. The slope of the small pH oscillations varies and according to this, the integrated sinus nerve activity (ISNA) shows different amplitudes. The tidal volumes show always high values if the maximal discharge of the receptor falls in the inspiratory period. The increase of the tidal volumes is nearly independent of the size of the ISNA

Figure 3 a–c shows the averaged signals of 30 oscillations with constant frequency and amplitude and the corresponding values of 30 ISNA's. The time between adjacent dots is 20 ms. The rise of ISNA followed the beginning of the decrease in pH with a delay of 800 ms. The ISNA curve is shifted to the left side by these 800 ms to let both signals start at the same time. The slope of the oscillations in Fig. 3a was determined with 0.04 pH units s^{-1}. This is 4–8 times higher than the slope of oscillations under resting conditions.

With decreasing pH the ISNA is increasing. The superposition shows that the peak of the receptor discharge was reached earlier than the peak of the pH decrease. In addition the change of activity within the 20 ms (adjacent dots) of the rising ISNA is greater than in the falling ISNA. The signal is not symmetric. This is a sign for a differential response characteristic of the glomus receptor. This result suggests that oscillations with low slopes will cause no differential response as shown in Fig. 3 c. With a slope 2 times higher than under resting conditions a differential response could be observed (Fig. 3 a, b).

Fig. 4 shows the influence of the arrival time of the oscillations in the respiratory cycle. If the maximal ISNA's occur during the inspiratory period, high tidal volumes are observed. The peaks of the periodic ISNA's reach different levels depending on different slopes of the pH-oscillations. The increase of the tidal volumes however, is influenced more by the timing than by the size of the ISNA.

Discussion

Influences of oscillations of pCO_2 or pH on respiration have been described by Black and Torrance (1967), Bernards and Sistermans (1969), Howard and Bromberger-Barnea (1969), Band et al. (1970). The importance of the phase relation of oscillations to the respiratory cycle was first described by Black and Torrance (1967) and Eldridge (1972). These results are confirmed in the present study. Since the recording of the ISNA in our experiments was done on intact nerves, the phase relation of oscillations to the respiratory cycle could be more correctly determined.

The glomus receptor response to pCO_2 changes showing adaptation was found by Black and Torrance 1967, Leitner et Dejours 1968, McCloskey 1968, Fitzgerald et al. 1969, Goodman 1974, and Lahiri et al. 1978.

No adaptation was shown by Gray 1968, Ponté and Purves 1974, and Purves 1966. The response of the aortic receptors shows no adaptation (Lahiri et al. 1979).

Our results corroborate the interpretation of Ponté and Purves (1974) who said that adaptation could only be observed with high amplitudes of changes of pCO_2. However, the present paper shows that the response of the peripheral chemoreceptor is already differential when the metabolism is increased only slightly more than under restring conditions. We suggest that the role of oscillations for the control of respiration may lie in the transition from rest to exercise. This is supported by the following reasons:

(1) The increase of the CO_2 production causes higher amplitudes of oscillations with an increase of maximal discharges of the glomus receptors. (2) In the transition the increase of heart minute volume causes an increase of the slope of oscillations even with constant amplitudes because the period lengths decrease (Doppler effect). (3) The time delay of the oscillations between their production in the lung and their arrival at the receptors is altered by changing heart minute volume. (4) With increasing oscillation amplitudes a more differential behaviour of the receptor response is preponderant. This causes an earlier stimulation of central neural structures. During resting conditions, the maximal stimulus of oscillations was found in the expiratory phase (Petersen et al. 1978). During exercise the arrival time of the oscillations is changed as stated earlier. The change from an expiratory phase relation to an inspiratory phase relation causes a rapid increase of ventilation.

We suggest that the peripheral chemoreceptors have a proportional response under resting conditions as long as the slope of oscillations is small. The slope of oscillations could be altered by CO_2 production, a change in the functional residual capacity and/or a decrease in the red cell fraction of the blood. With higher slopes of pCO_2 a more differential behaviour became preponderant.

References

Band DM, Cameron IR, Semple SJG (1969) Oscillations in arterial pH with breathing in the cat. J Appl Physiol 26:261–267

Bernards JA, Sistermans JF (1969) Transient changes in lung ventilation by brief stimulation of the carotid bodies in the dogs. Acta Physiol Pharmacol Ned 15:28-29

Black Ams, Torrance RW (1967) Chemoreceptor effects in the respiratory cycle. J Physiol 189: 59–61

Black AMS, McCloskey DI, Torrance RW (1971) The responses of carotid body chemoreceptors in the cat to sudden changes of hypercapnic and hypoxic stimuli. Respir Physiol 13:36–49

Eldridge FL (1972) The importance of timing on the respiratory effects of intermittent carotid sinus nerve stimulation. J Physiol (London) 222:297–318

Fitzgerald RS, Parks DC (1971) Effect of hypoxia on carotid chemoreceptor response to carbon dioxide in cats. Respir Physiol 12:218–229

Fitzgerald RS, Gross N, Dutton RE (1968) Ventilatory responses to transient acidic and hypercapnic vertebral artery infusions. Respir Physiol 4:387–395

Fitzgerald RS, Leitner LM (London, Liaubet MJ (1969) Carotid chemoreceptor response to intermittent or sustained stimulation in the cat. Respir Physiol 6:395–402

Folgering H, Smolders FDJ, Kreuzer F (1978) Respiratory oscillations of the arterial pO_2 and their effects on the ventilatory controlling system in the cat. Pflügers Arch 375:1–7

Gehrich JL, Moore GP (1973) Statistical analysis of cyclic variations in carotid body chemoreceptor activity. J Appl Physiol 35:642–648

Goodman NW (1974) Some observations on the homogeneity of response of single chemoreceptor fibres. Respir Physiol 20:27–28

Gray BA (1968) Response of the perfused carotid body to changes in pH and pCO_2. Respir Physiol 4:229–245

Howard P, Bromberger-Barnea B, Fitzgerald RS, Bane HN (1969) Ventilatory responses to peripheral nerve stimulation at different times in the respiratory cycle. Respir Physiol 7: 389–398

Lahiri S, Gelfand R, Mokashi A Nishino T (1978) Significance of peripheral chemoreceptor response and adaptation in the regulation of breathing. In: Fitzgeral RS, Gautier H, Lahiri S (eds) The Regulation of Respiration during Sleep and Anesthesia. Adv Exp Med Biol, vol 99. Plenum Press, New York, p 343

Lahiri S, Mulligan E, Nishino T, Mokashi A (1979) Aortic body chemoreceptor responses to changes in P_{CO2} and P_{O2} in the cat. J Appl Physiol: Respir Environ Exercise Physiol 47(4):858–866

Leitner LM, Dejours P (1968) The speed of response of chemoreceptors. In: Torrance RW (ed) Arterial chemoreceptors. Blackwell, Oxford, p 71

Majcherczyk S, Willshaw P (1973) Inhibition of peripheral chemoreceptor activity during superfusion with alkaline CSF of the ventral brainstem surface of the cat. J Physiol (London) 231:26–27

McCloskey DI (1968) Carbon dioxide and the carotid body. In: Torrance RW (ed) Arterial chemoreceptors. Blackwell, Oxford, p 279

Petersen ES, Whipp BJ, Drysdale DB, Cunningham DJC (1978) Carotid arterial blood gas oscillations and the phase of the respiratory cycle during exercise in man: Testing a model. In: Fitzgerald RS, Gautier H, Lahiri S (eds) The regulation of respiration during sleep and anesthesia. Adv Exp Med Biol, vol 99. Plenum Press, New York, London, p 335

Plaas-Link A, Mueller J, Mückenhoff K, Luttmann A, Loeschcke HH (1976) Measurements of arterial CO_2-oscillations and their dependence on respiratory frequency, tidal volume, CO_2 mean level and blood Pressure. Pflügers Arch 365: R 38

Ponté J, Purves MJ (1974) Frequency response of carotid body chemoreceptors in the cat to changes of P_{aCO2} and P_{aO2} and pH_a. J Appl Physiol 37(5):635–647

Purves MJ (1966) Fluctuations of arterial oxygen tension which have the same period as respiration. Respir Physiol 1:281–296

Purves MH, Ponté J (1977) Further studies on the fluctuation of chemoreceptor discharge in the cat. In: Acker H, Fidone S, Pallot D. Eyzaguirre C, Lübbers DW, Torrance RW (eds) Chemoreception in the carotid body. Springer, Berlin Heidelberg New York, p 175

The Significance of Carotid Chemoreceptor Stimulus-Impulse Transmission for the Respiratory Control System of the Rabbit

P. KIWULL and H. KIWULL-SCHÖNE[1]

Introduction

Since the findings of Nielsen and Smith (1952) it is well established that in humans hypoxia and hypercapnia, as ventilatory stimuli, act in a multiplicative manner. Based on results obtained by indirect methods in man, this mode of interaction was mainly attributed to the arterial chemoreceptors, as reviewed by (Cunningham 1974). This was also supported by recordings of chemoreceptor activity in animals, particularly in cats (Fitzgerald 1976, Lahiri and DeLaney 1975b). On the other hand, for rabbits, the overall reaction of ventilation to combined hypoxia and hypercapnia was found to be not multiplicative, but, approximately additive (Kiwull et al. 1976). Therefore the question arose, whether there were principal species differences in the stimulus-impulse transmission at peripheral chemoreceptors. Since aortic chemoreceptors do not play a functional role in the rabbit (Kiwull-Schöne and Kiwull 1979, Ott et al. 1971a), only the behaviour of carotid chemoreceptors would be of relevance for this question.

Methods

The experiments were performed in ten rabbits weighing 2.6 ± 0.2 kg. anaesthetized by sodium pentobarbital. All animals were paralyzed with Succinyl-Asta or Flaxedil (Abott), and artificially ventilated (Harvard respiratory pump). Blood coagulation was prevented by Heparin-Riker.

Surgery and Apparatus

The animals were fixed in a supine position. Rectal temperature was controlled at a mean of 38.0 ± 0.5°C. Tracheal CO_2-concentration was measured continuously by infrared absorption and tracheal O_2 concentration by polarography. Blood samples were drawn from a bypass of a femoral artery. The systemic arterial blood pressure

1 Institut für Physiologie, Ruhr-Universität, Postfach 102148, D-4630 Bochum, Fed. Rep. of Germany

Central Neurone Environment
(ed. by M.E. Schläfke, H.P. Koepchen and W.R. See)

was measured in the contralateral femoral artery by strain gauge. Correspondingly, the pressure of the carotid region was monitored in the external carotid artery. All signals were recorded on a direct polygraph. The arterial vessels of the carotid region were prepared for ligation. One carotid sinus nerve was carefully exposed and cut at its central end. The distal part was desheathed and small filaments were prepared for spike recordings by a conventional bipolar platinum electrode connected to a differential preamplifier. For further evaluation the recordings were stored on magnetic tape.

Experimental Protocol

For identification of chemoreceptor single fibre or few fibre activities, we tested their responsiveness to hypoxic gas mixtures and/or intravenous injection of 250 μg sodium cyanide. In order to eliminate baroreceptor activity, in some cases, the perfusion pressure of the carotid body was decreased to a range above ischemic activation. This was accomplished by clamping the common carotid artery and ligating most of the other arterial vessels of the carotid region. Subsequently, keeping the alveolar CO_2 pressure constant at normocapnia, the animal inhaled gas mixtures of different oxygen pressures, each for about 2 min. At each steady state level of blood gas conditions, the chemoreceptor activity was recorded for one minute. The procedure was repeated at one or two levels of constant hypercapnia (about 2 and 4 kPa above normocapnia), and finally, again during normocapnia.

Evaluation

For each condition of chemical stimulus, chemoreceptor activity was counted (window discriminator, electronic counter). On the basis of these data the fibre impulse rate per second (IPS) was calculated and plotted against P_AO_2 at constant P_ACO_2. For each fibre the best fitting of two nonlinear funcions, a power function ($IPS = a(P_AO_2)^{-b}$) or an exponential function ($IPS = a \cdot e^{-b(P_AO_2)}$), was selected. For the same fibre, CO_2 response curves of chemoreceptor activity at different levels of oxygen pressure were constructed using the nonlinear hypoxic response curves at different CO_2 levels (Fig. 2).

Results

General Aspects

In all experiments, spike heights of chemoreceptor fibres were found to be much smaller (20–30 μV) than those of baroreceptor fibres (>100 μV) under similar recording conditions. Therefore, in contrast to baroceceptor fibres, chemoreceptor fibres

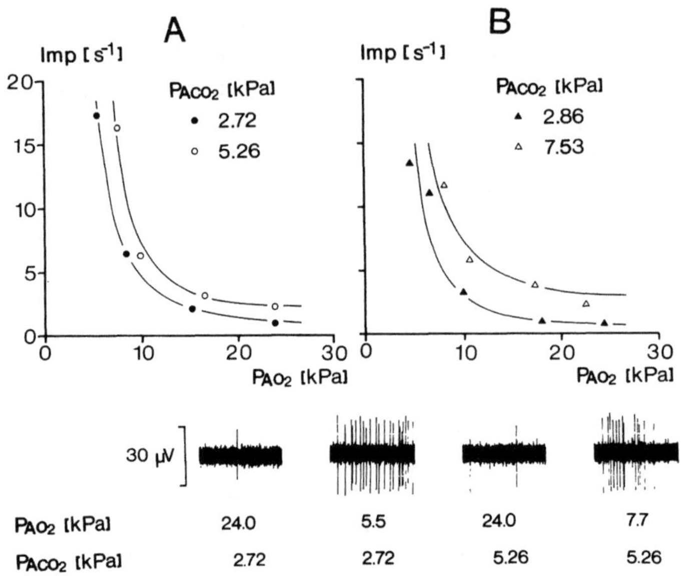

Fig. 1. Steady state O_2 responses of two carotid chemoreceptor fibres (**A, B**) at two different constant CO_2 levels. Each *curve* represents the best fit of the experimental data by a nonlinear function $y = ax^{-b}$. The coefficient b equals -2.37 (●), -3.17 (○), -2.70 (▲) and -2.21 (△)

could rarely be detected at all. Chemoreceptor discharge was much more irregular than baroreceptor activity. Artificial ventilation, at a rate of 20–30 breaths/min, never caused a rhythmical change of activity. Carotid chemoreceptors of the rabbit were found to be very insensitive to lowering of the perfusion pressure in the carotid region. Ischemic activation under normoxic conditions did not occur unless a perfusion pressure of about 1–3 kPa was reached.

Hypoxic Response

Lowering of the oxygen pressure at normocapnic constant PCO_2 led to a nonlinear increase of chemoreceptor activity (Fig. 1). This could be well described by nonlinear functions (see methods), as indicated by the correlation-coefficient: $r = 0.969 \pm 0.012$ ($n = 6$). In most of the fibres the best fitting function was the power function, in which factor b was found to range from -1.3 to -4.2. Based on these functions, the averaged absolute activity ($n = 6$) was 2.39 ± 0.56 Imp s^{-1} during hyperoxia ($P_AO_2 = 24$ kPa), 3.56 ± 0.51 Imp s^{-1} during normoxia ($P_AO_2 = 13.3$ kPa), and 11.4 ± 1.06 Imp s^{-1} during hypoxia ($P_AO_2 = 6.7$ kPa).

Interaction of Hypoxic and Hypercapnic Responses

The CO_2 sensitivity of a chemoreceptor fibre can be expressed as the increase of impulse activity versus the increase of CO_2 pressure (Δ Imp[s^{-1}]/ΔP_ACO_2 [kPa]) at

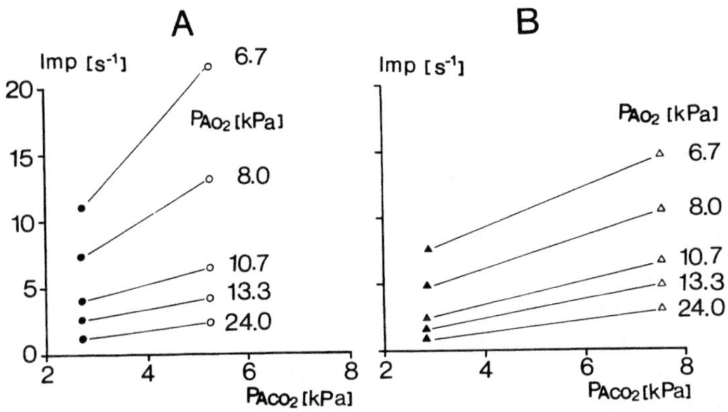

Fig. 2. Steady state CO_2 responses of two carotid chemoreceptor fibres (**A, B**) at different levels of alveolar oxygen pressure. Data are derived from nonlinear functions in Fig. 1

constant PO_2. These data, reconstructed from nonlinear functions (see above) are shown in Table 1 and in Fig. 2. It can be seen from the increase of the CO_2 sensitivity with hypoxia that there is a clear positive interaction between O_2 and CO_2 at the chemoreceptor level. In this population, the average CO_2 sensitivity during hypoxia (P_AO_2 = 6.7 kPa) is 6.1 ± 1.7 times greater than during hyperoxia (P_AO_2 = 24 kPa). In part of the experiments (n = 3 fibres) the arterial values for all CO_2 and O_2 conditions were obtained. The hypoxia induced augmentation of CO_2 sensitivity, which represents the degree of multiplicative interaction, became even more distinct when the data were brought in relation to blood gas tensions: it was twice as large as when brought in relation to alveolar gas tensions. This is mainly due to a diminution of the alveolar-arterial O_2 difference under the influence of CO_2.

Table 1. Influence of hypoxia on the CO_2 sensitivity in six different carotid chemoreceptor fibres as obtained by a CO_2 increase of about 4 kPa above normocapnia. The relative values refer to the absolute CO_2 response of each fibre during hyperoxia (P_AO_2 = 24 kPa)

CO_2 sensitivity = $\Delta Imp[s^{-1}]/\Delta P_ACO_2[kPa]$					
Fibre No.	At constant alveolar oxygen pressure [kPa]				
	24.0	13.3	10.6	8.0	6.7
1	0.44	0.58	0.93	2.22	4.20
2	0.47	0.70	0.87	1.21	1.50
3	0.66	0.71	0.83	1.14	1.46
4	0.47	0.79	0.96	1.34	1.87
5	0.23	0.46	0.71	1.00	1.04
6	0.25	0.42	0.73	1.77	3.27
Absolute means	0.42	0.61	0.84	1.45	2.22
(± S.E.M.)	± 0.07	± 0.06	± 0.04	± 0.19	± 0.50
Relative means	1.00	1.55	2.22	3.95	6.10
(± S.E.M.)		± 0.13	± 0.28	± 0.80	± 1.74

Discussion

The present data show that the functional properties of the carotid chemoreceptors in the rabbit, in several respects, are similar to those directly investigated in cats (Fitzgerald 1976, Lahiri and DeLaney 1975b) or indirectly investigated in man (Cunningham 1974). However, there are some differences too.

It was striking that the spike heights of chemoreceptor fibres in the rabbit were much smaller than those of the baroreceptor fibres. The evidence for this phenomenon is based on much more data than have been shown for the complete response curves in this presentation. The small spike heights indicate that most of the chemoreceptor fibres may be nonmyelinated. This might also explain why it is impossible to observe hypoxia or hypercapnia induced increases of the whole carotid sinus nerve activity as has been reported for cats (Plaas-Link, this volume). In rabbits, whole nerve chemoreceptor activity is completely concealed by baroreceptor activity. Furthermore, the threshold for ischemic activation due to lowering of the perfusion pressure is different from that in cats. Although we did not investigate this problem systematically, the threshold invariably was found to be in a very low range (about 1–2 kPa) of perfusion pressure. For cats, under comparable O_2 conditions the threshold range is at much higher perfusion pressures (Lahiri 1980). Both results support our earlier comparative studies in cats and rabbits (Ott et al. 1971b), in which the chemoreflex induced ventilatory reactions by lowered perfusion pressure were investigated.

The hypoxic response of chemoreceptor activity seems to be qualitatively similar to that in cats (Lahiri and DeLaney 1975a). There is also a good correspondence with the ventilatory response to hypoxia in rabbits (Kiwull-Schöne and Kiwull 1977, 1979).

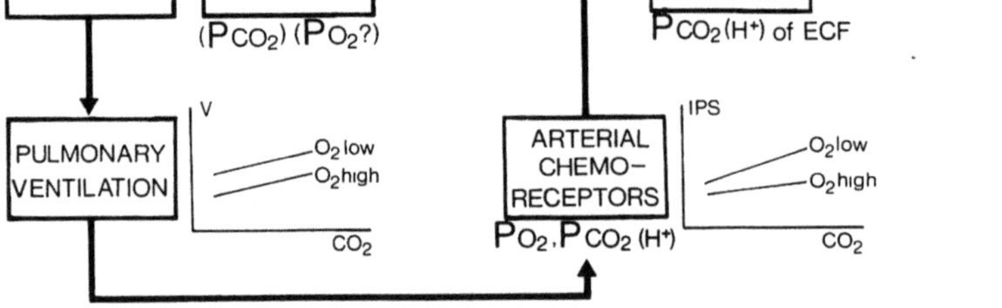

Fig. 3. Schematic diagram of the respiratory control system in the rabbit, including the different types of interaction of CO_2 and O_2 at different levels of nervous processing

The description by nonlinear functions makes it possible to compare the results of different investigations quantitatively. On the other hand, many mathematical models for the description of the hypoxic ventilatory response in humans (Cunningham 1974, Loeschcke and Gertz 1958), prefer a simple rectangular hyperbola, which for rabbits has been proved to be a less probable function (Kiwull-Schöne and Kiwull 1977).

The CO_2-response of chemoreceptor activity during hyperoxia is relatively small. This is in good agreement with our earlier results (Kiwull at al. 1972). In these studies we tried to separate the peripheral and the central portion of the CO_2 drive to ventilation by reversible cold-block of the carotid sinus nerves. The peripheral CO_2 drive was found to be practically negligible. In contrast, peripheral chemodenervation during hyperoxia causes a ventilatory depression of about 20% in cats (Katsaros 1968).

There is a distinct positive interaction of hypoxia and hypercapnia at the site of carotid chemoreactors in the rabbit. This agrees well with direct findings in cats (Lahiri and DeLaney 1975b) and indirect findings in humans (Cunningham 1974, Nielsen and Smith 1952). However, this kind of interaction is not reflected in the overall ventilatory reaction to combined hypoxia and hypercapnia in the rabbit, where it is purely additive (Kiwull et al. 1976, and Fig. 3). An explanation for this seeming difference may be found from our earlier investigations concerning the mode of interaction of different ventilatory afferent inputs at CNS level (Kiwull et al. 1974, 1976), which was found to be occlusive (Fig. 3). The same principle of central nervous integration was recently shown for cats (Eldridge et al. 1981). These and other results suggest that the overall reaction of ventilation to different chemical stimuli is a balance of several factors. We were able to demonstrate influences from peripheral chemoreceptors, from central chemosensitivity, and, not least of all, from pulmonary receptors (Kiwull-Schöne and Kiwull 1977, 1979). Special emphasis must be placed on vagal inputs from the lung since none of the classical response curves, either hypoxic or hypercapnic, can be achieved without vagal feedback. In humans, further influences of higher brain functions have to be considered.

In summary, for a given mode of stimulus-impulse-transmission at the peripheral chemoreceptors, the overall reaction of ventilation is subject to a variety of facilitatory and inhibitory factors quite competitive with each other

Acknowledgement. We would like to thank Mrs. P. Wollenberg for her expert technical assistance, Mrs. M.Th. Franckson for the drawings, Mrs. H. Spiethoff for typing the manuscript and Mr. I Ukat for reading and discussing the manuscript. Supported by the Deutsche Forschungsgemeinschaft (SFB 114 Bionach).

References

Cunningham DJC (1974) Integrative aspects of the regulation of breathing: a personal view. In: Guyton AC, Widdicombe JG (eds) MTP international review of science. Respir Physiol Ser 1, vol 2. Butterworth Univ Park, London, p 303

Eldridge FL, Gill-Kumar P, Milhorn DE (1981) Input-output relationships of central neural circuits involved in respiration in cats. J Physiol (London) 311:81–95

Fitzgerald RS (1976) Single fiber chemoreceptor responses of carotid and aortic bodies. In: Paintal AS (ed) Morphology and mechanisms of chemoreceptors. Navchetan Press (P), New Delhi, p 27

Katsaros B (1968) Evidence for the existence of a respiratory drive of unknown origin conducted in the carotid sinus nerves. In: Torrance RW (ed) Arterial chemoreceptors. Blackwell, Oxford, p 357

Kiwull P, Kiwull-Schöne H, Klatt H (1976) Interaction of central and peripheral respiratory drives: Differentiation between the role of stimuli and afferents. In: Loeschcke HH (ed) Acid base homeostasis of the brain extracellular fluid and the respiratory control system. Thieme, Stuttgart, p 146

Kiwull P, Schöne H, Wiemer W, Klatt H (1974) Quantitation of interaction of respiratory drives. In: Umbach W, Koepchen HP (eds) Central rhythmic and regulation. Hippokrates, Stuttgart, p 137

Kiwull P, Wiemer W, Schöne H (1972) The role of the carotid chemoreceptors in the CO_2-hyperpnea under hyperoxia. Pflügers Arch 336:171–186

Kiwull-Schöne H, Kiwull P (1977) Multifactor influences on the functional relationship between ventilation and arterial oxygen pressure. In: Acker H, Eyzaguirre C, Fidone SJ, Lübbers DW, Pallot DJ, Torrance RW (eds) Chemoreception in the carotid body. Springer, Berlin Heidelberg New York, p 190

Kiwull-Schöne H, Kiwull P (1979) The role of the vagus nerves in the ventilatory response to lowered PaO_2 with intact and eliminated carotid chemoreflexes. Pflügers Arch 381:1–9

Lahiri S (1980) Role of arterial O_2 flow in peripheral chemoreceptor excitation. Fed Proc Fed Am Soc Exp 39:2648–2652

Lahiri S, DeLaney RG (1975a) Relationship between carotid chemoreceptor activity and ventilation in the cat. Respir Physiol 24:267–286

Lahiri S, DeLaney RG (1975b) Stimulus interaction in the responses of carotid body chemoreceptor single afferent fibres. Respir Physiol 24:249–266

Loeschcke HH, Gertz KH (1958) Einfluß des O_2-Druckes in der Einatmungsluft auf die Atemtätigkeit des Menschen, geprüft unter Konstanthaltung des alveolaren CO_2-Druckes. Pflügers Arch 267:460–477

Nielsen M, Smith H (1952) Studies on the regulation of respiration in acute hypoxia. Acta Physiol Scand 24:293–313

Ott N, Kiwull P, Wiemer W (1971a) Zur Bedeutung der Chemoreflexe des Karotis- und Aortengebietes für die Atmungs- und Blutdruckregulation bei herabgesetztem Blutdruck. Z Kreislaufforsch 60:648–660

Ott N, Kiwull P, Wiemer W (1971b) Zur Aktivierung der Chemorezeptoren des Glomus caroticum durch lokale arterielle Drucksenkung bei Kaninchen und Katze. Pflügers Arch 325:28–39

The Effect of Acute Hypocapnia on the Response to Inhaled CO_2 in Man

E.S. PETERSEN[1], R. MUANGDIT[1,2] and D.J.C. CUNNINGHAM[1]

Rebreathing is a commonly used procedure by which man's ventilatory response to a ramp rise of Pco_2 and / or decrease of Po_2 may be studied. It was first described by Haldane in the early part of this century; the modification mostly used to day is that first described by Read (1967). Loeschcke and Gertz's important paper from 1958 describes an elegant modification of the rebreathing technique in which the effect of a falling Po_2 was studied while Pco_2 was maintained constant at different levels. This paper with the work of Nielsen and Smith (1952) and that of Lloyd et al. (1958) represent major early attempts at a quantitative analysis of the relationship and interaction between a rise of Pco_2 and lack of oxygen as ventilatory stimuli in man.

In euoxia, as demonstrated by Dripps and Comroe (1947), there is still a definite contribution of hypoxic drive mediated through the arterial chemoreceptors. When the alveolar Po_2 exceeds 200 torr (26.6 kPa), however, the absolute magnitude of this peripheral drive is negligible in resting man with normal blood acid base conditions. In this situation the central chemical drive is dominant and according to Loeschcke (e.g. this symposium; Loeschcke 1980) it is basically determined by the $[H^+]$ in the brain extracellular fluid at the site of the chemosensitive regions in the medulla.

Michel (1964) studied the time course of changes of pH and $[HCO_3^-]$ in the cerebrospinal fluid of dogs when respiratory acidosis and alkalosis were acutely induced by the inhalation of a CO_2-rich gas mixture or by hyperventilation. His results plotted in the HCO_3^-/pH diagram are shown in Fig. 1: the early changes followed the *in vitro* buffer line, but within 30–60 min significant changes in $[HCO_3^-]$ occurred, allowing some compensation for the changes in pH. Ahmad et al. (1978, and this symp) using direct pH measurements on the surface of the brain have reported much more rapid changes of ECF $[HCO_3^-]$ in response to step or nearly-step changes of Pco_2. They report time constants for $[HCO_3^-]$ changes in the order of 1–2 min.

The purpose of the present studies was to analyze man's response to rebreathing initiated at a reduced Pco_2 level. By studying the rebreathing responses at different levels of alveolar Po_2 it should be possible to obtain information about the magnitude of the respiratory drive both when arterial chemoreceptor drive is virtually absent, and when it is large. In the former case the rising ventilation would reflect the changing $[H^+]$ at the site of the intracranial chemosensitivity, and possible changes in the

1 The University Laboratory of Physiology, Oxford, England

2 *Present address:* Faculty of Medicine, Khonkaen University, Thailand

Central Neurone Environment
(ed. by M.E. Schläfke, H.P. Koepchen and W.R. See)

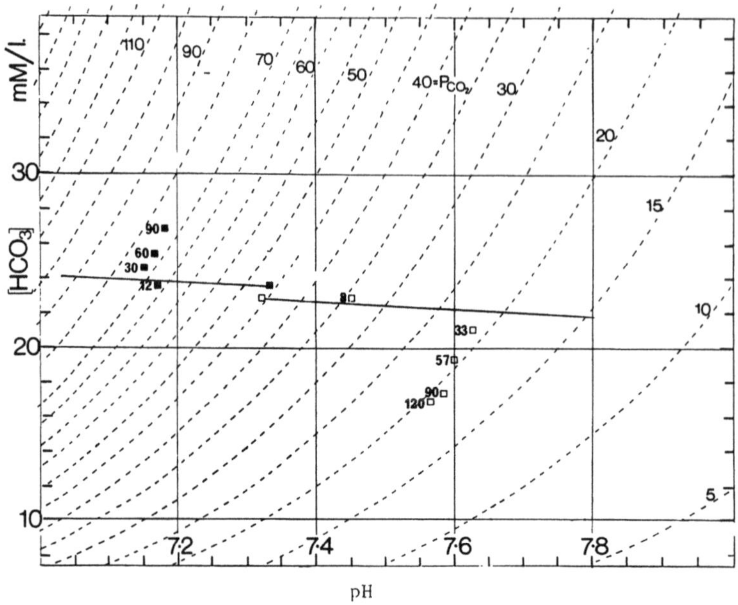

Fig. 1. Time course of changes of cerebrospinal fluid pH and HCO_3^- in dogs during hyperventilation (□) and CO_2-breathing (■). The figures at each point indicate the time in min after the onset of hyperventilation or CO_2-inhalation. The buffer lines showing the relation when CSF is equilibrated with CO_2 in vitro are shown (Data from Michel 1964)

ventilatory response might reflect changes in intracranial $[HCO_3^-]$. The other type should provide information about the CO_2–O_2 stimulus interaction.

Methods and Experimental Protocol

Rebreathing was performed in a thin-walled rubber balloon, initially containing ca. 7 L air or an O_2 -enriched gas mixture (30–40% O_2 in N_2). The balloon was enclosed in a rigid box connected to a spirometer and, through a Lloyd valve, to a dry gas meter which signalled every 10 litres. During inspiration atmospheric air passed into the box surrounding the balloon through the inspiratory port of the Lloyd valve, while during expiration air displaced from the box was directed through the expiratory port and from there to the gas meter. End-tidal gas was sampled at the mouth and analyzed for CO_2 by an infrared CO_2 analyzer and for O_2 by a paramagnetic O_2 analyzer, and the sample returned to the circuit. The analyzers were calibrated with standard gases immediately before and after each rebreathing run. The composition of the standard gases was checked by chemical gas analysis on the Lloyd-Haldane apparatus.

Tidal volume, Po_2, Pco_2 and the output from the gas meter were recorded on an U–V recorder.

The records were analyzed as follows: The time for each 10 L (at high ventilations 20 L) mark on the record was measured , and the minute ventilation computed. The mean Po_2 and Pco_2 for each of these segments were recorded. For the subsequent analysis the Po_2 and Pco_2 at each of four isopnoeic values ($\dot{V}_E$ = 20, 30, 40 and 50 L/min) were found by interpolation.

Protocol: Four healthy subjects aged 20–23 years were studied. Rebreathing was initiated in four different conditions.

1. Control. After 15 min of quiet air breathing the subjects were connected to the rebreathing bag by turning of a three-way tap.

2. Hypocapnia induced by *voluntary hyperventillation.* The subjects were asked to breathe at an increased tidal volume following a bleeper (set at 20/min) maintaining a P_{ET,CO_2} of 20–25 torr (2.7–3.3 kPa) for 30 min. After the period of voluntary hyperpnoea rebreathing was initiated without disconnecting the subjects from the apparatus.

3. Hypocapnia induced by the inhalation of an *hypoxic* gas mixture (10% O_2 in N_2) for 30 min. As in (2) rebreathing was initiated without disconnecting the subjects from the apparatus. Recorded P_{ET,O_2} was never below 50 torr (6.7 kPa).

4. Hypocapnia induced by metabolic acidosis consequent to a brief bout of exhausting *exercise* of the treadmill. In these runs venous blood samples were taken immediately before and after each rebreathing run. The plasma lactate concentrations in samples taken approximately 10 min after exercise, just before rebreathing commenced, ranged from 6.9 mM in subject 530 to 10.9 mM in subject 593 (mean 9.5), which might be expected to lower arterial pH by 0.15–0.2 units.

Only one of these conditions was studied on any one day.

The order of the two rebreathing runs (air or O_2 rich mixture) was randomized except for post-exercise where the air-run was always done first. At least 1 h was allowed between the two runs. The two post-exercise rebreathing runs were therefore initiated at different levels of metabolic acidosis and thence Pco_2 (see Table 1).

Results

The P_{ET,CO_2} values immediately before the start of rebreathing are shown in Table 1.

The recorded values for $\dot{V}_E$ plotted in the Pco_2/Po_2 diagram are shown in Fig. 2. A given level of ventilation was achieved at a lower Pco_2 when rebreathing was initiated on a background of hypocapnia than it was in control runs however the hypocapnia was induced, by voluntary hyperventilation or, less severely, by the inhalation of an hypoxic gas mixture. This was the case for every single isopnoeic comparison with the exception of 20 L/min for subject 530.

Table 2 shows for the high oxygen runs the mean Pco_2 required to produce a ventilation of 20, 30, 40 and 50 L/min. Although the scatter was large the values were lower in all three test conditions than in control runs. The differences were significant at the 0.05 level (for post-exercise 0.001).

Table 1. End-tidal Pco_2 (torr) at start of rebreathing runs

Subject	Control Air	Control $+O_2$	Post-Hyperventil Air	Post-Hyperventil $+O_2$	Post-Hypoxia Air	Post-Hypoxia $+O_2$	Post-Exercise Air	Post-Exercise $+O_2$
531	35	38	18	23	27	24	24	32
530	40	40	22	26	35	39	37	40
546	37	33	20	22	32	25	23	29
593	39	40	19	18	30	26	27	34
Mean	38	38	20	22	31	29	28	34

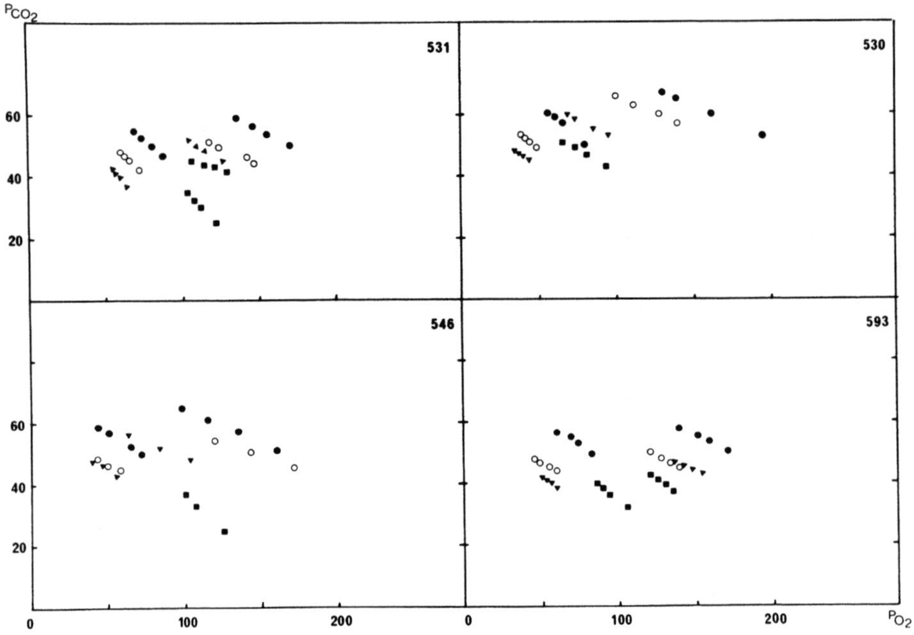

Fig. 2. Pco_2 and Po_2 (torr) at each of four isopnoeic ventilations (20, 30, 40 and 50 L/min) during rebreathing either air or an O_2 rich gas mixture in each of four conditions: Control (●), Post-hyperventilation (○), Post–hypoxia (▲) and Post-exercise (■)

For the analysis of the ventilatory effect of progressive hypoxia during rebreathing of air, the hyperbolic relationship between ventilation and Po_2 was linearized by replotting the data in a diagram showing Pco_2 against an hypoxic index = $1000/(Po_2 - C)$ (cf. Lloyd and Cunningham 1963) assuming a value for C of 30 torr (Fig. 3). The gradients of straight lines joining isopnoeic points from the two rebreathing runs provide an estimate of the hypoxic sensitivity in terms of change of Pco_2 per unit change of hypoxia, a high value indicating high hypoxic sensitivity and vice versa. The mean values of the hypoxic sensitivities thus expressed are shown in Table 3. It was significantly

Table 2. End-tidal P_{CO_2} (torr) at four levels of ventilation during rebreathing of an O_2 rich gas mixture, pooled data

Ventilation L/min	Control	Post-Hyperventil.	Post-Hypoxia	Post-Exercise
20	51	49	47	31
30	56	51	50	36
40	59	53	52	39
50	62	56	53	42
Mean ± SD	57.2 ± 5.4	51.9 ± 6.7	50.3 ± 5.3	36.7 ± 7.9

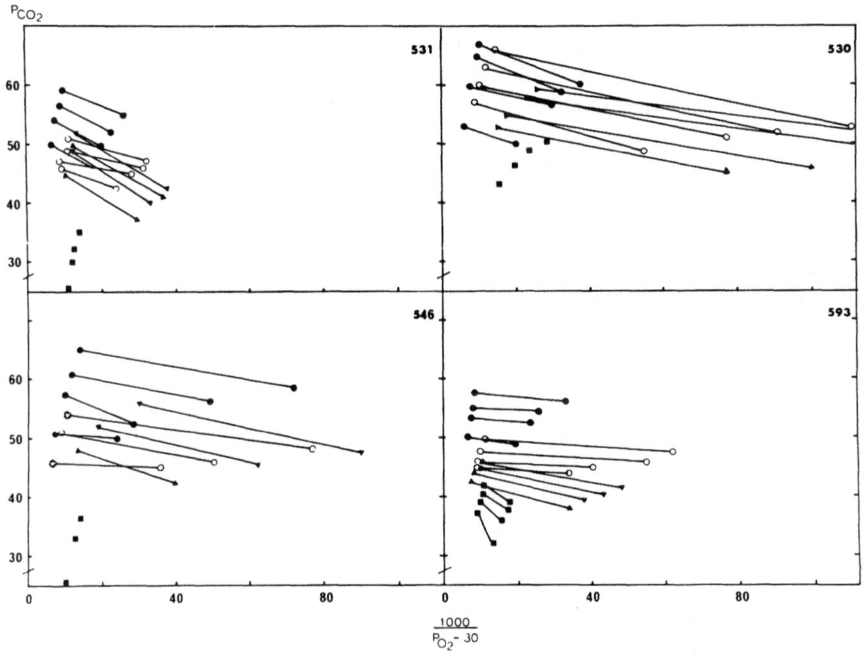

Fig. 3. P_{CO_2} (torr) and $1000/(P_{O_2} - 30)$ at each of four isopnoeic ventilations (20, 30, 40 and 50 L/min) during rebreathing either air (*right*) or an O_2 rich mixture (*left*) in each of four conditions. *Straight lines* joining isopnoeic points at the same conditions have been drawn. Symbols as in Fig. 2

($P < 0.05$) depressed when rebreathing was initiated on a background of hypocapnia induced by hyperventilation, but largely unaffected by the less intense hypocapnia induced by a preceding period of hypoxia.

Although a true comparison of the two post exercise rebreathing runs was not possible, a rough estimate of the hypoxic sensitivity in this condition (e.g. subject 593, Fig. 3) gives values manifestly larger than those seen in the other conditions.

Table 3. Hypoxic sensitivity, ΔPco_2 per unit hypoxic drive (1000/(Po_2-30). Pooled data, n = 16

	Mean	SD
Control	0.19	0.11
Post-Hyperventilation	0.11	0.06
Post-Hypoxia	0.20	0.12

Discussion

The experiments described above have provided a number of consistent findings:

1. Any given level of ventilation during rebreathing a high O_2 mixture was achieved at a considerably lower P_{ET,CO_2} when rebreathing was initiated on a background of hypocapnia of 30 min duration than it was in control runs starting at a normal resting Pco_2. This observation may be explained with reference to Fig. 1 on the assumption that the HCO_3^- concentration in the brain ECF at the site of intracranial chemosensitivity was lowered significantly at the time rebreathing started, whereas the ca. 4 minute ramp rise of Pco_2 during the actual rebreathing run was associated with little further change of [HCO_3^-]. Michel's data on the dog are in basic agreement with this interpretation since ECF [HCO_3^-] would be expected to to change earlier than CSF [HCO_3^-]. It can for the present data be estimated that 30 min hypocapnia was associated with decreases of [HCO_3^-] ranging from about 1 mM in subject 530 to 4 mM in subject 593.
2. The hypoxic sensivity was depressed following a 30-min period of hyperventilation with hypocapnia, whereas it was unaffected by a 30 minute period of preceding hypoxia with hypocapnia (cf. Cormack et al. 1956) The reasons for this difference are not quite clear. It might be related to the different levels of hypocapnia induced by the two conditions (Table 1). This and possibly different levels of plasma lactate in the two conditions might account for a different rate of adaptation of peripheral and central chemosensitivity. Plasma lactate was not measured in these parts of the experiments. However, from measurements performed after similar periods of voluntary hyperventilation in our laboratory we would expect increments in the order of 1 mM. It is probable that larger increments occurred during the hypoxic periods, and that an increased degree of metabolic acidaemia might account for a larger peripheral chemoreceptor response.
3. The ventilatory responses to rebreathing following exhausting exercise were greater than in any of the other two conditions. This observation is most likely explained through the intense metabolic acidosis, which would be expected to provide a considerably increased drive from the peripheral chemoreceptors. The greatly intensified hypoxic drive post exercise may be related to interaction with the increased CO_2 transport (cf. Yamamoto 1960).

Acknowledgements. R. Muangdit supported by a grant awarded by the WHO.

References

Ahmad HR, Loeschcke HH, Woidtke HH (1978) Three compartments model for the bicarbonate exchange of the brain extracellular fluid with blood and cells. In: Fitzgerald RS, Gautier H, Lahiri S (eds) The regulation of respiration during sleep and anesthesia. Plenum Press, New York, p 195

Cormack RS, Cunningham DJC, Gee JBL (1956) The effects of hypercapnia and acapnia on the respiratory response to acute want of oxygen in man. J Physiol (London) 133:47–48

Dripps RD, Comroe JH (1947) The effect of the inhalation of high and low oxygen concentrations on respiration, pulse rate, ballistocardiogram and arterial oxygen saturation (oximeter) of normal individuals. Am J Physiol 149:277–291

Lloyd BB, Cunningham DJC (1963) A quantitative approach to the regulation of human respiration. In: Cunningham DJC, Lloyd BB (eds) The regulation of human respiration. Blackwell Oxford, p 331

Lloyd BB, Jukes MGM, Cunningham DJC (1958) The relation between alveolar oxygen pressure and the respiratory response to carbon dioxide in man. Q J Exp Physiol 43:214–227

Loeschcke HH (1980) Chemical alterations of cerebrospinal fluid acting on respiratory and circulatory control systems. In: Wood JH (ed) Neurobiology of cerebrospinal fluid. Plenum Press, New York, London, p 29

Loeschcke HH, Gertz KH (1958) Einfluß des O_2-Druckes in der Einatmungsluft auf die Atemtätigkeit des Menschen, geprüft unter Konstanthaltung des alveolaren CO_2-Druckes. Pflügers Arch 267:460–477

Michel CC (1964) C.s.f. [HCO_3^-] during respiratory acid-base disturbance. J Physiol (London) 170:66–67

Nielsen M, Smith H (1952) Studies on the regulation of respiration in acute hypoxia. Acta Physiol Scand 24:293–313

Read DJC (1967) A clinical method for assessing the respiratory response to carbon dioxide. Australas Ann Med 16:20–32

Yamamoto WS (1960) Mathematical analysis of the time course of alveolar CO_2. J Appl Physiol 15:215–219

Patterns of Reflex Responses to Dynamic Stimulation of the Human Respiratory System

D.J.C. CUNNINGHAM[1], E.F. METIAS[1,2], M.G. HOWSON[1], and E.S. PETERSEN[1]

For more than 10 years we have been interested in the reflex effects in man of changing the time profile of the alveolar p_{CO_2}. I hope to show you today that the precise pattern of the reflex response may be influenced by the detail of the forcing pattern.

When a human subject is rendered hypercapnic by adding CO_2 to the inspired air the small swings of alveolar p_{CO_2} with each respiratory cycle are as represented at the top left of Fig. 1, and give rise to small oscillations of arterial p_{CO_2}. Just above is a diagrammatic spirogram. When the subject becomes hypercapnic as a result of adding CO_2 to only the first part of the inspirate, the alveolar p_{CO_2} pattern is slightly different (Fig. 1, top middle): the small fall of p_{CO_2} is delayed and when it occurs it is steeper than normal. If the subject is also hypoxic the breathing is stimulated by more than can be accounted for by the hypoxia and the raised p_{CO_2}. If the time pattern of inspiratory CO_2 is reversed so that the small fall of p_{CO_2} occurs early rather than late in inspiration (Fig. 1, top right), breathing is depressed. If these two patterns are given in alternate breaths we should expect to find alternate-breath stimulation and depression of breathing (Fig. 1, bottom right), and indeed, this is what actually occurs. Note that the end-tidal gases do not change if the experiment is done properly (Metias et al. 1981). We call this pattern Within Breath Changes, WBC.

We reported on the responses to another pattern of alternating stimulus when I last visited Bochum (see Ward and Cunningham 1977): whole breaths of CO_2-rich or CO_2-free gas were given and alveolar p_{CO_2} swung by 7–10 torr (Between Breath Changes, BBC). In this case, too, there was an alternating reflex response (Fig. 1, bottom left).

Figure 2 shows an experimental trace of each kind. The test experiments show obvious reflex alternation of tidal volume superimposed on 'noise', whereas the control trace shows only 'noise'. On the right a large expiratory tidal volume follows a large inspiratory tidal volume, and small follows small. On the left, however, the phase relations are reversed, and the difference between the tidal volumes with each breath implies an alternation of FRC. Five out of 88 WBC runs and 4 out of 30 BBC runs showed a significant I–E phase difference, like this one.

The latencies of the alternating reflexes are about the same as the lung-to-ear circulation time; this, together with their dependence on the presence of hypoxia, indica-

1 The University Laboratory of Physiology, Oxford, England

2 Present address, University of Mansoura, Egypt

Central Neurone Environment
(ed. by M.E. Schläfke, H.P. Koepchen and W.R. See)

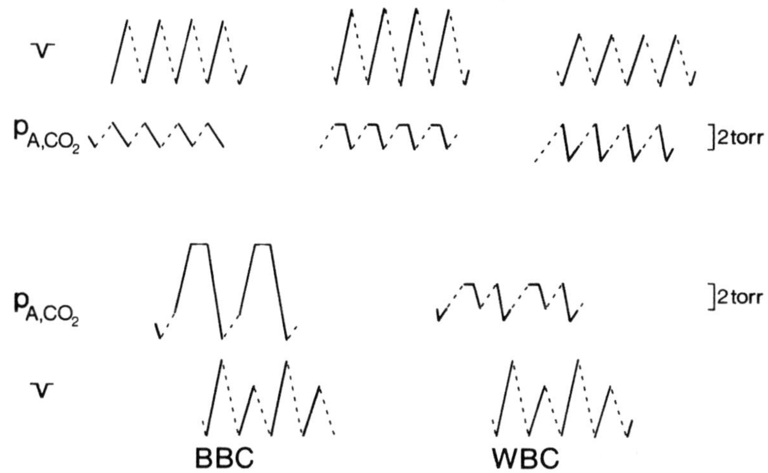

Fig. 1. Diagram of alternating stimuli and responses. *Full lines,* inspiration; *dashed lines,* expiration. For explanation see text

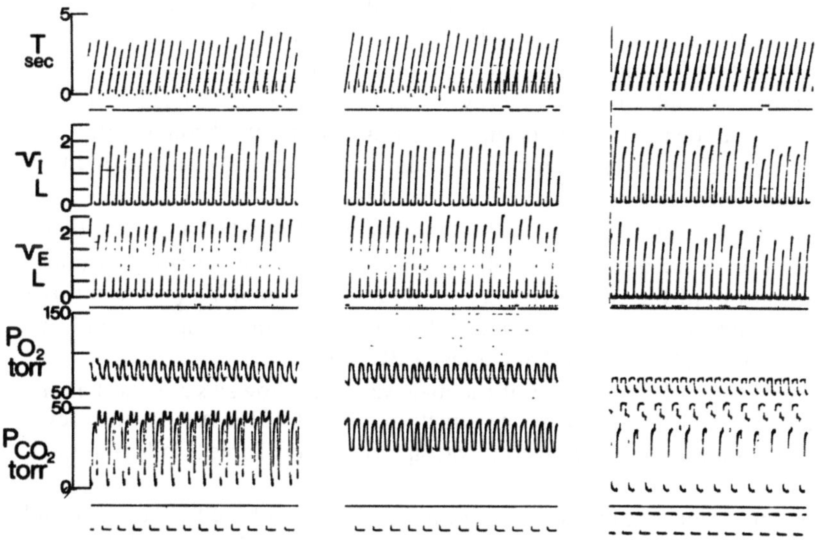

Fig. 2. Three experimental recors. *Left*, Within-Breath Changes, WBC, of alveolar Pco_2. *Centre*, control. *Right* Between-Breath Changes, BBC, of alveolar Pco_2. *From below upwards,* Pco_2 at the mouth; Po_2 at the mouth (note hypoxia in all three); expiratory volume; inspiratory volume; time from start of inspiration, with break at end of inspiration (height to break is T_I, heigh to top is cycle duration, $T_I + T_E$). (Metias et al. 1981; courtesy of Springer-Verlag)

tes that they are mediated by the arterial chemoreceptors. We think it is reasonable to regard the existence of the reflexes as fully established, and the rest of this paper is concerned with the small differences between the patterns of reflex response to WBC and BBC.

The details of the runs have been analysed as shown diagrammatically in Fig. 3. At the top a variable x, which might be inspiratory tidal volume, is represented by the

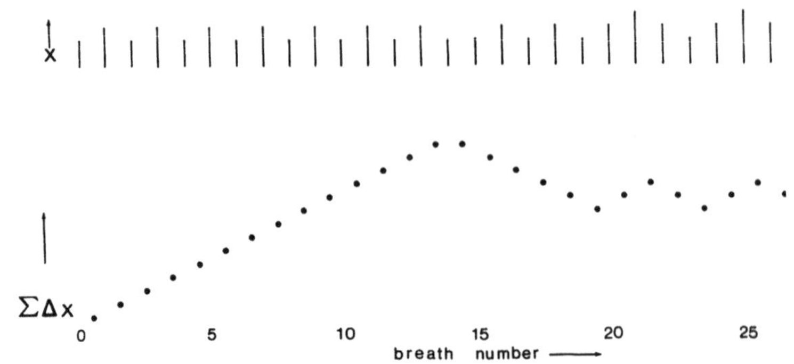

$$\Sigma\Delta x = \Sigma(x_1 - x_0) - (x_2 - x_1) + (x_3 - x_2) - (x_4 - x_3) \ldots (x_n - x_{n-1})$$

Fig. 3. Method of analysing runs: see text

vertical strokes. Breaths 0–20 follow the alternating pattern we are studying; at breath 14 the phase of the alternation is reversed. Breaths 21–27 follow a different, 2-up 2-down pattern. Below, the differences in heights of the vertical strokes are summed, the signs of alternate differences being reversed; the resulting quantity $\Sigma\Delta x$ is the cumulative alternation amplitude. Though the rhythm of breaths 21–27 is regular, it is not the one that this treatment detects, and $\Sigma\Delta x$ wanders around. From breath 0 to breath 14, however, $\Sigma\Delta x$ moves steadily in one direction. Reversal of phase of alternation leads to movement in the opposite direction (breaths 14–20). In control runs, and when the alternating CO_2 signal occurs against a background of high Po_2, these alternating patterns occur quite commonly for short periods, but are rarely sustained. Figure 4 is a printout of the cumulative Δx for volumes $V_{T,I}$ and $V_{T,E}$, mean flows $\bar{v}_I$ and $\bar{v}_E$ and durations T_I and T_E of inspiration and expiration, together with ΔFRC against breath number, for a control and a test WBC run.

In the control trace of Fig. 4 none of the main variables showed significant alternation, though $\Sigma\Delta T_E$ moved mainly in one direction for part of the run. As in a few other control runs mean ΔFRC, however, was significant (cf. Table 2 in Metias et al. 1981).

In the test run several features deserve comment. $\Sigma\Delta\bar{v}_E$ and $\Sigma\Delta T_E$ moved steadily downwards and upwards, respectively, over the whole run. $\Sigma\Delta V_{T,I}$ moved downwards over the first two-thirds, but showed no further consistent change. The inspiratory variables showed no consistent change over the tirst half; $\Sigma\Delta V_{T,I}$ and $\Sigma\Delta\bar{v}_I$ then move steadily upwards. The trace illustrates that reflex expiratory alternation can occur without being preceded by inspiratory alternation (first half) and that reflex inspiratory alternation does not always provoke expiratory alternation which, when it occurs, may be out of phase with inspiration (slopes are opposite; cf. Fig. 2). Finally, alternation of $V_{T,E}$; if not exactly balanced by alternation of $V_{T,I}$, must give rise to alternation of FRC (Ward and Cunningham 1977). This independence of expiratory from inspirator alternation of V_T and so the lack of tight control of FRC under these conditions receive strong statistical support from analysis of the distribution of significant alternation of these two variables in the whole series. There was no significant association of alternatio of either with alternation of the other, though both alternated significantly in a majorit of runs. Equally surprisingly, mean alternation of T_E (by some 90–120 ms) was independent of alternation of T_I.

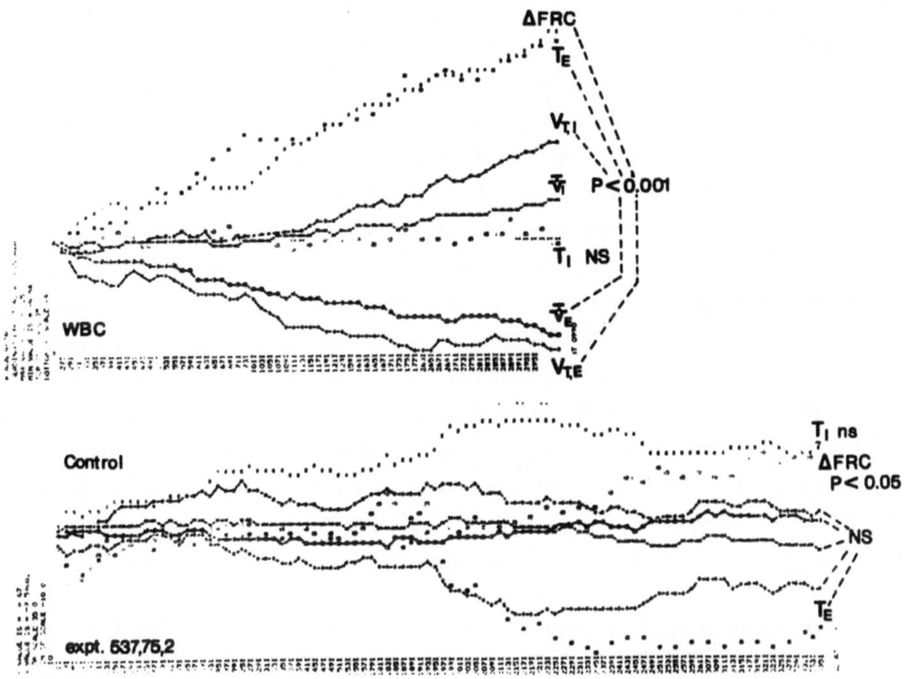

Fig. 4. Control and test runs from expt. 537.75. Print-out of cumulative Δx as run proceeds. On right, significance of differences from zero of mean alternation amplitude. In control print-out, most of the means for the individual variables are not significant and are unlabelled for lack of space. In the print-out of the test experiment all are labelled, but volumes and mean flows are emphasised

We have used data like these to determine the mean $\overline{\Delta x}$ ($=\Sigma\Delta x/N$) and the statistical significance of its difference from zero (t-test) for each of the variables over each of 88 runs of WBC and 30 runs of BBC, both in hypoxia, interspersed with many controls and runs in high O_2. Every test run in hypoxia included at least one and usually several significant ($p < 0.05$) values of mean Δx, but such values occurred only sporadically in the control runs and the runs in high O_2 (table 2 of Metias et al. 1981).

Out of the many results of this analysis we shall describe some of the similarities and differences of the responses to BBC and WBC as they are probably of most interest.

In the first place, although the swings of mean alveolar p_{CO_2} from one breath to the next were 7–10 torr in BBC but were nearly always less than 2 torr in WBC, the amplitudes of the reflex alternations were not very different, and the overall incidences of significant responses were also about the same (Metias et al. 1981). We therefore think that the reflex effects of WBC are due to something other than the small swings of mean p_{CO_2} in gas and blood.

The differences of pattern are conveniently illustrated on mean V_T, T_I, T_E diagrams (Fig. 5). As we use them, T_E is plotted leftward from the origin, and in the present context significant alternation is made to stand out by great exaggeration. In these diagrams the slope of a line joining a point to the origin represents the mean flow

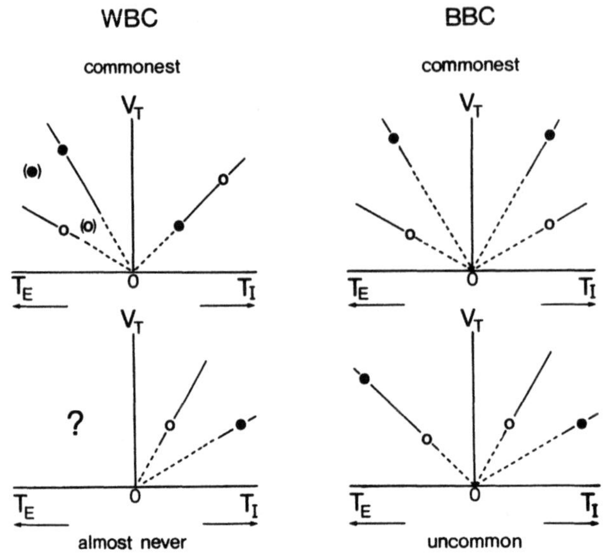

Fig. 5. The commonest and least common patterns of inspiratory alternation on the tidal volume, time diagram, with T_E plotted leftwards from the origin. Not to scale. See text for detail

($\bar{v} = V_T/T$ for both inspiration and expiration); the presence of two such lines or two points in a half-diagram indicates significant alternation of flow, tidal volume and/or half-cycle duration.

Figure 5 shows the commonest and least common patterns. The two preferred inspiratory patterns occured significantly more often than average, and the inspiratory pat tern (flow + time) was significantly rare in WBC. The (volume + time) pattern, preferred in WBC but not in BBC, carries implications for models of the mechanism which terminates inspiration (e.g., Bradley et al. 1975, Cunningham et al. 1980).

With regard to expiration, reflex alternation of expiratory flow took part in a higher proportion of the multiple alternations in BBC than in WBC. Apart from differentiating between the effects of the two methods of forcing, it is difficult in the present state of knowledge to interpret this finding in physiological terms. The somewhat scanty literature on expiration originally left us with the impression that an influence acting DIRECTLY on expiration, rather than secondarily, through inspiration, might have affected expiratory muscle activity or airway calibre, i.e., factors contributing to flow, rather than volumes, but our own evidence is not always consistent with this view.

Inspiratory-expiratory Linkage

We will now consider two differences between the inspiratory-expiratory combinations of alternating patterns encountered during WBC and BBC.

Table 1. The inspiratory-expiratory linkage. The association of two inspiratory patterns of alternation with expiratory alternation

		WBC			BBC	
		$V_{T,I} + \bar{\dot{v}}_I$	$V_{T,I} + T_I$		$V_{T,I} + \bar{\dot{v}}_I$	$V_{T,I} + T_I$
Any expiratory alternation	+	16	15		7	0
	–	3	8		4	5
		NS		P	= 0.034	

+ present; – absent

Table 2. The inspiratory-expiratory linkage. Occurence of alternation of inspiratory and expiratory mean flows separately and together.

		WBC			BBC	
Alternation of		$\bar{\dot{v}}_I$			$\bar{\dot{v}}_I$	
		+	–		+	–
$\bar{\dot{v}}_E$	+	9	27		8(27)	7(15)
	–	24	28		8(22)	7(20)
		< 0.05		P	NS	

Bracketed figures include data of S.A. Ward

Table 1 is a pair of 2 x 2 contingency tables, one for WBC and the other for BBC, showing the presence or absence of significant alternation of any combination of expiratory variables in the same run as either of two common patterns of inspiratory alternation. In WBC alternation of ($V_{T,I} + T_I$), the commonest inspiratory pattern in WBC, was accompanied more often than not by some form of expiratory alternation. In our small BBC series it was not so common, and when present it was NEVER accompanied by any kind of expiratory alternation ($P<0.05$). We are unable to offer any explanation for this difference.

Table 2 is another pair of 2 x 2 contingency tables. It shows the numbers of runs in which significant alternation of mean inspiratory or mean expiratory flows did or did not occur, separately or together. The distribution in BBC is symmètrical, both for the present series alone and when combined with the larger series of Ward and Cunningham (1977), shown in brackets in the table; that is to say, the presence or absence of alternation of one of the flows had no influence on whether the other would alternate or not. In WBC, on the other hand, the distribution is not symmetrical ($P = 0.048$) : it seems that alternation of one flow was less likely to be present it the other was alternating. The table describes the situations for the means of whole runs. The 9 in the top left cell of the WBC table includes some runs in which one flow or the other, but not both, showed alternation at any one time. This situation is illustrated for another WBC run in Fig. 6: here values of $\Sigma\Delta\bar{\dot{v}}_E$ are plotted

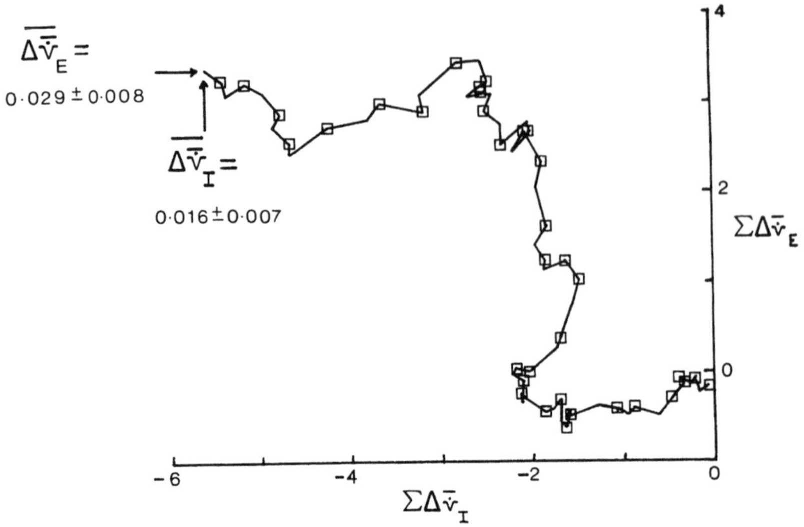

Fig. 6. Cumulative alternation amplitudes of inspiratory and expiratory mean flows plotted against each other for another run from expt. 537.75. this is from one of the nine WBC runs counted in the top left cell of Table 2. Mean alternation amplitude ($\overline{\Delta\dot{v}}_I$; $\overline{\Delta\dot{v}}_E$) was significant in both, but scarcely occurred simultaneously throughout the run. The segments of the line join the points for alternate breaths and there were 75 pairs of breaths in the run

against the corresponding values of $\Sigma\Delta\bar{\dot{v}}_I$ as the run developed, starting at the origin. Large sections of the line are either vertical or horizontal; thus even within this run the preference was for "either/or" rather than "both". Two more runs examined in this way gave similar results. It seems that the use of run means in Table 2 has led us to underestimate the tendency of two flow alternations in WBC to occur separately rather than simultaneously.

It is not difficult to suggest an underlying mechanism for this last observation. In WBC the distinction between the two parts of the alternating signal appears to depend upon events occurring, or not occurring, in the first 2/3 of the inspiratory 1/3 of the whole cycle (Fig. 2). The results of Black and Torrance (1967) and several more recent authors agree in showing that timing within the cycle is important in determining whether a brief stimulus from the arterial chemoreceptors is effective (for references see Metias et al. 1981, and Plaas-Link, this symp). An event that occupies only a minority of the cycle time may be unable to influence both inspiratory and expiratory outputs within the same cycle. In BBC, on the other hand, an important part of the stimulus may last throughout each cycle, and thus be more easily able to influence events in both parts at once.

In summary, these two kinds of forcing stimulate breathing through the same peripheral pathway and at effective intensities that are not very different. Nevertheless, the patterns of reflex response to them, though overlapping, show qualitative differences. It is suggested that some time-dependent component of each determines to some extent which out of more than one possible pathway within the CNS the excitation is likely to traverse.

Acknowledgements. This work was supported by a grant from the Medical Research Council. E.F. Metias was supported by a grant from the Egyptian Government.

References

Black AMS, Torrance RW (1967) Chemoreceptor effects in the respiratory cycle. J Physiol (London) 189:59–61

Bradley GW, Euler Cvon, Marttila I, Roos B (1975) A model of the central and reflex inhibition of inspiration in the cat. Biol. Cybern 19:105–116

Cunningham DJC, Howson MG, Metias EF, Petersen ES (1980) Dynamic changes in the time profile of the alveolar oscillations of p_{CO_2} and their effects on the pattern of breathing in man. Bull Eur Physiopath Respir 16:218–219

Metias EF, Cunningham DJC, Howson MG, Petersen ES, Wolff CB (1981) Reflex effects on human breathing of breath-by-breath changes of the time profile of alveolar p_{CO_2} during steady hypoxia. Pflügers Arch 389:243–250

Ward SA, Cunningham DJC (1977) Separartion of the inspiratory and expiratory reflex effects of alternate-breath oscillation of p_A,CO_2 during hypoxia. Respir Physiol 29:379–390

Central Hypoxic-Hypercapnic Interaction in Mild Hypoxia in Man

Y. HONDA[1]

Introcudtion

Whether hypoxic-hypercapnic interaction is exclusively generated in the peripheral chemoreceptors or further integration takes place in the respiratory centers is a matter of controversy (Torrance 1968, Cunningham 1974). Furthermore, an additional factor calles 'medullary depression' became recognized by a number of investigations (Severinghaus et al. 1966, Weiskopf and Gabel 1975). Thus, evaluation of hypoxic-hypercapnic interaction now appears to be more complicated.

Method

Subjects and Experimental Setup

Twelve healthy male subjects aged 23–50 years were studied twice on different days. Figure 1 shows experimantal setup used in the study, which is a closed circuit including a 9 l respirometer of about 15 l capacity. Breath by breath end-tidal P_{O_2} and P_{CO_2} were continuously monitored and adjusted by an oxygen electrode and an infrared CO_2 analyzer, respectively. Breath by breath/minute volume was also calculated from the output of a potentiometer attached to the respirometer and digitally continously displayed. Just in front of the inlet of the respiratory valve, a rubber bag of 5 l capacity was attached by a three-way stopcock and was filled by humidified O_2.

Withdrawal Test

To evaluate the ventilatory activity of the peripheral chemoreceptors, we used the withdrawal test advocated by Miller et al. (1974). The subject was first brought to a steady state of hypoxia (P_{ETO_2} = 55 mmHg) with hypercapnia (P_{ETCO_2} = 5 mmHg higher than the control level), then two breaths of oxygen were inhaled from the rubber

1 Department of Physiology, Chiba University School of Medicine, Chiba, Japan

Central Neurone Environment
(ed. by M.E. Schläfke, H.P. Koepchen and W.R. See)

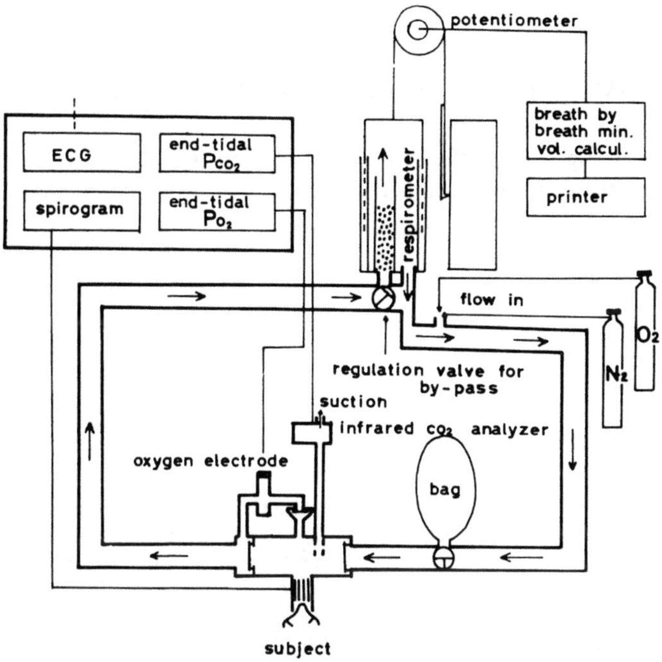

Fig. 1. Experimental setup

bag by turning the three-way stopcock during the preceding expiratory period. The average fall of breath by breath / minute ventilation beginning from 5 s till 20 s after the first O_2 breath was defined as the withdrawal response [$\Delta V_{(w)}$]. This trial was repeated five times in a l-day experiment, so that in total, 10 withdrawal tests were conducted on each subject.

Steady State Hypercapnic-Hypoxia Response

The difference in steady state ventilation between hypoxia (P_{ETO_2} = 55 mmHg) and hyperoxia (P_{ETO_2} > 200 mmHg) with the same degree of hypercapnia (P_{ETCO_2} = 5 mmHg higher than the control) was obtained for each individual and defined as steady state hypercapnic-hypoxia response [$\Delta V_{(s)}$]. Table 1 summarizes the physical characteristics of the subjects and alveolar as well as arterial blood gas levels at steady hypercapnic-hypoxia. The details of the present method are reported elsewhere (Honda et al. 1981).

Table 1. Subjects andsstate of hypoxic hypercapnia (n = 12)

	Age yr	Weight kg	Height cm	P_{ETO_2} mmHg	Pa_{O_2} mmHg	P_{ETCO_2} mmHg
Mean	31	62	167	55.2	53.6	45.9
(SD)	(9.4)	(5.6)	(4.1)	(1.5)	(3.3)	(1.8)

Results

Comparison of steady state [$\Delta V_{(s)}$] and withdrawal hypercapnic-hypoxia response [$\Delta V_{(w)}$].

The two left hand colums in Fig. 2 show the comparison of averaged $\Delta V_{(s)}$ vs $\Delta V_{(w)}$. The difference was statistically significant ($P < 0.01$).

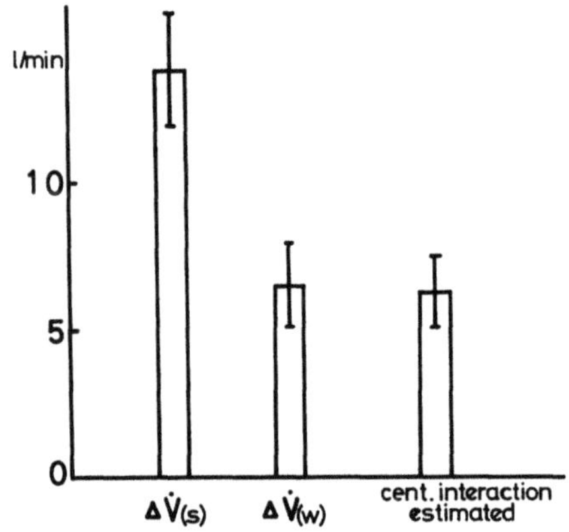

Fig. 2. Steady state [$\Delta V_{(s)}$] and withdrawal response [$\Delta V_{(w)}$] to hypoxic hypercapnia, compared with estimated central hypercapnic interaction. Each *column* indicates average value and *vertical bar* two times SE. The difference between [$\Delta V_{(s)}$] and [$\Delta V_{(w)}$] is significant ($P < 0.01$)

Indications Suggesting the Presence of Hypoxic Depression

The magnitudes of ventilation in hyperoxia of subject Y.S. were mostly higher than in hypoxia with the same degree of hypercapnia, although his withdrawal responses were consistently, positive. Therefore, his [$\Delta V_{(s)} - \Delta V_{(w)}$] values were all negative. Such negative responses were seen on 28 of a total 120 trials in the 12 subjects studied. These results were considered indicative of the presence of hypoxic depression.

Discussion

Comparison Between Steady State and Transient Withdrawal Response to Hypercapnic Hypoxia

The studies , so far conducted, examining hypoxic-hypercapnic interaction are inconsistent in regard to the central or peripheral origin of interaction. Above all Miller et al. (1974), using a similar technique to our, did not see a significant difference between steady state and withdrwal response. The discrepancy between their results and ours was considered to be due to the following two reasons: (1) Miller et al. (1974) used fixed N_2 and CO_2 concentrations, whereas we monitored breath by breath end-tidal P_{O_2} and P_{CO_2}. (2) Among the three subjects examined by Miller et al. (1974) one showed a clear hypoxic depression, just as seen in our subject Y.S. Since the existence of hypoxic depression resulted in an underestimation of steady ventilation, the higher rate of its occurrence, i.e., 1/3 in Miller et al. (1974) vs 1/12 in our's experiments, would have obscured the difference $\Delta V_{(s)}$ from $\Delta V_{(w)}$.

Figure 3 schematically shows the ventilatory drives concerned with the present experiments. In steady state hypercapnic hypoxia the factors concerned were assumed to be hypoxic and hypercapnic stimulation, central as well as peripheral interaction, and in addition hypoxic depression, which was demonstrated in the previous section. On the other hand, in steady state hyperoxic hypercapnia, hypercapnic stimulation is the only ventilatory drive. Therefore, $\Delta V_{(s)}$ may have been derived from hypoxic stimulation, central and peripheral hypercapnic-hypoxic interaction, and hypoxic depression. Since the withdrawal maneuver only eliminates the peripheral chemoreceptor drives with the humoral environment of the central respiratory mechanisms kept unchanged, the response $\Delta V_{(w)}$ should be due to hypoxic stimulation plus peripheral interaction. Therefore, $[\Delta V_{(s)} - \Delta V_{(w)}]$ should express the magnitude of ventilation resulting from central interaction plus hypoxic depression.

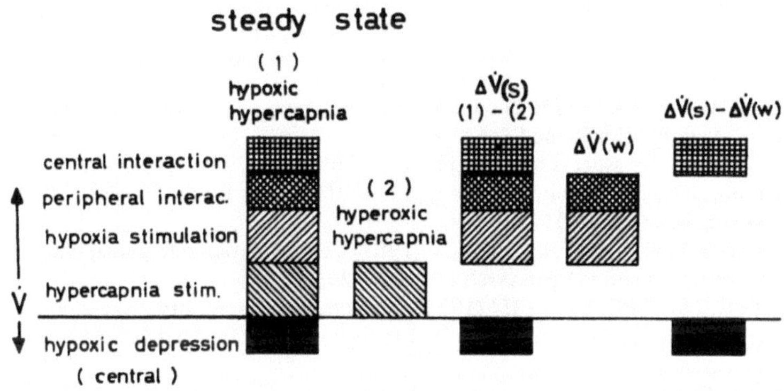

Fig. 3. Schematic representations of the ventilatory drives concerned with the present experiment. Explanations in the text

Estimation of Central Hypoxic-Hypercapnic Interaction

As to the cause of hypoxic depression, we considered a direct inhibitory effect on the central neuronal mechanisms unlikely because the experiments were conducted in mild hypoxia, and even subject Y.S. who exhibited clear hypoxic ventilatory depression, responded much more strongly in hypoxia than in hyperoxia to a Doxapram injection. Therefore, ventilatory depression due to increased Cerebral Blood Flow (CBF) was assumed to be the major part of hypoxic depression. Its magnitude was calculated referring to the data of Ponten and Siesjö (1966) and Borgström et al. (1975). By adding this amount to $[\Delta V_{(s)} - \Delta V_{(w)}]$ we estimated the magnitude of the central hypoxic-hypercapnic interaction. As shown in Fig. 2, this value was nearly as large as the withdrawal response in the mean.

Summary

Hypoxic-hypercapnic interaction in mild hypoxia was studied in twelve healthy males. Steady state ventilatory responses to hypercapnic-hypoxia [$\Delta V_{(s)}$] were obtained as the difference in ventilation between hypoxia (P_{ETO_2} = 55 mmHg) and hyperoxia, with the same degree of hypercapnia (P_{ETCO_2} = 45 mmHg). On the other hand, withdrawal responses [$\Delta V_{(w)}$] were obtained as the magnitude of depression of ventilation caused by two O_2 breaths strarting from the above mentioned hypoxic hypercapnia. Averaged $\Delta V_{(s)}$ and $\Delta V_{(w)}$ were 9.57 ± 5.45 and 6.45 ± 4.90 (mean ± SD) l/min, respectively, the difference being statistically significant. Furthermore, if we assume the presence of ventilatory depression to be due to a fall of tissue P_{CO_2} resulting from an increase in CBF caused by hypoxia, the magnitude of central hypoxic-hypercapnic interaction was estimated to amount to 6.25 ± 4.26 l/min.

References

Borgström L, Johannsson H, Siesjö K (1975) The relationship between arterial P_{O_2} and cerebral blood flow in hypoxia. Acta Physiol Scand 93:423–432

Cunningham DJC (1974) Integrative aspects of the regulation of breathing, a personal view. In: Guyton AC, Widdicombe JG (eds) MTP international review of science. Respir Physiol, ser 1, vol II. Butterworth, London, p 303

Honda Y, Hata N, Sakakibara Y, Nishino T, Satomura Y (1981) Central hypoxic-hypercapnic interaction in mild hypoxia in man. Pflügers Arch 391:289–295

Miller JP, Cunningham DJC, Lloyd BB, Young JM (1974) The transient respiratory effects in man of sudden changes in alveolar CO_2 in hypoxia and in high oxygen. Respir Physiol 20:17–31

Ponten U, Siesjö KB (1966) Gradients of CO_2 tention in the brain. Acta Physiol Scand 67:129–140

Severinghaus JW, Bainton CR, Carcelen A (1966) Respiratory insensitivity to hypoxia in chronically hypoxic man. Respir Physiol 1:308–334

Torrance RW (1968) Prolegomena. In: Torrance RW (ed) Arterial chemoreceptors. Blackwell Sci Pub Oxford, p 1

Weiskopf RB, Gabel RA (1975) Depression of ventilation during hypoxia in man. J Appl Physiol 39:911–915

Neural Control of the Bronchomotor-System

W.T. ULMER[1]

Investigations on neural control of breathing have mostly concentrated on the relationship between the ventilation of the lung and the in- and output of the centers responsible for the regulation. Neural control of ventilation is also influenced by the ability of the "ventilation machine" to function. This function is dependent on the compliance of the lung and on the airway resistance. Both lung compliance and airway resistance (R_{aw}) are of great interest from the clinical point of view. Patients with certain lung diseases die mostly because of disturbances in the mechanical behaviour of the "ventilation machine." Therefore, it seems worthwhile also to consider the regulation of the bronchomotor-system as an important part of the regulation of breathing.

Little is known about any regulation of the compliance of the lung, therefore the lung compliance is always believed to be constant. Does this perhaps imply that there is no neural regulation, no neural control?

The airway resistance, which mainly is controlled by the bronchial muscles, shows spontaneous changes during the day (De Millas and Ulmer 1971) and somestimes during seconds.

Therefore, we can presume that there are some changes of the activity of the bronchial muscles which could be under neural control.

A few yaers ago we believed that the tremendous increase of the R_{aw} which could be measured on patients with obstructive airway disease, was predominantly caused by the release of mediators like histamine, serotonin, SRS-A or acetylcholine, which work directly on the bronchial muscles.

However, investigations by different groups clearly show that there is also a neural control of the bronchomotor-system, which may, under clinical conditions, be more important than the direct effects of mediators on the muscles. De Kock et al. (1966), Simonsson et al. (1967), Widdicombe et al. (1962), Widdicombe (1977), Nadel (1980), Islam et al. (1972, 1975), Islam and Ulmer (1973), Ulmer (1979) clearly showed that the N. vagus is involved in controlling the tone of the bronchial muscles probably via a reflex (Fig. 1).

By chemical block (or section) of the N. vagus we blocked the motor and the sensory part of this nerve. Is this control of the bronchial muscles a specific reflex, or is it only

1 Medizinische Universitätsklinik und Poliklinik der Berufsgenossenschaftlichen Krankenanstalten „Bergmannsheil Bochum", Hunscheidtstraße 1, D-4630 Bochum, Fed. Rep. of Germany

Central Neurone Environment
(ed. by M.E. Schläfke, H.P. Koepchen and W.R. See)

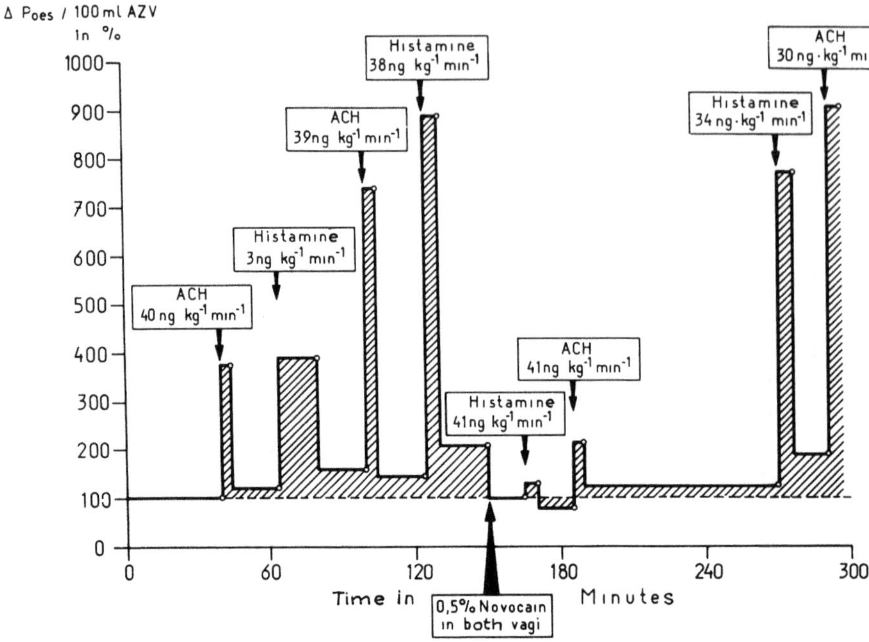

Fig. 1. Δ Poes/100 ml tidal volume (TV) as measurement of airway resistance in % of the initial v during histamine and acetylcholine (*ACH*) inhalation before, during and after vagus blockade by novocain on dogs. (Islam and Ulmer 1973)

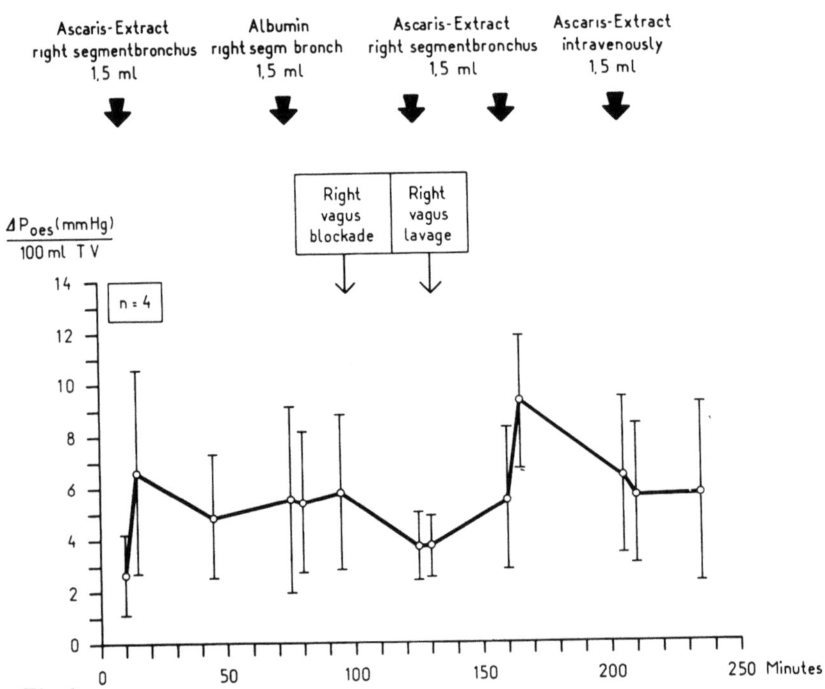

Fig. 2.

the basic activity of the N. vagus which we block thereby preventing any activity of the bronchial muscles? Challenge by histamine or allergens of local areas restricted to the trachea or one segmental bronchus produces a constriction of all muscles of the bronchial tree (Fig. 2).

From the experiments performed by Zimmermann et al. (1979 a,b,c) we also know that there are sensory receptors in the upper part of the trachea, and that by cutting the N. hypoglossus, glossopharyngicus or laryngicus cranialis, we can prevent the bronchoconstriction which otherwise developes by challenge of this restricted part of the trachea (Fig. 3).

Patients with chronic obstructive airway disease are frequently hyperreactive to bronchoconstrictive agents. The question that should be answered is where this increased sensitivity is located.

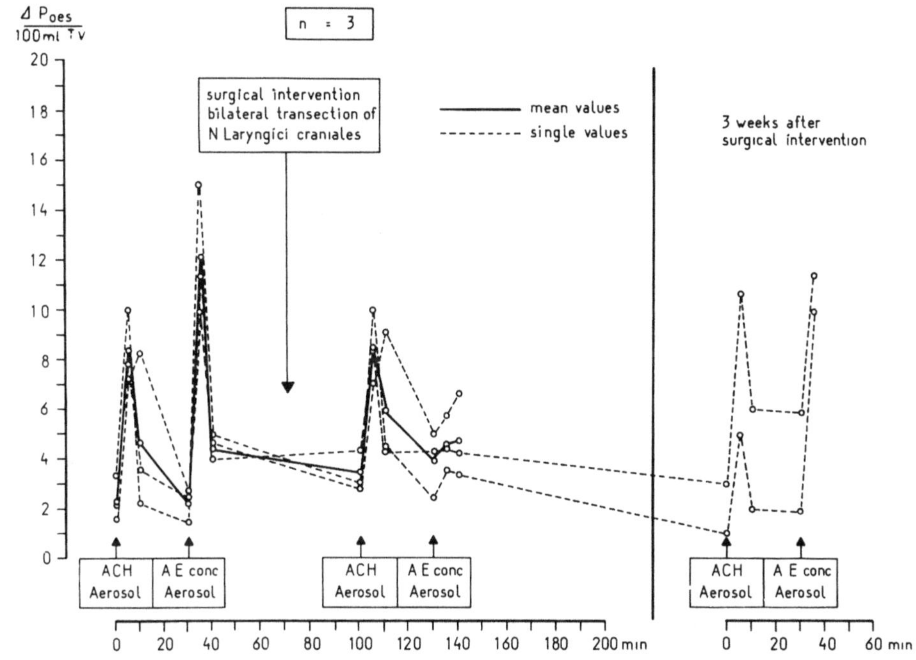

Fig. 3. Δ Poes/100 ml TV as measurement of airway resistance during inhalation of acetylcholine (*ACH*) and Ascaris-suum-extract. (*AE*) aerosols before and after cutting of the N. laryngicus cranialis

---

Fig. 2. Δ Poes/100 ml TV as measurement of airway resistance during installation of Ascaris-suum-extract locally in one segmental bronchus before, during and following block of the N. vagus. Albumin solution was given once as control

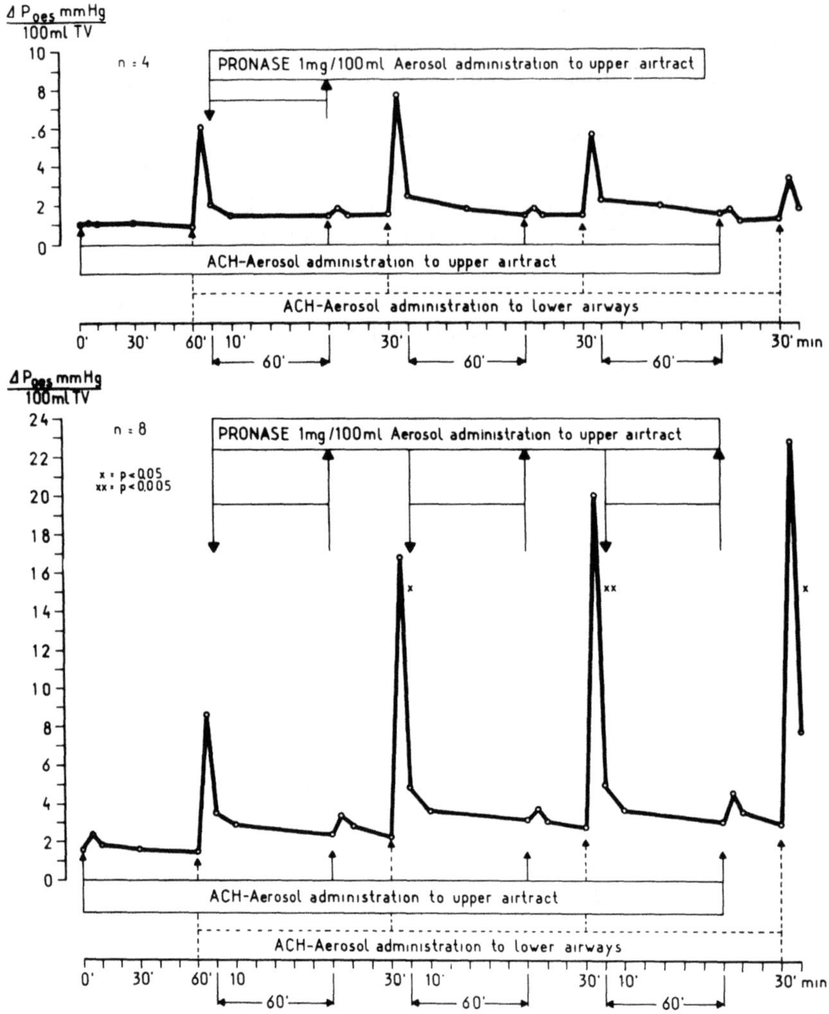

Fig. 4. Δ Poes/100 ml TV as measurement of airway resistance during inhalation of acetylcholine (*ACH*) to upper and lower airways before and following the inhalation of pronase to the upper airway. **a** Pronase was given over 60'; **b** Pronase was given three times, each time over 60'. The sensitivity to ACH in the lower part of the airways increased significantly

Histamine, allergens (Ascaris-suum-extract) as well as proteolytic enzymes when given in very low dosages, increase the sensitivity of this reflex arc (Ulmer et al. 1971, Zimmermann et al. 1978, 1979 a,b,c) (Fig. 4).

However, when we gave pronase only to the upper part of the airways, a statistically significant increase in airway resistance, after challenge with ACH, occurred only during exposure of the peripheral (lower) part of the airways to ACH (Zimmermann et al. 1979 a,b,c).

This and other experiments point in the same direction (Josenhans et al. 1969) namely that there may be some central integration of inputs from different areas. The change of the "peripheral sensitivity" could be caused by an increased output due to altered inputs from different areas to the CNS.

Still there are many open questions on this clinically important problem of neural control of bronchoconstriction. Where are the sensory receptors really located and how can these receptors be influenced? Where in the CNS are the areas of regulation located and what are the interactions with other centers? Are there interactions for reflex mucus secretion or reflex bronchoconstriction and how do they work? For the present we may be sure that there exists a neural control of the bronchomotor-system which may be of primary importance for most of the patients with so-called obstructive airway disease.

References

Islam MS, Ulmer WT (1973) Der Wirkungsmechanismus von Serotonin (5-Hydroxitryptamin) und Histamin bei der Atemwegsobstruction. Respiration 30:360–377

Islam MS, Ulmer WT (1975) Lokale Überempfindlichkeit sensorischer Rezeptoren als Ursache reflektorischer Atemwegsobstruktion. Respiration 32:445

Islam MS, Rasche B, Vastag E, Ulmer WT (1972) Über den Wirkungsmechanismus von Histamin in den Atemwegen. Respiration 29:538

Josenhans WT, Melville GN, Ulmer WT (1969) The effect of facial cold stimulation on airway conductance in healthy man. Can J Physiol Pharmacol 475:vol 47

Kock De MA, Nadel JA, Zwi S, Colebatch HJH, Olsen CR (1966) New method for perfusing bronchial arteries: histamine bronchoconstriction and apnea. J Appl Physiol 21:185–194

Millas De H, Ulmer WT (1971) Der Tagesrhythmus der Strömungswiderstände in den Atemwegen und deren Beeinflußbarkeit. Pneumonologie 144:237

Nadel JA (1980) Pathophysiology of Asthma: Neural effects on airway smooth muscle, submucosal glands, breathing and cough. Prog Respir Res 14:1–8

Simonsson BG, Jacobs FM, Nadel JA (1967) Role of autonomic nervous system and the cough reflex in the increased responsiveness of airways in patients with obstrutive airway disease. J Clin Invest 46:1812

Ulmer WT (1979) Die obstruktiven Atemwegserkrankungen: Bronchitis, Asthma, Emphysem. Handbuch der Inneren Medizin, vol IV/2. Springer, Berlin, Heidelberg, New York, pp 449–541

Ulmer WT, Islam MS, Bakran I Jr (1971) Untersuchungen zur Ursache der Atemwegsobstruktion und des überempfindlichen Bronchialsystems. Dtsch med Wochenschr 96:1759

Widdicombe JG (1977) Some experimental models of acute asthma. JR Coll Physicians London 11: 141-155

Widdicombe JG, Kent DC, Nadel JA (1962) Mechanism of bronchoconstriction during inhalation of dust. J Appl Physiol 17:613–616

Zimmermann I, Ulmer WT (1978) Influence of low concentrations of allergens on bronchial system. Respiration 35:87

Zimmermann I, Islam MS, Ulmer WT (1976) Effect of ascaris extract applied intravenously on segment bronchus and the influence of ipsilateral vagus blockade. Respiration 33: 270

Zimmermann I, Walkenhorst M, Ulmer WT (1979a) The location of sensoric bronchoconstricting receptors in the upper airway. Respiration 38:1–11

Zimmermann I, Ulmer WT, Weller W (1979b) The role of upper airways and of sensoric receptors on reflex bronchoconstriction. Res Exp Med 174: 253-265

Zimmermann I, Haxhiu MA, Bugalho De Almeida AA, Ulmer WT (1979c) Localisation of the sensitivity of the airways be proteolytic enzymes (Pronase) against acetylcholine. Respiration 38: 249–256

Reflex Effects of J Receptors

A. S. PAINTAL[1]

At a symposium held in Philadelphia in 1976 evidence was provided to show that the J receptors, which are normally stimulated during exercise, were apparently the main pulmonary mechanism responsible for producing reflex tachypnoea, bronchoconstriction, laryngeal constriction, (and also cough in man), as there was no firm evidence that the rapidly adapting stretch receptors specifically excited inspiration. Indeed one could asssume that the latter produced the same reflex effects as the slowly adapting stretch receptors (Paintal 1977b). There is now further evidence to show that the two (slowly and rapidly adapting stretch receptors) constitute a functionally homogenous group. In fact a review of the available information [detailed elsewhere (Paintal 1982 in press) including some recent observations by Paintal and Ravi (1980) and Sant'Ambrogio and Sant'Ambrogio (1981)] has revealed that apart from differences in the volume threshold of the receptors, the properties and responses of the slowly and the rapidly adapting receptors to 14 different stimuli are similar – in fact almost identical in several instances as detailed in Table 1. In view of this and earlier information one can safely assume that these pulmonary stretch receptors together produce bronchodilation, tachycardia, and excitation of expiratory muscles (i.e., effects opposite to those produced by J receptors) (see Paintal 1973) although, they can under certain experimental situations excite inspiratory muscles at eupnoeic levels of lung inflation (see Figs. 2 and 3 in Larrabbee and Knowlton, 1946). Using a 'Servo-respirator', Dimarco et al. (1980) have also observed that at low volumes pulmonary stretch receptors can facilitate inspiratory muscle activity during eupnoeic breathing.

Role of J Receptors in Bronchoconstriction

It has been established (see Paintal 1977a,b) that stimulation of J receptors produces reflex bronchoconstriction in rabbits (Karczweski and Widdicombe 1969) and cats (Barer and Nusser 1953). Recently, Russell and Lai-Fook (1979) have shown that stimulation of the J receptors of dogs with capsaicine, leads to narrowing of the airways. In the case of man the evidence is also clear for it was shown by Bruderman et al. (1966) that

1 Department of Physiology and DST, Centre for Visceral Mechanisms, Vallabhbhai Patel Chest Institute, Delhi – 110 007, India

Central Neurone Environment
(ed. by M.E. Schläfke, H.P. Koepchen and W.R. See)

Table 1. Comparison of the characteristics and responses of slowly and rapidly adapting pulmonary stretch receptors of cats; data from other animals is specified. Numbers in parenthesis relate to source of information given in the list of references

Sl. No.	Stimulus or characteristic of receptors	Slowly adapting stretch receptor	Rapidly adapting stretch receptor	Inference
1.	Conduction velocity of fibres (mean)	36 m/s (20) (dogs) 31.6 (27)	25 m/s (20) (dogs) 28.8 (27)	Different Similar
2.	Inflation threshold	Below 40 ml in 91–100% of receptors (17, 32)	Below 40 ml in 50–68% of receptors (17, 32)	Different
3.	Effect of deflation	33% stimulated (17)	66% stimulated (17)	Different
4.	Location	83% in lung parenchyma (25) (dogs) 74% intrapulmonary (29)	...... (dogs) 86% intrapulmonary (29)	Simlilar
5.	Accessibility to chemical substances through pulmonary circulation	83% of higher–threshold and 25% of low-threshold accessible (25) (dogs) 90% of intrapulmonary receptors accessible (29)	..... (dogs) 100% of intrapulmonary receptors accessible (29)	Similar
6.	Effect of CO_2			
	a) Effect of lowering CO_2 from 32 to 3 mm Hg.	(dogs) Activity increased (9.)	(dogs) Activity increased (9.)	Similar
	b) Increasing CO_2 from low levels	(dogs) Activity decreased (9.)	(dogs) Activity decreased (9.)	Similar
	c) Raising CO_2 from 32 to 50 mm Hg	(dogs) Reduction of activity by 4.5% (9.)	(dogs) No change (9.)	Similar
7.	Histamine			
	a) intravenous	Sensitized ? (33)	Sensitized (2)	Similar
	b) aerosol	Stimulated (34)	(dogs) Stimulated (28)	Similar
8.	Pulmonary congestion	Stimulated (19)	(Rabbits) Stimulated (30)	Similar
9.	Pulmonary microembolism	Stimulated (3)	Stimulated (3)	Similar
10.	Sulphur dioxide	15% sensitized (32)	11% sensitized (32)	Similar
11.	Insufflation of Powders	Sensitization present (32)	60% sensitized (32)	Similar
12.	Ammonia vapour	66% stimulated (32)	82% stimulated (32)	Similar
13.	Ether vapour	60% partially depressed (32)	75% depressed (32)	Similar
14.	Halothane	Sensitized (31,8)	Sensitized (2)	Similar
15.	Veratridine into right atrium	(dogs) Stimulated (29)	(Dogs) stimulated (29)	Similar
16.	Bradykinin	(dogs) Variable stimulation/sensitization (16)	(Dogs) Stimulation/sensitization in 33% (16)	Similar
17.	Benzonatate (right atrium)	(dogs) Blocked (29)	(Dogs) Blocked (29)	Similar
18.	Bupivacaine aerosol	Depressed-blocked (12)	Depressed-Blocked (12)	Similar

injection of lobeline, which stimulates the J receptors of man (Jain et al. 1972) increased airway resistance. This is of significance since these receptors are stimulated by increase in pulmonary blood flow (see below) and so could play an important role in the genesis of exercise induced asthma.

Effect of Blood Flow on J Receptors

Perhaps the most significant recent observation is that the J receptors are stimulated by increases in pulmonary blood flow, that normally occur during moderate exercise (Anand and Paintal 1980). This information was obtained through experiments carried out on anaesthetized cats in which impulses from J receptors in the upper two right lung lobes through which blood flow was varied, were recorded. The variation in blood flow was achieved through the experimental arrangement shown in Fig.1. In order to obtain sufficient increases in blood flow above the initial levels, the third lobe of the right lung was removed. The resting discharge (usually near zero) from a J receptor in a right lung lobe was recorded before occluding the left pulmonary artery. Thereafter the left pulmonary artery was occluded so that the entire cardiac output (measured by the Fick principle) flowed through the right lung as shown in Fig. 1B. Thereafter the lungs of both sides were removed and weighed so that the blood flow through the right lung before occlusion of the left pulmonary artery could be determined as indicated in Fig. 1A.

The results obtained in the case of six (of ten) receptors, in which the activity was increased by increasing blood flow, are shown in Fig. 2. The variation in the responses

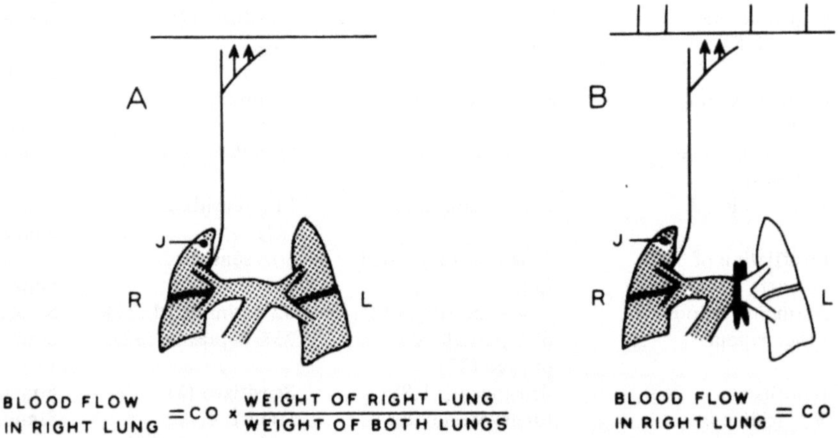

Fig. 1. Schematic representation of the experimental arrangement for studying the effect, of varying blood flow through the upper two lobes of the right lung (third lobe removed), on the activity of J receptors located in these lobes. Variation in blood flow was achieved by occluding the left pulmonary artery in cats with open chests. The cardiac output was measured using the Fick principle. With left pulmonary artery occluded, the entire cardiac output flowed through the right lung lobes. The amount of blood flow before occlusion was estimated using the formula at the bottom of A

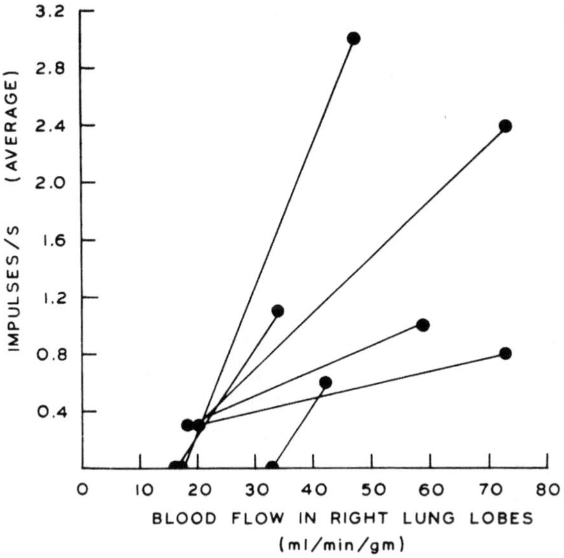

Fig. 2. Effect of increase in blood flow on the activity of six J receptors in the right lung showing variation in responses. There are two points for each receptor, the *lower* one shows the initial activity before occlusion of the left pulmonary artery and the *upper* one the activity during increased blood flow produced by occluding the left pulmonary artery. Abscissa, blood flow, ml/min/gm of lung tissue. Ordinate, frequency of discharge averaged over means from 0 to 10 s

was considerable; indeed in four of the ten receptors there was no increase in activity following increase in blood flow (not shown in Fig. 2). On the other hand it is clear from Fig. 2 that there must be some receptors (e.g., the case on which blood flow increased from 33 to 42 ml/min/gm which are stimulated by even small increases in blood flow (of the order of 27%). On averaging the activity in the whole sample of ten receptors it was found that increasing the blood flow by an average of 133% resulted in a discharge of 0.75 impulses/s. Although this might seem negligible activity, it became clear that this order of input from J receptors (provided by injecting 18 μg/kg of phenyl diguanide into the right atrium) produced powerful visceral and somatic reflex effects (Anand and Paintal 1980). The significance of these findings lies in the fact that increases of 133% in blood flow are produced by moderate exercise and that, since J receptor activity of about 0.75 impulses/s, produced by such increases, yields powerful reflex effects, much smaller increases in blood flow would be effective in producing noticeable reflex effects.

Reflexly Effective 'Threshold' Increases in Blood Flow

An estimate of the reflexly effective 'threshold' increase in blood flow can be made in the light of the fact that the threshold dose for producing barely visible reduction in the size (or Ti) of one breath, was found to average 6 μg/kg in 9 cats. Thus if 18 μg/kg of phenyl diguanide produces activity equivalent to that produced by a 133% increase in blood flow, it follows that a 6 μg/kg dose will be equivalent to about a 44% increase in blood flow, i.e., an increase that would be produced by mild exercise. In fact the actual reflexly effective threshold increase in blood flow in the absence of anaesthesia would be expected to be even less.

Reflex Potency of J Impulses

In the cat (and possibly also in the rabbit) the reflex potency of J receptor impulses is considerable, since the reflexly effective 'threshold' dose of phenyl diguanide is 6 μg/kg – a dose that would be expected to yield about 0.3 impulses/s since 12 μg/kg yields 0.5 impulses/s (see Fig. 5 in Anand and Paintal, 1980). In contrast, the reflex potency of J receptor impulses in anaesthetized dogs is comparatively much lower – a fact that has emerged from a consideration of some observations of Coleridge and Coleridge (1977 b) on the effects of phenyl diguanide on J receptors as well as of some new observations described below.

Effect of Phenyl Diguanide on J Receptors of Dogs

Coleridge and Coleridge (1977 b) have reported that phenyl diguanide stimulated about 40% (10 of 25) of the J receptors of dogs but its excitatory effect was much less than that of capsaicine. The discharge produced in the 10 receptors by 10 μg/kg of phenyl diguanide averaged 3.3 impulses/s. However, for the whole sample of 25 receptors the average would be about 1.3 impulses/s – a response that is greater than that produced by the same dose in cats which, according to the data published by Anand and Paintal (1980), would yield less than 0.5 impulses/s (see above). In those dogs that yielded activity of this order Coleridge and Coleridge (1977b) apparently observed no reflex effects (of injecting 10 μg/kg phenyl diguanide) that could be attributed to receptors. However, in some dogs with open chests the same dose yielded marked activity of the order of 17 impulses/s, and in such dogs they observed typical cardiovascular reflex effects of J receptor stimulation. However, as mentioned above, on the whole the excitatory effects of phenyl diguanide are much smaller than those produced by capsaicine, and perhaps for this reason and also the fact that phenyl diguanide does not produce any reflex effects attributable to the J receptors, it is possible that the excitatory effects of phenyl diguanide on the J receptors of dogs have been ignored so far (see Paintal 1982 in press).

Why Phenyl Diguanide Does not Produce Reflex Effects Through J Receptor Impulses in Dogs

In view of the fact that activity of the order of 0.3 impulses/s (see above) in cats can produce reflex effects the question arises, 'why does activity of the order of 1.3 impulses produced by phenyl diguanide not produce noticeable reflex effects in dogs?" One reason could be the presence of a relatively high level of "spontaneous" activity of about 1.9 impulses/s in dogs with intact chests (Coleridge and Coleridge 1977 a). For comparison one has to exclude the exceptional endings found by Coleridge and Coleridge (1977 b) that yielded a discharge of 17 impulses/s following injection of phenyl diguanide. After doing so the average increase in activity in the remaining 24 receptors produced by 10 μg/kg phenyl diguanide would be about 0.7 impulses/s. This is less than the resting activity of 1.9 impulses/s in dogs with intact chests (Coleridge and Coleridge 1977 a). Perhaps it is this already existing "spontaneous" activity that renders the small increase in activity, produced by phenyl diguanide, ineffective for the manifestation of reflex effects.

Reflexly Effective Level af Activity in J Receptors of Dogs.

The above conclusion raises the question "At what level of activity in J receptors are reflex effects produced in dogs?" In order to answer this question experiments were carried out on dogs and the reflex effect, of injecting graded doses of capsaicine (starting with 1 μg/kg) into the right atrium, was recorded with the purpose of finding the reflexly effective dose of capsaicine. As expected the strength of the reflex effects depended on the type of anaesthetic used and the level of anaesthesia. However, it was surprising to find that in some dogs, anaesthetized with sodium pentobarbital, no reflex effects could be elicited even by 10 μg/kg of capsaicine. The best response was obtained in a dog anesthetized with inactin in which we observed that the threshold dose for producing a reflex respiratory effect was 3 μg/kg injected into the right atrium (Islam, Paintal, Ravi and Ulmer 1981 – unpublished observations).

There is no relationship available, between the dose of capsaicine and the activity produced, in the J receptors of dogs, from which one could determine the level of activity generated by 3 μg/kg capsaicine. However, since 10 μg/kg yields an average peak activity of about 34 impulses/s (Coleridge and Coleridge 1977 a) it is likely [assuming that the activity of J receptors is roughly proportional to the dose of capsaicine as in the case of the phenyl diguanide (see Fig. 5 in Anand and Paintal 1980) [that the average peak activity generated by 3 μg/kg will be about 10 impulses/s. Although, when compared to the responses in cats, this is considerable activity, it does appear to be the threshold activity for producing reflex effects (apnoea and bradycardia) in anaesthetized dogs. However, it is possible that using other kinds of inputs from J receptors the reflexly effective "threshold" activity might be less since Kalia (1977) observed that the J reflex (inhibition of the knee jerk) was elicited in dogs when the lungs were congested to apparently sub-oedema levels and the average activity during pulmonary oedema is 7.3 impulses/s (Coleridge and Coleridge 1977a). Alternatively it is conceivable that the reflex effects of low level activity in J receptors take time to manifest and are concealed by opposing reflex effects produced by other receptors further downstream e.g., chemoreceptors, in the case of phenyl diguanide, which accelerate respiration and produce a rise in blood pressure (Dawes et al. 1952). In any case it is obvious that the anaesthetized dog is not a suitable animal for the study of respiratory and cardiovascular reflex effects of J receptors.

Summary

1. A review of past and recent information shows that the slowly and rapidly adapting pulmonary stretch receptors constitute a functionally homogenenous group.
2. Available evidence shows that the J receptors produce bronchoconstriction in rabbits, cats dogs and man.
3. In anaesthetized cats the threshold activity in J receptors for producing reflex effects is probably about 0.3 impulses/sec. It is estimated that such activity would be produced by about 44% increase in pulmonary blood flow.
4. The anaesthetized dog apparently needs a relatively large input from J receptors for reflex effects to be elicited by the injection of chemical substances.

References

1. Anand A, Paintal AS (1980) Reflex effects following selective stimulation of J receptors in the cat. J Physiol (London) 299:553–572
2. Armstrong DJ, Luck JC (1974) A comparative study of irritant and type J receptors in the cat. Respir Physiol 21:47–60
3. Armstrong DJ, Luck JC, Martin VM (1976) The effect of emboli upon intrapulmonary receptors in the cat. Respir Physiol 26:41–54
4. Barer GR, Nusser E (1953) The part played by bronchial muscles in pulmonary reflexes. Br J Pharmacol Chemother 8:315–320
5. Bruderman I, Stern S, Braun K (1966) Respiratory effects of intravenous lobeline in normal subjects and in patients with mitral stenosis. Br Heart J 28:740–745
6. Coleridge HM, Coleridge JCG (1977a) Afferent vagal C-fibers in the dog lung: their discharge during spontaneous breathing, and their stimulation by alloxan and pulmonary congestion. In: Paintal AS, Gill-Kumar P (eds) Respiratory adaption, capillary exchange and reflex mechanisms. Vallabhbhai Patel Chest Inst, Delhi, p 396–405
7. Coleridge HM, Coleridge JCG (1977b) Impulse activity in afferent vagal C-fibres with endings in the intrapulmonary airways of dogs. Respir Physiol 29:125–142
8. Coleridge HM, Coleridge JCG, Luck JC, Norman J (1968) The effect of four volatile anaesthetic agents on the impulse activity of two types of pulmonary receptor. Br J Anaeth 40:484–492
9. Coleridge HM, Coleridge JCG, Banzett RB (1978) II Effect of CO_2 on afferent vagal endings in the canine lung. Respir Physiol 34:135–151
10. Dawes GS, Mott JC, Widdicombe JG (1952) Chemoreceptor reflexes in the dog and the action of phenyl diguanide. Arch Int Pharmacodyn Thér 90:203–222
11. Dimarco AF, Euler C von, Romaniuk JR, Yamamoto Y (1980) Low threshold facilitation of inspiration by lung volume increments. Acta Physiol Scand 109:343–344
12. Fahim M, Jain SK (1979) The effect of bupivacaine aerosol on the activity of pulmonary stretch and 'irritant' receptors. J Physiol (London) 288:367–378
13. Jain SK, Subramanian S, Julka DB, Guz A (1972) Search for evidence of lung chemoreflexes in man: study of respiratory and circulatory effects of phenyldiguanide and lobeline. Clin Sci 42: 163–177
14. Kalia M (1977) The J reflex produced by pulmonary congestion in the cat. Proc Int Un Physiol Sci 13:364
15. Karczewski W, Widdicombe JG (1969) The role of the vagus nerves in the respiratory and circulatory responses to intravenous histamine and phenyl diguanide in rabbits. J Physiol (London) 201: 271–291

16. Kaufman MP, Coleridge HM, Coleridge JCG, Baker DG (1980) Bradykinin stimulates afferent vagal C-fibers in intrapulmonary airways of dogs. J Appl Physiol 48:511–517
17. Knowlton GC, Larrabee MG (1946) A unitary analysis of pulmonary volume receptors. Am J Physiol 147:100–114
18. Larrabee MG, Knowlton GC (1946) Excitation and inhibition of phrenic motoneurones by inflation of the lungs. Am J Physiol 147:90–99
19. Marshall R, Widdicombe JG (1958) The activity of pulmonary stretch receptors during congestion of the lungs. Q J Exp Physiol 43:320–330
20. Paintal AS (1953) The conduction velocities of respiratory and cardiovascular afferent fibres in the vagus nerve. J Physiol (London) 121:341–359
21. Paintal AS (1973) Vagal sensory receptors and their reflex effects. Physiol Rev 53:159–227
22. Paintal AS (1977a) Thoracic receptors connected with sensation. Br Med Bull 33:169–174
23. Paintal AS (1977b) The nature and effects of sensory inputs into the respiratory centres. Fed Proc Fed Am Soc Exp Biol 30:2428–2432
24. Paintal AS (1982) Lung and ariway receptors. In: Pallot DJ (ed) Some aspects of the control of respiration. Croom Helm, London (in press)
25. Paintal AS, Ravi K (1980) The relative location of low – and higher threshold pulmonary stretch receptors. J Physiol (London) 307:50–51P
26. Russell JA, Lai-Fook SJ (1979) Reflex bronchoconstriction induced by capsaicin in the dog. J Appl Physiol 47:961–967
27. Sampson SR (1977) Sensory neurophysiology of airways. Am Rev Resp Dis 115 Suppl:107–115
28. Sampson SR, Vidruk EH (1975) Properties of 'irritant' receptors in canine lung. Respir Physiol 25:9–22
29. Sant'Ambrogio FB, Sant'Ambrogio G (1981) Circulatory accessibility of afferent endings in the tracheo-bronchial tree of the dog. J Physiol (London) 318:52–53P
30. Sellick H, Widdicombe JG (1969) The activity of lung irritant receptors during pneumothorax, hyperpnoea and pulmonary vascular congestion. J Physiol (London) 203:359–381
31. Whitteridge D (1958) Effects of anaesthetics on mechanical receptors. Br Med Bull 14:5–7
32. Widdicombe JG (1954a) Receptors in the trachea and bronchi of the cat. J Physiol (London) 123:71–104
33. Widdicombe JG (1954b) The site of pulmonary stretch receptors in the cat. J Physiol (London) 125:336–351
34. Widdicombe JG (1961) The activity of pulmonary stretch receptors during bronchoconstriction, pulmonary oedema, atelectasis and breathing against a resistance. J Physiol (London) 159:436–450

Respiratory and Cardiovascular Reflexes from the Afferent Sympathetic Nerves

N.R. PRABHAKAR and W. MAREK[1]

The vagal (parasympathetic) sensory receptors giving rise to various repiratory and cardiovascular reflexes have been extensively studied and were dealt with in many papers and reviews (Paintal 1973, 1977). Less well known are the reflexes from the afferent nerves passing through the sympathetic system. The "sympathetic afferents" are of considerable importance because the sensations, such as pain and discomfort (Nociception), were assumed to be mediated by these afferents.

In the present investigation the afferent mesenterial nerves were studied, because their stimulation is known to increase ventilation, especially the tidal volume (Chernigovsky 1976, Frank 1975). With regard to the neuronal regulation of respiration, it was of interest to analyse the response of the phrenic nerve and the intercostal muscles (as an index of respiratory neuronal output) to the stimulation of the mesenterial afferents.

Stimulation of the mesenterial afferents in addition to the respiratory changes leads to an increase in the arterial pressure (Fank 1975, Saphir and Rapaport 1969). Recently it was shown that the renin-angiotensin system plays an important role in the arterial pressure increase induced by bladder distension (Prabhakar and Marek 1979, Prabhakar et al. 1978). Therefore we felt it worthwhile to investigate whether the same meachanism would also play a role in the cardiovascular changes observed with stimulation of the afferent mesenterial nerves.

Methods and Results

The experiments were performed on chloralose-urethane anaesthetized, spontaneously breathing cats. The tidal volume was recorded, with Krogh's spirometer connected to the tracheal cannula, along with the integrated electrical activities of the phrenic nerve and inspiratory intercostal muscles. The arterial pressure was recorded from a catheter connected to one femoral artery. Expiratory p_{CO_2} was analysed with an infra-red analyser. After an abdominal incision the mesenteric plexus running along the superior mesenteric artery was isolated, the nerve bundles were cut and the central ends of these nerves were used for electrical stimulation. Statistical evaluation was performed using the paired or unpaired t-test.

1 Institut für Physiologie, Ruhr-Universität Bochum, D–4630 Bochum, Fed. Rep. of Germany

Central Neurone Environment
(ed. by M.E. Schläfke, H.P. Koepchen and W.R. See)

Respiratory Responses

A typical example of an experiment showing the respiratory response to continuous (repetitive) electrical stimulation of the mesenteric plexus is presented in Fig. 1A. In general, stimulation of the mesenteric plexus for about 0.5 to 1 min (4.0–6.0 V; 10–20 Imp · s^{-1} and 2 ms duration) resulted in an initial inspiratory arrest of breathing, accompanied by a tonic excitation of the inspiratory intercostal muscles and a total inhibition of the phrenic nerve activity (Fig. 1A). Following the inhibition, during maintained stimulation, the phrenic nerve activity and breathing movements reappeared and slowly returned to prestimulus values. The tonic activity of the inspiratory intercostal muscles became rhythmic. The amplitude of the rhythmic inspiratory intercostal muscle activity was above the prestimulus value. Similar respiratory responses were also obtained with distension of the small intestine through a balloon with intraluminal pressures between 5–8 kPa.

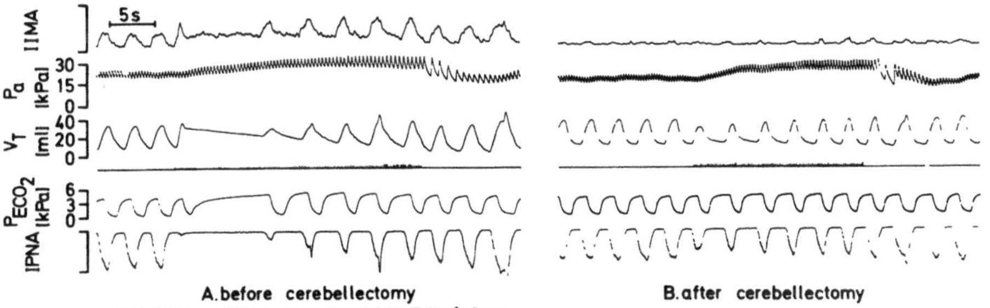

a

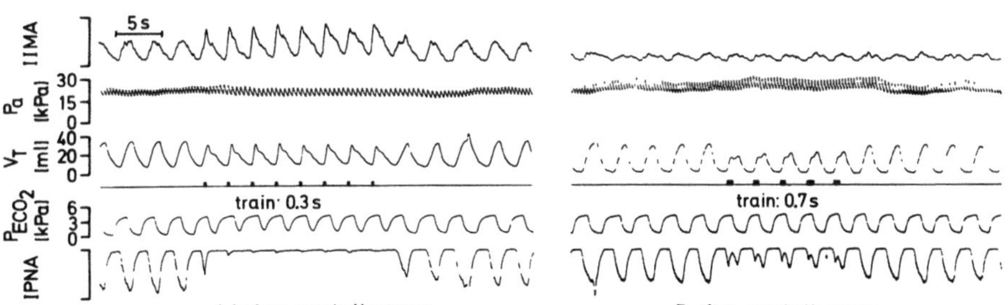

b

Fig. 1 a, b. Example of an experiment showing the effect of continuous stimulation (**a**); brief inspiratory electrical stimulation (**b**) of the mesenteric plexus, before (*left*) and after cerebellectomy, (*right*). IIMA = Integrated inspiratory intercostal muscle activity (inspiration upwards); P_a = arterial pressure; V_T = spirometric registration of the tidal volume (inspiration upwards); $P_{E\,CO_2}$ = expiratory CO_2 partial pressure; IPNA = Integrated phrenic nerve activity (inspiration downwards). After cerebellectomy, with stimulation of the mesenteric plexus, there was only a partial inhibition of the phrenic nerve activity and no excitation in the inspiratory intercostal muscles. The increase in the arterial pressure during the stimulation of the mesenteric plexus was unaffected after cerebellectomy

On the other hand, brief electrical stimulation of the mesenteric plexus (0.3–0.7 s train duration) at the beginning of inspiration led to a complete and maintained inhibition of the phrenic nerve activity and a marked increase in the amplitude of the rhythmic inspiratory intercostal muscles activity. The amplitude of the tidal volume, however, was not appreciably decreased (Fig. 1B). As soon as the stimulation was stopped both, the phrenic and the inspiratory intercostal muscle activities, returned to their prestimulus values.

To study the role of the cerebellum in the respiratory responses, in some experiments, after an occipital craniotomy the cerebellum was removed by suction during the course of the experiment. After cerebellectomy, continuous as well as brief inspiratory electrical stimulation of the mesenteric plexus resulted only in a partial inhibition of the phrenic nerve activity, without excitation of the inspiratory intercostal muscles. The stimulation of the mesenteric plexus after cerebellectomy led to a decrease in the tidal volume (Fig. 1B).

Cardiovascular Responses

An increase in the arterial pressure was consistently observed with electrical stimulation of the mesenteric plexus (Fig. 1A). The increase in arterial pressure was significantly enhanced after bilateral sino-aortic denervation ($P = 0.0008$ systolic; $P = 0.0002$ diastolic, $n = 11$).

To analyse the role of the renin-angiotensin system in the observed pressor responses, the renal venous plasma renin activity was measured by radio-immunoassay (Haber et al. 1969). In sino-aortic denervated cats, during electrical stimulation of the mesenteric plexus, the plasma renin activity increased from 68.9 ± 9.1 to 130 ± 12.6 A_T I ng $\cdot$ ml^{-1} $\cdot$ h^{-1} ($P = 0.0002$, $n = 7$). After infusion of 1 Sar-8-Ala-Angiotensin II (Sarlasin, Beckman) a specific angiotensin II antagonist, the pressor response with stimulation of the mesenteric plexus was significantly reduced ($P = 0.0008$ systolic pressure; $P = 0.018$ diastolic pressure, $n = 6$). The pressor responses to intravenous injection of 1 ml of norepinephrine (2 $\mu g \cdot ml^{-1}$), however were unaltered after infusion of Sarlasin.

Bilateral cervical vagotomy did not alter the respiratory as well as the cardiovascular changes with stimulation of the mesenteric plexus, whereas bilateral splanchnicotomy abolished them.

Discussion

An inspiratory arrest of respiration is designated as "apneusis" and accompanied with tonic activity of the diaphragm (Sears 1977). Electrical stimulation of the mesenteric plexus and also the distension of the small intestine led to an inspiratory arrest of respiration, but accompanied with an inhibition of the phrenic nerve activity. Simultaneously there was a tonic excitation of the inspiratory intercostal muscles. So, the inspiratory arrest was due to an excitation of the intercostal muscles and not due to the tonic excitation of the phrenic nerve activity as it is the case with "apneusis". During the brief inspiratory phasic stimulation of the mesenteric plexus the phrenic nerve activity was completely inhibited with a simultaneous increase in the rhythmic activity of the inspiratory intercostal muscles. It is important to note that the amplitude of the tidal volume was not diminished, although the phrenic nerve activity was absent. This would mean that during abdominal afferent nerve stimulation there is a switching of the abdominal breathing to the thoracic one. Furthermore, these results suggest a preferential participation of intercostal motoneurones to maintain the tidal volume in the absence of phrenic nerve activity. Bilateral vagotomy did not affect this paradoxical pattern of breathing, while bilateral splanchnicotomy abolished it, implying that the sympathetic afferents constitute the major afferent pathway.

In cerebellectomized animals, there was only a partial inhibition in the phrenic nerve activity in response to mesenteric plexus stimulation as compared to the total inhibition observed with intact cerebellum. Moreover, the excitation of the intercostal muscles was either markedly reduced or absent. It was shown that afferent fibers of the splanchnic nerve project more specifically to the anterior lobe of the cerebellum (Newman and Paul 1969). Furthermore, Loeschcke and Trouth (1974) reported marked changes in respiration with electrical stimulation of the cerebellum. It appears that the cerebellum is to some extent responsible for the altered breathing pattern observed with abdominal sympathetic afferent stimulation.

An increase of the arterial pressure was seen with electrical stimulation of the mesenteric plexus. This increase can be due to an activation of the efferent sympathetic system i.e. increased release of catecholamines. Using adrenergic blocking agents, the contribution of catecholamines to the observed pressor response can be studied. However, the adrenergic blocking substances not only affect the pressor reaction to catecholamines, but also markedly diminish the release of other vasoactive substances, for example renin from the kidney (Coote et al. 1972). It means that the relative contribution of other vasoactive substances cannot be studied by using adrenergic antagonists. For these reasons, in the present investigation Sarlasin, a specific angiotensin II antagonist, was used to study the role of the renin-angiotensin system in the pressor responses. After Sarlasin, the pressor response to mesenteric plexus stimulation was markedly reduced. On the other hand, the increase of arterial pressure in response to intravenous injection of norepinephrine was unaltered, suggesting the intactness of the noradrenergic system. Moreover, there was a significant increase in the plasma renin activity during the stimulation of the mesenteric plexus. From these results it can be concluded that the renin-angiotensin system also plays an important role in the increased arterial pressure induced by the stimulation of the afferent mesenterial nerves.

References

Chernigovsky VN (1976) Tissue receptors. Historical scope. Modern view. Perspectives. In: Iggo A, Illyinsky OB (eds) Somatosensory and visceral receptor mechanisms. Prog Brain Res 43:3–14

Coote JH, Johns EJ, McLeod VH, Singer B (1972) Effect of renal nerve stimulation, renal blood flow and adrenergic blockade on plasma renin activity in the cat. J Physiol (London) 226:15–36

Frank MH (1975) Cardiovascular chemoreflexes from the perfused innervated ileum of the cat. Am J Physiol 228:944–953
Haber E, Koerner T, Page LB, Kliman B, Purnode A (1969) Application of radioimmuno-assay for angiotensin I to the physiologic measurements of plasma renin activity in normal human subjects. J Clin Endocrinol 29:1349–1355
Loeschcke HH, Trouth CO (1974) Effects on ventilation and circulation of topical electrical stimulation of the cerebellum. Pflügers Arch 374:R 56
Newman PP, Paul DH (1969) The projection of splanchnic afferents on the cerebellum of the cat. J Physiol (London) 202:223–237
Paintal AS (1973) Vagal sensory receptors and their reflex effects. Physiol Rev 53:159–227
Paintal AS (1977) Thoracic receptors connected with sensation. Br Med Bull 33:169–174
Prabhakar NR, Marek W (1979) Involvement of renin-angiotensin system in the pressor responses during the distension of the urinary bladder. Pflügers Arch 377:R 9
Prabhakar NR, Marek W, Loeschcke HH (1978) Role of the kidney in the pressor responses to stimulation of the abdominal autonomic afferents. Fed Proc Fed Am Soc Exp Biol 3:352
Saphir R, Rapaport E (1969) Cardiovascular responses of the cat to mesenteric intraarterial administration of nicotine, cyanide and venous blood. Circ Res 25:713–724
Sears TA (1977) The respiratory motoneuron and apneusis. Fed Proc Fed Am Soc Exp Biol 36:3412 3420

Further Evidence for the Existence of a Pulmonary Respiratory "Oscillator" in Early Postnatal Life

P. JOHNSON and J.E. FEWELL[1]

The newborn actively maintains its lung volume above functional residual capacity (Olinsky et al. 1974). Inspiratory intercostal muscle tone has been suggested as an important mechanism (Lopes et al. 1981) for this and there is a reduction in functional residual capacity during active sleep when the intercostal muscles are generally inactive (Henderson-Smart and Read 1979).

However the other mechanisms that maintain lung volume in the newborn have not been seriously considered. First there is good clinical evidence for reconsidering the facts. The audible expiratory grunt and the positive expiratory subglottic pressure that occurs with the respiratory distress syndrome in the preterm infant, has been shown to be beneficial to ventilation, presumptively regulated by vagal reflex mechanisms (Harrison et al. 1968). The prominence of this mechanism forms the physiological basis for using continuous expiratory distending pressure (Spiedel and Dunn 1975) in the treatment of respiratory distress. The true significance of positive expiratory distending pressure in reflexly sustaining effective breathing in the newborn is observed in the prevention of apnoea by the use of continuous distending pressure (Spiedel and Dunn 1975), which in practice is often only applied during expiration. Maintaining lung volume in such infants by negative pressure applied to the chest wall can also be effective in reducing or abolishing apnoea (Thibeault et al. 1967). Thus it could be inferred that there must be a "threshold" expiratory vagal afferent tone above which there is a phasic expiratory input which sustains effective rhythmic breathing. The lack of a "peripheral" rhythm generator is an important component of neonatal respiratory failure. The clinical evidence then is against simple pathological abnormality but points to a physiological basis for many of the postnatal respiratory disorders which depends on a loss of expiratory "drive" to breathing.

Secondly there is increasing experimental evidence that expiratory laryngeal adduction causes a substantial increase in resistance to expiratory airflow and positive subglottic pressure in unanaesthetized newborn animals. This has been shown to be a normal feature of breathing during quiet sleep in lambs (Johnson 1979, Harding et al. 1980). It becomes especially prominent as breathing frequencies fall between the third and fourth weeks of life, when glottic closure with total arrest of early expiratory airflow occurs with subglottic pressures rising to +10 cms H_2O. Opening a tracheal window,

---

[1]Nuffield Institute for Medical Research, University of Oxford, Oxford, England

Central Neurone Environment
(ed. by M.E. Schläfke, H.P. Koepchen and W.R. See)

thus removing the resistance of the upper airway and preventing any rise in subglottic pressure causes a further *decrease* in expiratory laryngeal calibre. There was also a marke(lengthening of expiratory time with a fall in breathing frequency and in some lambs a grossly irregular breathing pattern occurred. This pattern of breathing would persist during quiet sleep for many hours and was reversible simply by closing the tracheal window. Further studies demonstrated that pulmonary vagal afferents regulate both expiratory laryngeal calibre and respiratory frequency by sensing expiratory pressure. This pathway actually forms a negative feedback loop, since adduction occurs when the expiratory sensory information from the lungs is interrupted by intrathoracic vagal block. Conversely, preventing this reflexly controlled laryngeal constriction, or its effectiveness in elevating expiratory pressures in the lung by opening a tracheostomy, opens another negative feedback loop: breathing frequency falls and becomes irregular.

We have also shown that pulmonary vagal afferents, chiefly pressure sensitive, control the very considerable expiratory activity of the diaphragm that occurs in the newborn lamb during sleep (Henderson-Smart et al. 1982, in press). This mechanism, which can be transiently evoked in adult cats (Remmers and Bartlett 1977) by suddenly opening a tracheal window during expiration, is therefore also normally operative in the newborn.

The expiratory activity of the abdominal muscles is also under vagal control (Bishop and Bachofen 1972) and is overt in the infant with respiratory distress causing the audibl expiratory grunt as air is forced through the adducted glottis. This would seem to be an extreme method of maintaining lung expansion, again controlled by pressure sensitive lung afferents. The vagal recruitment of the abdominal muscles in expiration can be demonstrated simply by artificially increasing expiratory resistance in cats (Bishop and Bachofen 1972) or lambs.

Although these results provide good evidence for the importance of pulmonary vagal afferents in respiratory rhythm generation during quiet sleep in young lambs it was equally striking that in most instances arterial blood gases were maintained in the normal range, even when breathing frequency fell and became irregular with a large reduction in minute ventilation. Clearly minute ventilation, which is generally used as a measure of ventilatory drive, gives a poor guide to effective ventilation in this situation.

Nevertheless we wished to know whether there were circumstances in which this reduction of vagal drive, as indicated by a loss in frequency and rhythmicity of breathing could significantly compromise effective ventilation. We therefore decided to test the effects of hypoxia during sleep with and without the intact upper airway. Hypoxia is a stimulus to breathing in the newborn, probably at the carotid body, although the ventilatory response as measured by minute ventilation is sometimes not sustained (Rigatto et al. 1975). Hypoxia has also generally been believed to dilate the larynx, the chief effect being exerted by pulmonary vagal afferents. Could it be the loss of vagal sensory input that is the reason why the newborn sometimes develops respiratory failure during hypoxia?

Methods

Nine lambs were investigated on several occasions between 7 and 45 days of age. The lambs were separated from their ewes at 24 h of age after receiving colostrum and established on an ad-lib supply of artificial milk from a reservoir. Electrodes (Cooner Wire Co.) recording eeg, oculogram, ecg, thyroarytenoid (TA), intercostal and diaphragm emgs, and an arterial catheter were implanted under halothane/N_2O anaesthesia and led subcutaneously to connectors on the back of the lamb.

For the studies, the lamb was connected to a miniature first stage amplifier harnessed on the back and then by a heavy duty cable to amplifiers in an adjacent room where the signals were recorded on a Beckman 411 dynograph and a 14-channel analogue tape recorder (Ampex 2000) for offline computer analysis. The lamb always remained in a perspex box, with which it was familiar, in the room in which it lived with several other lambs.

Control recordings of several hours duration, encompassing several sleep cycles, were made before a tracheostomy was carried out under halothane/N_2O anaesthesia. A cuffed fenestrated (Shiley) plastic tracheostomy tube was inserted which allowed the animal to breathe through either; an intact upper airway (inner cannulae out and external plug in); both the tracheostomy and the upper airway (inner cannula and plug out); or the lower airway (inner cannula in, thus occluding the fenestration to the upper airway). After recovery the lambs were recorded for several hours before the inspired oxygen was rapidly reduced from air to 0.06–0.08%, either with the upper airway intact, or isolated (tracheostomy open) with a resistive valve providing an expiratory resistance to stimulate the intact larynx. The hypoxia was maintained for 60–90 min before returning to room air for a recovery period of several hours. In some studies carbon dioxide 1–4% was added during the hypoxia. Arterial pO_2, pCO_2, pH and Base deficit were measured frequently throughout the study period. Only one hypoxic episode was performed each day iand the sequence randomised over successive days. In five lambs, studies were repeated at different ages. Between the studies the tracheotomy was closed with the upper airway intact.

In four lambs the behavioural and ventilatory responses to acute hypoxia (2 min) was repeatedly tested. These results are not all presented in detail but will be compared qualitatively with the chronic results described here.

Data Analysis

The breathing frequency, amplitude (TV or integrated Diaphragm emg), Integrated laryngeal adductor (TA), emg; behavioural state – awake (active or quiet), quiet sleep (high voltage ECoG – no eye movements), rapid eye movement sleep (low voltage ECoG – no eye movements) – were measured for each minute before being averaged for each 10 min or so during control and hypoxic sleep periods. In four lambs subglottic pressure was measured from an indwelling transtracheal catheter.

Results

The lambs were usually asleep initially and all responded to hypoxia (FIO_2 0.07) first by increasing ventilation, frequency and tidal volume, blood pressure and heart rate before

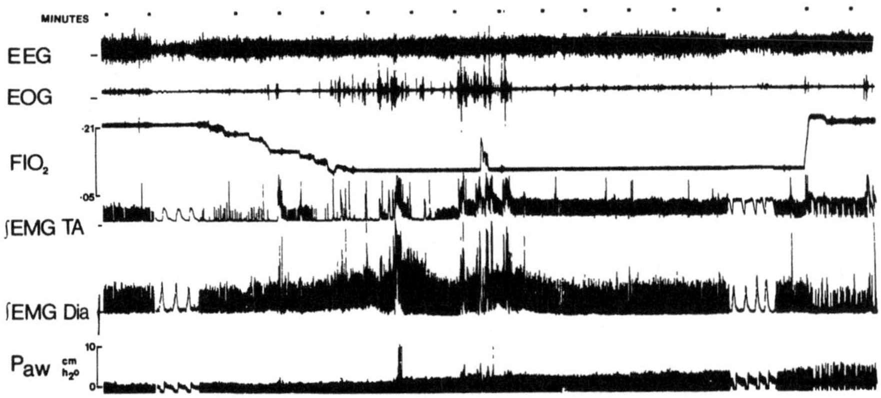

Fig. 1. This is a typical record from a 21 day old lamb made hypoxic during quiet sleep with upper airway intact. Note the faster record during the initial air breathing period in which TA activity peaks in early expiration. Hyperpnoea (tachycardia and hypertension not shown) with loss of TA activity occurs during progressive hypoxia followed by arousal (eye movements) and gross activity. The lamb settles again into quiet sleep (high voltage – no eye movements) and the TA activity plateaus throughout expiration and subglottic pressure is increased. The increased TA activity did not disappear with the increase in oxygen suggesting that it was chiefly vagal and not chemoreceptor in origin

arousing after 45–75 s depending on the behavioural state. TA activity, if present initially, disappeared during this phase. The lambs usually stood up for a few minutes before lying down again and passing into a quiet sleep-like state as judged by high voltage ECoG, absent eye movements and behavioural observations. The PaO_2 at this stage was usually approximately 20 mmHg. The initial tachypnoea, tachycardia, and hypertension gradually fell and laryngeal adductor activity became prominent, both in amplitude and the proportion of expiratory duration, whether or not the upper airway was intact. If the upper airway was intact, or positive expiratory pressure of +8–+10 cm H_2O was applied to the tracheotomy, breathing frequency was maintained above control with considerable expiratory TA activity for the length of the hypoxia (60–90 min). However if the upper airway was open at the start of the hypoxia, or opened during this latter phase of the hypoxia, breathing frequency would fall below control and usuall become irregular. Conversely TA activity increased so that the glottis was closed through out expiration. Significantly arterial PaO_2 and $PaCO_2$ were not different between the slow irregular breathing pattern with the upper airway open and the regular pattern with the upper airway intact.

In younger lambs less than 30 days of age, when breathing frequency fell below control, an associated bradycardia occurred and in two of nine lambs expiratory apnoea followed with glottic closure. During apnoea the PaO_2 fell further, $PaCO_2$ rose, heart rate fell and the eeg became isoelectric and convulsions with gasping occurred. This was followed by a small burst of breathing followed by recurrent apnoea. These cycles were repeated until either the upper airway was reintroduced; positive expiratory pressure ap applied; or the hypoxia was relieved. Repeated strong cutaneous stimulation (pinching) briefly restored regular breathing patterns when the breathing became irregular but had

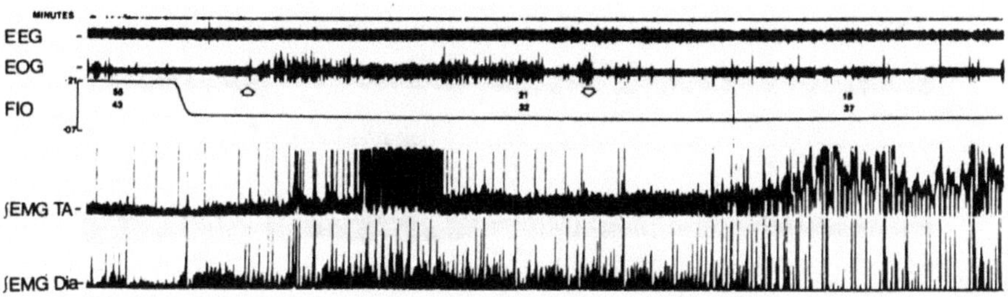

Fig. 2. This is a similar record from an 11 day old lamb, initially asleep with a tracheostomy open. The initial responses are similar to the lambs with upper airway intact. ↑ indicates where the lamb aroused and stood up before feeding from ad-lib milk supply (note swallowing on TA record). The lamb lay down at ↓ and entered quiet sleep with slow breathing and increased TA activity. Within a few minutes the breathing had become irregular with recurrent expiratory apnoea occurring with the glottis tightly adducted

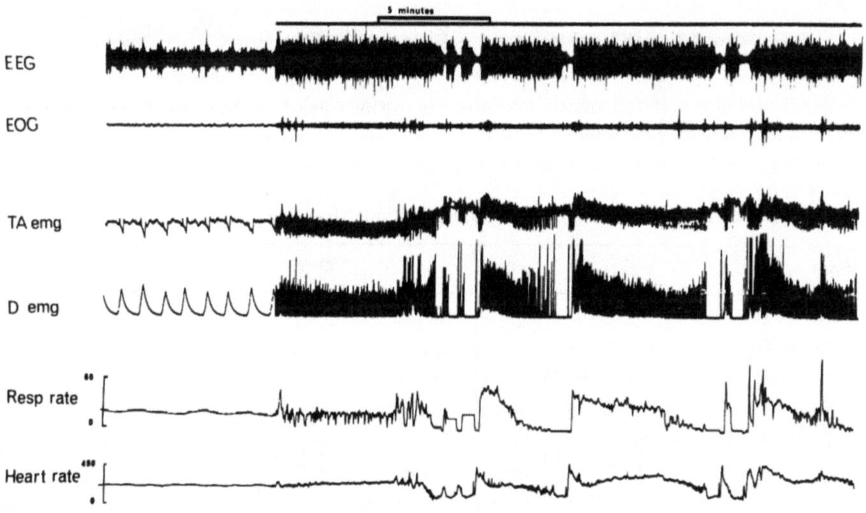

Fig. 3. Shows a faster record of a 9 day old lamb exhibiting recurrent expiratory apnoea during hypoxia with the tracheostomy open. Note the isoelectric periods followed by arousal, gasping and rapid breathing. Bradycardia occurred during the apnoea

little effect once apnoea had occurred. Except where recurrent apnoea occurred, and when PaO_2 fell below 15 torr with an increase in $PaCO_2$ (up to 50 torr), a significant acidemia seldom occurred during these severe hypoxic periods.

There was a wide range of results which are summarized in Table 1. However it can be seen that the principle determinants of the response to 30 minutes of hypoxia were age and whether or not the upper airway was intact (or simulated by positive expiratory pressure). Only in lambs under 30 days of age did a significant reduction in breathing frequency and/or irregular breathing occur. Two of nine lambs exhibited prolonged recurrent expiratory apnoea. In older lambs the increased breathing frequency was sustained above control throughout the hypoxic period and was little influenced by the presence

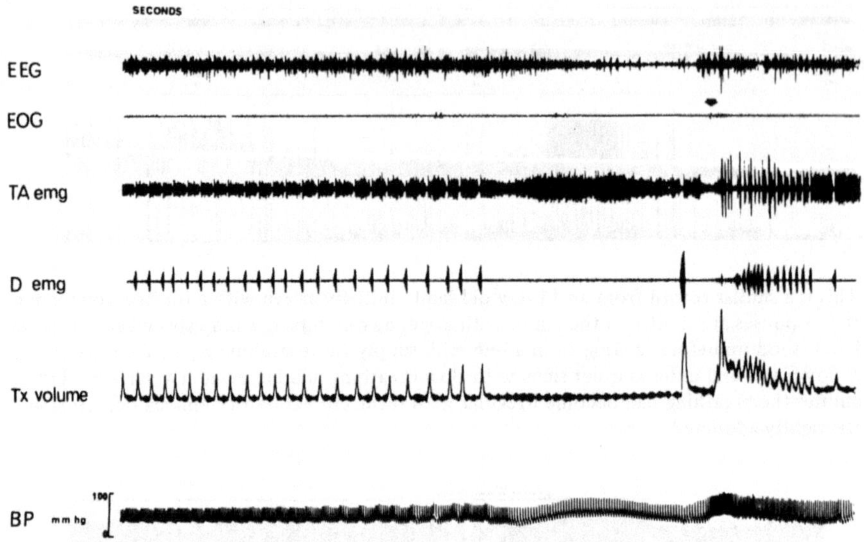

Fig. 4. A faster record shows the sustained TA actitivity during one of the hypoxic apnoeic episodes in the lamb with its tracheostomy open

Table 1. Breathing pattern after 30 min of chronic hypoxemia (F_1O_1.12 – .08)

Age (days)	3-21	30-55
UPPER AIRWAY INTACT	(12 trials) All Regular f ▲ 35 - 70% 1/12 REM	(3 trials) All regular f ▲ 55 - 74% no REM
WITHOUT UPPER AIRWAY OR WITH TRACHEOTOMY OPEN	(9 trials) All irregular f ▼ 15 - 75% 1/9 REM	(3 trials) All regular f ▲ 50 - 73% no REM

f = respiratory frequency as a percentage of control; REM = rapid eye movement sleep. This only occured in one lamb during hypoxia

or absence of the upper airway. In several lambs 2–3% carbon dioxide was added during the hypoxia; this immediately restored regular breathing during the irregular breathing periods with or without the upper airway being intact. There was however no significant change in arterial $PaCO_2$.

Discussion

In addition to generating rhythmic breathing during quiet sleep in young lambs breathing air, it is clear from these studies that the development of positive expiratory pressure by laryngeal constriction provides an important stimulus to breathing during hypoxia, after the initial acute responses have subsided (Fewell and Johnson 1981). It seems likely that on the afferent limb of this mechanism are pressure sensitive vagal receptors in the lung, as was suggested by Dixon et al. (1974); however, with the opposite response to their studies, namely laryngeal constriction in response to hypoxia. Until recently hypoxia, hypercapnia, or hypoxic hypercapnia were considered similarly as causing laryngeal dilatation during asphyxia. These studies were conducted under anaesthesia with the upper airway isolated, usually for calibre and resistance measurements. Recently Megirian and Sherrey (1981) in unanaesthetized rats, and England and Bartlett (1981) in unanaesthetized man have shown that hypoxia constricts the larynx. Bartlett, however, shwowed that inspired carbon dioxide dilates the larynx. These results accord entirely with our own in these unanaesthetized lambs. However it is also clear that low level inspired carbon dioxide (0.02–0.03) is a stimulus to breathing during hypoxia, even with an isolated upper airway or no positive expiratory pressure, and without elevating the arterial CO_2. CO_2 administration was just as effective in sustaining rhythmic breathing as was the addition of positive expiratory pressure. Thus hypoxia and hypercapnia (by CO_2 inhalation) have quite opposite effects on laryngeal calibre and breathing pattern. This makes teleological sense in that glottic closure with lung distension during expiration could enhance O_2 extraction from a diminishing reservoir. Such a mechanism would have no such benefit to the organism in the case of carbon dioxide. We do not believe that this possibility will be elucidated by the addition of CO_2 to the inspired gas in order to simulate hypercapnia as is traditionally done. Endogenous hypercapnia, except where it occurs with exercise, is not usually a consequence of increased carbon dioxide production, but is a failure to eliminate what may often be a reduction in carbon dioxide production.

There remains an intriguing possibility that it is the afferent sensory information from the lung, during expiration in particular, which largely sustains rhythmic breathing in the newborn during sleep (except in REM sleep where central mechanisms predominate) for many days or even weeks after birth. In certain situationsthe larynx is reflexly recruited to reinforce this drive to breathing. Lung collapse and hypoxia clearly can be prevented or limited by this vagal mechanism. Conversely, failure of this pathway (afferents in the lung, the synapses in the brainstem, or the motor control of the larynx) could lead to incipient respiratory failure or death. The conventional testing of ventilatory response to acute hypoxia i.e. to arousal (Phillipson et al. 1978) (usually in less than 1 min) or to determine whether a biphasic response exists (5–10 min) (Rigatto et al. 1975) completely fails to test this possibility. It seems likely, from what is known of the characteristics of carotid body single fibre activity, that a sustained discharge would occur during both the initial acute phase (tachypnoea) and the later phase (slow irregular breathing). Thus the changing pattern of breathing, some might say depression of breathing, must be due either to peripheral effects, e.g. mechanical changes in the airways, lung, or chestwall; loss of vagal afferent sensory drive; or to central effects of the hypoxia on medullary or higher centre neurones. Since the effects of removing or restoring the upper airway or the application of expiratory positive pressure causes instantaneous changes in breathing frequency with imperceptible changes in PaO_2 or $PaCO_2$ or behavioural state, it seems reasonable to conclude that phasic expiratory vagal afferent sensory information is a major factor in sustaining breathing even when the carotid body afferent traffic is increased.

These results dispute the legitimacy of the biphasic response of the newborn to hypoxia as being one of respiratory depression (Rigatto et al. 1975). That interpretation depends on changes in minute ventilation, which in our studies bear little relation to blood gas changes or alveolar ventilation. In other words, except for the two instances of apnoea, ventilation was equally effective whether rapid and regular or slow and irregular. Thus the effectiveness of breathing was not depressed. Whether regular breathing during hypoxia is less vulnerable to a further withdrawal of respiratory stimuli remains to be answered. It seems possible that the change with age reflects a change in the central integration of the carotid body and vagal afferents on the drive to breathing. Whether the vagal input modifies efferents to the carotid body, thus changing its gain, is a possibility which is difficult to answer experimentally. Again evidence for a change with age would be necessary. An, as yet, unquantified contribution to the change with age must be the increasing rigidity and stability of the chest wall which prevents passive lung collapse.

Of course, bypassing the upper airway deprives the animal of important phasic flow and pressure information sensed in the upper airways which can greatly influence the pattern of breathing (Yousef et al. 1981). That the effect we are seeing largely emanates from the lower airway vagal pressure receptors is supported by the fact that the application of expiratory pressure through the tracheotomy tube alone restores the pattern of breathing observed when the upper airway is intact and endogenous positive subglottic pressure can be developed.

It is worth mentioning here that the post-ramp inspiratory neurones described by Richter and Ballantyne (this symp) correspond closely with the normal emg activity of thyroarytenoid muscle which occurs exclusively during expiration (Harding et al. 1980).

Since expiratory vagal afferents regulate both the "post inspiratory" activity of the diaphragm and the expiratory calibre of the larynx, should these neurones be more appropriately considered as expiratory? While the postinspiratory activity of the diaphragm may appear analogous to the use of flexors in the control of extension during locomotion, can the same argument hold for an immutable expiratory muscle such as the thyro-arytenoid? On the other hand it could be debated that any muscle activity that aids what is traditionally considered an inspiratory function, namely gas exchange, should be labelled inspiratory.

The loss of breathing frequency and rhythm which follows the removal of expiratory vagal sensory input during quiet sleep follows from the observation that apnoea occurs during high voltage electrocortical periods in late fetal life in the lamb (Dowes et al. 1972). There is now good evidence that an active inhibition of breathing exists in this state in that mid-brain transection and "releases" an apnoeic response to hypoxia to one of stimulated breathing (Dawes et al. 1980). This latent inhibition carries over into the newborn in that spontaneous breathing activity is abolished by eupnoic artificial ventilation exclusively during quiet sleep (Henderson-Smart et al. 1980).

The reduction in the spontaneous use of expiratory laryngeal adduction and positive subglottic pressure that occurs with increasing age during normal breathing in lambs probably reflects the increase in rigidity and stability of the chest wall. Nonetheless the control of this mechanism which is normally evident in young lambs and recruited in older lambs during hypoxia probably lies in central neural organisation. In normal breathing, and now hypoxia, the importance of lung afferents in sustaining breathing has been well demonstrated and represents a significant peripheral rhythm generator or oscillator. Failure of this system is being investigated as a possible cause of sudden infant death.

Acknowledgements. Support for this work from the Foundation for Study of Sudden Infant Death, Wellcome Trust and NIH Pulmonary Faculty Training Prrogram (JEF) is gratefully acknowledged. The assistance of Sarah Kitt with the typescript is also greatly appreciated.

References

Bishop B, Bachofen H (1972) Comparison of neural control of diaphragm and abdominal muscle activities in the cat. J Appl Physiol 32:798–805

Dawes GS, Fox HE, Leduc BM, Liggins GC, Richards RT (1972) Respiratory movements and rapid eye movement sleep in the fetal lamb. J Physiol (London) 220:119–143

Dawes GS, Gardner WN, Johnston BM, Walker DW (1980) Breathing patterns in fetal lambs after mid-brain transection. J Physiol (London) 208:29

Dixon M, Szereda-Przestaszewska M, Widdicombe JG, Wise JCM (1974) Studies on laryngeal calibre during stimulation of peripheral and central chemoreceptors, pneumothorax and increased respiratory loads. J Physiol (London) 239:347–363

England SJ, Bartlett D (1981) Human vocal cord movements during hypoxia differ from those during hypercapnia. Physiologist 24 (4):100

Fewell JE, Johnson P (1981) The role of the larynx on the regulation of breathing during hypoxia in sleeping lambs. J Physiol (London) 320:57–58

Harding R, Johnson P, McClelland ME (1980) The expiratory role of the larynx during development and the influence of behavioural states. Respir Physiol 40:165–179

Harrison VC, Heese HB, Klein M (1968) The significance of grunting in hyaline membrane disease. Pediatrics 41:549

Henderson-Smart DJ, Read DJC (1979) Reduced lung volume durial behavioural active sleep in the newborn. J Appl Physiol 46:1081–85

Henderson-Smart DJ, Johnson P, McClelland ME (1980) The abolition of spontaneous breathing during quiet sleep by eupnic positive pressure ventilation in lambs. J Physiol (London) 307: 48–49

Henderson-Smart DJ, Johnson P, McClelland ME (1982) A syndronous respiratory activity of the diaphragm during spontaneous breathing in the lamb. J Physiol (London) (in press)

Johnson P (1979) Comparative aspects on the control of breathing during development. In: Euler C, Lagerkrantz H (eds) Proceedings of the Wenner-Gren Symposium. Central nervous control mechanisms in regular, periodic and irregular breathing. Pergamon Press, Oxford, p 337

Lopes J, Muller NL, Bryan MH, Bryan AC (1981) Importance of inspiratory muscle tone in maintenance of FRC in the newborn. J Appl Physiol Respir Environ Exercise Physiol 51 (4):830–834

Megirian D, Sherrey JH (1981) Respiratory functions of the laryngeal muscles during sleep. Sleep 3 (3/4):289–298

Olinsky A, Bryan MH, Bryan AC (1974) Responses of newborn infants to added respiratory loads. J Appl Physiol 37:190–193

Phillipson EG, Sullivan CE, Read DJC, Murphy E, Kozar LF (1978) Ventilatory and waking responses to hypoxia in sleeping dogs. J Appl Physiol 44:512–520

Remmers JE, Bartlett D (1977) Reflex control of expiratory air flow and duration. J Appl Physiol Respir Environ Exercise Physiol 43 (1):80–87

Rigatto H, Brady JP, Torre Verduzco de la R (1975) Chemoreceptor reflexes in preterm infants, I. The effects of postnatal age on the ventilatory response to inhalation of 100% and 15% oxygen. Pediatrics 55:604–613

Spiedel BD, Dunn PM (1975) Effect of continuous positive airway pressure on breathing pattern of infants with Respiratory-Distress snydrome. Lancet 302–304

Thibeault DW, Wong MM, Auld PAM (1967) Thoracic gas volume changes in premature infants. Pediatrics 40:403

Yousef KAO, Oommen PM, Thach BT (1981) An animal model for airway sensory deprivation producing obstructive apnea with postmortem findings of sudden infant death syndrome. Pediatrics 68:6796

On the Functional Organization of the Generators of Rhythmic Motor Synergy in Breathing

C. von EULER [1]

It is obvious to most people (at least when they are reminded of it) that during wakefulness, the breathing apparatus and its neural control mechanisms, in addition to its homeostatic functions, are being utilized for many important non-homeostatic behavioural tasks (see e.g., von Euler 1981) which may, on a short term scale, be competing with, or intruding on, the homeostatic regulation of the blood gases and extracellular pH. At rest the organism would seem to be able to 'afford' to give some behavioural functions a higher 'priority' than the homeostatic functions with respect to the usage of the breathing apparatus (von Euler 1981, 1982, Purves 1979). Thus in the awake state and at relative rest, respiratory regulation is characterized by the complex interactions and integrations of the control mechanisms subserving both the homeostatic and the various behavioural demands on the respiratory apparatus according to the priorities which in each condition are given to the different functions.

Experimental results obtained in studies of human respiratory functions are often discussed only in terms of the operation of the automatic metabolic bulbo-pontine mechanisms, neglecting the possibility that they might depend partly, or exclusively, on forebrain mechanisms. Some divergencies in opinion concerning the operation of the automatic, metabolic regulator may have been caused by taking for granted that observed reactions and reflexes were part of the metabolic control system rather than depending on forebrain influences (von Euler 1977). When utilizing breathing for the purpose of speech, for example, the rate and volume of inhalation and the rate of exhalation during the utterance seem to be governed largely by the speech controlling mechanisms according to their requirements for phrasing, loudness, and articulation (Bunn and Mead 1971, Mead and Bunn 1974, Phillipson et al 1978). Thus, no inferences concerning the neural regulation of automatic breathing patterns can be drawn from observations on the control of 'speech breathing', or vice versa, as some authors have been tempted to do (e.g., Phillipson et al. 1978).

The presentation to follow will be restricted to problems concerning the generation and control of the pattern of automatic metabolic breathing.

1 Nobel Institute for Neurophysiology, Karolinska Institutet, 10401 Stockholm, Sweden

Central Neurone Environment
(ed. by M.E. Schläfke, H.P. Koepchen and W.R. See)

The 'Central Pattern Generator' (CPG) for Automatic Breathing

At relatively low drive for ventilation, expiration is passive in terms of engagement, or lack of engagement of the expiratory muscles, driven as it is, by the recoil forces of lungs and chest walls. Some inspiratory muscle activity, however is present during about the first half of the expiratory phase, performing negative work to slow the rate of exhalation (Clark and von Euler 1972, von Euler 1977, Richter and Ballantyne 1981 and this symp.) Thus, the most basic rhythm generation only requires that the main part of the inspiratory activity driving inspiration should be inhibited at a suitable level and kept inhibited during the phase of expiration, which should be long enough for the lungs to empty by the recoil forces impeded by the on-lingering post-inspiration activity in the inspiratory muscles and later, if drive for ventilation is sufficiently strong, supported by active expiratory motor activity.

At present, there seems to be general agreement that the generation of the breathing rhythmicity does not depend on a reciprocal inhibition between the inspiratory and expiratory motoneurons nor between inspiratory and expiratory bulbospinal premotor neurons (Bradley et al. 1975, von Euler 1977, Cohen 1979, Merrill 1979). Thus it appears that the 'half-center' hypothesis of Graham Brown is not immediately applicable to rhythm generation for breathing (von Euler 1977, Merrill 1979, 1981). Instead a picture has emerged suggesting an asymmetrical neural network which may be described in terms of three main neural mechanisms: One for the integration of tonic 'drive' inputs from central and peripheral chemoreceptors as well as other central and reflex 'drive' inputs (Bruce et al. 1979) to form the progressively increasing central inspiratory activity, CIA. When this activity has grown to a critical level it is inhibited and terminated by a second set of mechanisms i.e., the inspiratory 'off-switch' mechanisms. The third mechanisms is keeping the CIA integrator inhibited during this phase and controlling the duration of the expiratory phase. The non-linearities necessary for a rhythmic, oscillating operation of a neural network are provided for by the sharp threshold functions of the 'off-switch' mechanisms and the non-linear decay rate of the activity during the course of expiration. Of great importance in this respect are also the mechanisms for 'gating' and 'phase dependent reflex reversals' which at the moments of phase transitions close or open the transmission of the different drive inputs. These mechanisms add to the non-linearities and thus 'sharpen' the phase transitions (von Euler 1981).

All three of the above mentioned operational mechanisms seem to be drive dependent: The rate of rise of inspiratory activity increases with increasing drive; the threshold for the inspiratory 'off-switch' is raised with increased chemical drive, but not with increased temperature, and not initially with locomotion; the rate of decay of inspiratory inhibition during expiration is faster with increasing drives and the expiratory phase is correspondingly shorter (Bruce et al. 1979, von Euler and Trippenbach 1976a, b, von Euler et al. 1976).

With respect to the neural correlates to these operationally defined submechanisms we have to admit that our knowledge still is quite limited, despite the recent very rapid progress concerning the properties of the different types of respiration-related neurons, their membrane properties, synaptic connections, and projections (see. e.g., the papers in this volume by Richter and Ballantyne, and by Cohen et al. as well as the recent reviews by Cohen 1979, 1981, Kalia 1981, Merrill 1979, 1981).

For instance, it is not yet known where the CIA integrator or generator is located. It is tempting to speculate that its location is related to the nucleus paragigantocellularis lateralis, nPGCL (Andrezik et al. 1981, Dermietzel 1976, Schlaefke 1976, Schlaefke et al. 1979, Dev and Loeschcke 1979, King and Knox 1977) which includes the areas on the ventral surface of the medulla which, especially by the Bochum School have been shown to be of primary importance for the central chemoceptive drive and its integration with other ventilatory drive inputs (see several papers in this volume; cf. also Cherniack et al. 1979a, b). This nucleus has been reported to receive afferents from many central and peripheral sources which can influence breathing including strong bilateral cross connections with the nPGCL of the opposite side. It is of particular interest that the afferent input from nTS originates in the part of the nTS complex which receives the primary afferents from the carotid body (Janscó and Király 1980).

With respect to the neural correlate to the "off-switch" mechanisms it seems likely that the function of its "stage 1", the graded, reversible inhibition (Younes et al. 1978) is performed by those neurons in nTS described by Feldman and Cohen (1978) as late, high threshold R_β neurons. Remmers et al. (1979) have shown that the activity pattern of these neurons correlates very closely to the dynamic properties of the inhibition of "stage 1" of the inspiratory "off-switch" mechanism. The neurons maintaining the inhibition of the CIA-integrator during expiration and of the bulbo-spinal output neurons during the latter part of expiration have however, yet to be defined (cf. Richter and Ballantyne 1981 and this symp.).

Are There One or More CPGs for the Motor Synergy in Breathing?

Whereas locomotor patterns are generated in the spinal cord by separate CPGs for each limb, possibly even with a "generating unit" for each muscle group (Grillner 1981), the CPG for the basic automatic breathing rhythmicity seems to be governed by one CPG located in the lower brainstem. This probably reflects the difference in complexity between the motor synergy in breathing and in locomotion; the breathing movements being fairly symmetrical bilaterally and with but small variations in phase relationships between the different muscle groups under different conditions. The more complex locomotor movements, on the other hand, exhibit great variations in the phase relationships between the limbs, and between the different muscle groups within the limbs, as the locomotor activity changes e.g., from slow walk to trot or to gallop and during locomotion in a circle or around a corner when the "outer" leg(s) is (are) walking faster than the "inner" leg(s).

The relatively small changes in phase relationships which do occur within the breathing apparatus between different synergistic muscle groups might be explained purely on the basis of differences, and variations, in thresholds for recruitment. The inspiratory intercostals, for instance, often exhibit a somewhat later onset than the diaphragm. This difference decreases with increasing drive. Further, the upper airway muscles controlling the airway patency usually have a somewhat earlier onset than the main air "pumping" muscles. Thus our present knowledge concerning the motor synergy in breathing does not make it necessary to postulate the existence of more than one CPG for the breathing motor synergy.

The question as to whether there are two functionally complete, bilaterally symmetrical CPGs synchronized only by cross connections or, alternatively, whether the interconnections between the bilateral structures are necessary links in the network constituting one functional CPG has only very recently found an answer. Karczweski and Gromysz (1981) have shown in the rabbit that by a sagital midline section the phrenic motoneurone output on the two sides can exhibit independent rhythmic discharges each with a pattern of typical respiratory character. A similar demonstration of two independent CPGs for breathing movements following a sagital midline section splitting the medulla into two halves had been made in lamprey by Kawasaki (1979). According to Karzewski this has now been shown also for the baboon (personal communication). Since the crossing axons of the bulbo-spinal premotor neurons are cut by midline transections the maintenance of rhythmic efferent activity in the phrenic nerves following such a section implies that in these species the proportion of uncrossed bulbo-spinal axons should be relatively high. However, there is no available knowledge on this matter for the species. In the cat it has been much more difficult to demonstrate bilaterally independent rhythmicity even by recording from laryngeal motoneurons which seem to receive their input from uncrossed premotor neurons or by recording directly from the bulbo-spinal inspiratory neurons in medulla (Budzinska, Euler and Yamamoto, unpublished). It would seem possible, however, that the main differences between rabbit and cat in this respect are of quantitative nature.

Differential Control of the Phrenic and External Intercostal Motoneuron Pools

As mentioned above the inspiratory bulbo-spinal neurons conveying the output from the CIA generator to spinal motoneuron pools seem to form two separate pools on each side, one in the dorsal part of medulla, in the solitary tract nuclear complex, nTS, and one in the more ventrolateral part of medulla in the nucleus retro- or rather, paraambigualis, nPA (Merrill 1970, 1979, 1981, Kalia 1981) The results of Merrill (1974) indicate that the communication between these two pools of inspiratory premotor neurons is a "one w traffic" from the nTS group to the nPA group. It has been suggested by several investigators that the nTS group of inspiratory neurons may project predominantly to the phrenic motoneuron pool whereas the inspiratory bulbo-spinal nPA neurons may project mainly to the inspiratory intercostal motoneuron pools. This is an interesting proposition which, if it were true, might provide the basis for a differential control of the diaphragm and inspiratory intercostals and of their relative contributions to the work of inspiration which can be very different in different conditions (cf. von Euler 1979). However, there seems to be very little conclusive evidence for such a view. It seems fairly well established that the nTS group projects *both* onto thoracic *and* onto phrenic motoneuron pools and that the nPA group of inspiratory bulbo-spinal neurons project *both* to thoracic *and* phrenic pools, though possibly with different proportions.

At least part, but possibly all, of the differental control on phrenic and external intercostal inspiratory activity is exerted at the segmental levels, including the differental control exerted by cerebellum (Aminoff and Sears 1971, Corda et al. 1966, Decima and von Euler 1969a, b, Prabhakar et al. 1979). So far we don't know whether any of the groups of bulbo-spinal premotor neurons can be influenced from cerebellum or other structures from which differential effects on phrenic and intercostal motoneuron pools can be exerted.

A very dramatic change in the proportional contribution of the various respiratory muscles to the work of breathing occurs at the onset of locomotion in unanaesthetized 'thalamic' cats walking or running on a variable speed treadmill (DiMarco et al. 1981). It has been shown that in parallel with spontaneous or experimental (electrical stimulation in the subthalamic locomotor region) induction of locomotion ventilation increases by a corollary activation of the respiratory mechanisms (Eldridge et al. 1981, DiMarco et al. 1981). This increase in ventilation is characterized by an increase in respiratory rate, a near normal or only slightly increased peak phrenic amplitude, a *decreased* inspiratory intercostal activity and a pronounced *increase* in expiratory intercostal motor activity (DiMarco et al. 1981). Thus the pattern of breathing in this diencephalic (or mesencephalic) corollary activation of ventilation is quite different from that produced by hypoxic or hypercapnic stimulation. Of special interest with respect to the theme of this symposium is the further finding that all pattern parameters associated with induction of locomotion show a mainly additative 'threshold-lowering' (Cherniack et al. 1979a) interaction with central chemoceptive stimulation (DiMarco et al. 1981).

Acknowledgements. Quoted work from this laboratory was supported by the Swedish Medical Research Council (project no. 14X–544), Magn. Bergwalls Stiftelse and Knut och Alice Wallenbergs Stiftelse.

References

Aminoff MJ, Sears TA (1971) Spinal integration of segmental cortical and breathing inputs to thoracic respiratory motoneurones. J Physiol (London) 215:557–575

Andrezik JA, Chan-Palay V, Palay SL (1981) The nucleus paragigantocellularis lateralis in the rat. Anat Histol Embryol 161: 373–390

Bradley GW, Euler von C, Martilla I, Roos B (1975) A model of the central and reflex inhibition of inspiration in the cat. Biol Cybernet 19: 105–116

Bruce EN, Euler von C, Yamashiro SM (1979) Reflex and central chemoceptive control of the time course of inspiratory activity. In: Euler von C, Lagercrantz H (eds) Central nervous control mechanisms in breathing. Pergamon Press, Oxford New York, p 177

Bunn JC, Mead J (1971) Control of ventilation during speech. J App Physiol 31:870–872

Cherniack NS, Euler von C, Homma I, Kao FF (1979a) Graded changes in central chemoceptor input by local temperature changes on the ventral surface of medulla. J Physiol (London) 287: 191–211

Cherniack NS, Euler von C, Homma I, Kao FF (1979b) Experimentally induced Cheyne-Stokes breathing. Respir Physiol 37: 185–200

Clark FJ, Euler von C (1972) On the regulation of depth and rate of breathing. J Physiol (London) 222: 267–295

Cohen MI, (1979) Neurogenesis of respiratory rhythm in the mammal. Physiol Rev 59: 1105–1173
Cohen MI (1981) Central determinants of respiratory rhythm. Ann Rev Physiol 43: 91–104
Corda M, Euler von C, Lennerstrand G (1966) Reflex and cerebellar influences on α and on 'rhythmic' and 'tonic' γ activity in the intercostal muscle. J Physiol (London) 184: 898–923
Decima EE, Euler von C (1969a) Excitability of phrenic motoneurones to afferent input from lower intercostal nerves in the spinal cat. Acta Physiol Scand 75: 580–591
Decima EE, Euler von C (1969b) Intercostal and cerebellar influences on afferent phrenic activity in the decerebrate cat. Acta Physiol Scand 76: 148–158
Dermietzel R (1976) Central chemosensitivity, morphological studies. In: Loeschcke HH (ed) Acid base homeostasis of the brain extracellular fluid and the respiratory control system. Thieme, Stuttgart, p 52
Dev NB, Loeschcke HH (1979) Topography of the respiratory and circulatory responses to acetylcholine and nicotine on the ventral surface of the medulla oblongata. Pflügers Arch 379: 19–27
DiMarco AF, Euler von C, Romaniuk R, Yamamoto R (1981) Changes in respiratory activity associated with locomotion. Fed Proc Fed Am Soc Exp 40: No 3/ I: 379
Eldridge FL, Millhorn DE, Waldrop TG (1981) Exercise hyperpnea and locomotion: Parallel activation from the hypothalamus. Science 211: 844–846
Euler von C (1977) The functional organization of the respiratory phase-switching mechanisms. Fed Proc Fed Am Soc Exp 36: 2375–2380
Euler von C (1979) On the neural organization of the motor control of the diaphragm. Am Rev Respir Dis 119: Suppl Part 2: 45–50
Euler von C (1981) The contribution of sensory inputs to the pattern generation of breathing. Can J Physiol Pharmacol 59: 700–706
Euler von C (1982) Some aspects on speech breathing physiology. In: Grillner S et al. (ed) Speech Motor control. Pergamon Press. Oxford, New York, p 95
Euler von C, Trippenbach T (1976a) Temperature effects on the inflation reflex during expiratory time in the cat. Acta Physiol Scand 96: 338–350
Euler von C, Trippenbach T (1976b) Excitability changes of the inspiratory 'off-switch' mechanism tested by electrical stimulation in nucleus parabrachialis in the cat. Acta Physiol Scand 97: 175–188
Euler von C, Martilla I, Remmers JE, Trippenbach T (1976) Effects of lesions in the parabrachial nucleus on the mechanisms for central and reflex termination of inspiration in the cat. Acta Physiol Scand 96: 324–337
Feldman JL, Cohen MI, (1978) Relation between expiratory duration and rostral medullary expiratory neuronal discharge. Brain Res 141: 172–178
Grillner S (1981) Control of locomotion in bipeds, tetrapods, and fish. In: Brooks V (ed) Handbook of physiology, Section 1. The Nervous System, vol II. Motorcontrol. Am Physiol Soc Bethesda, Md, p 1179
Jansco G, Kiraly E (1980) Distribution of chemosensitive primary sensory afferents in the central nervous system of the rat. J Comp Neurol 190: 781–792
Kalia MP (1981) Anatomical organization of central respiratory neurons. Ann Rev Physiol 43: 105–120
Karczewski WA, Gromysz H (1981) The 'split respiratory centre'. Lessons from brainstem transections. In: Hutas I, Debreczeni LA (eds) Advances in physiological sciences, vol X. Respiration. Pergamon Press, Oxford, New York, p 587
Kawasaki R (1979) Breathing rhythm-generation in the adult lamprey, Entosphenus japonicus. Jpn J Physiol 29: 327–338
King GW, Knox CK (1977) Neuroanatomical substructure of medullar respiratory neurons. Neurosci Abstr 3:22
Mead J, Bunn JC (1974) Speech as breathing. In: Wyke B (ed) Ventilatory and phonatory control systems. Oxford Univ Press, London, p 33
Merrill EG (1970) The lateral respiratory neurones of the medulla: their associations with nucleus ambiguus, nucleus retroambigualis, the spinal accessory nucleus and the spinal cord. Brain Res 24: 11–28

Merrill EG (1974) Finding a respiratory function for the medullary respiratory neurones. In: Bellairs R, Gray EG (eds) Essays on the nervous system. Clarendon, Oxford, p 451

Merrill EG (1979) Is there reciprocal inhibition between medullary inspiratory and expiratory neurones? In: Euler von C, Lagercrantz H (eds) Central nervous control mechanisms in breathing. Pergamon Press, New York, p 239

Merrill EG (1981) Where are the real respiratory neurons? Fed Proc Fed Am Soc Exp 40: 2389–2394

Phillipson EA, McClean PA, Sullivan CE, Zamel N (1978) Interaction of metabolic and behavioral respiratory control during hypercapnia and speech. Am Rev Respir Dis 117: 903–909

Prabhakar NR, Marek W, Loeschcke HH (1979) Changing from diaphragmatic to thoracic respiration during stimulation of sympathetic afferents. Pflügers Arch 382: Suppl R20

Purves MJ (1979) What do we breathe for? In: Euler von C, Lagercrantz H (eds) Central nervous control mechanisms in breathing. Pergamon Press, New York, pp 7

Remmers JE, Baker JP Jr, Younes MK (1979) Graded inspiratory inhibition: The first stage of inspiratory 'off-switching'. In: Euler von C, Lagercrantz H (eds) Central nervous control mechanisms in breathing Pergamon Press, New York, p 195

Richter DW, Ballantyne D (1981) Central rhythm generation by inspiratory neurones in the anaesthe-t tised cat. Neurosci Lett Suppl 7: S117

Schlaefke ME (1976) Central chemosensitivity: Neurophysiology and contribution to regulation. In: Loeschcke HH (ed) Acid base homeostasis of the brain extracellular fluid and the respiratory control system. Thieme, Stuttgart, p 66

Schlaefke ME, See WR, Kille JR (1979) Origin and afferent modification of respiratory drive from ventral medullary areas. In: Euler von C, Lagercrantz H (eds) Central nervous control mechanisms in breathing. Pergamon Press, New York, p 25

Younes MK, Remmers JE, Baker J (1978) Characteristics of inspiratory inhibition by phasic volume feedback in cats. J Appl Physiol 45: 80-86

A Three Phase Theory About the Basic Respiratory Pattern Generator

D.W. RICHTER and D. BALLANTYNE[1]

It is generally accepted that the neurones which generate rhythmic respiratory activity are located in the lower brainstem and that these neurones are not dependent for their rhythmicity on phasic sensory inputs. Until recent years the rhythmicity of respiration was widely held to be the result of antagonistic connections between two populations of inspiratory and expiratory bulbar neurones. Although antagonistic inhibition has been demonstrated between these neurones (Mitchell and Herbert 1974, Richter et al. 1979) the significance of such inhibition for rhythmicity remains doubtful since in eupnoeic breathing expiratory intercostal and abdominal muscles are usually not active and expiration is 'passive', i.e., bulbar expiratory activity may not be essential to the maintenance of a normal inspiratory rhythm. One approach then to the problem of central rhythmicity would be to concentrate on the mechanisms generating the inspiratory rhythm. However, an equally important aspect of this problem concerns the nature of the central mechanisms, which enable *controlled* 'passive' expiration to occur, and how they are distinguished from mechanisms producing 'active' expiration. These mechanisms can all be defined with reference to events occurring in bulbar inspiratory neurones. Thus, what are the mechanisms which are responsible for: (1) *Terminating* the ramp-like activity of inspiratory neurones, i.e., what is the mechanism of the inspiratory *'off-switch'* (von Euler et al. 1973)? (2) *Relaxing* inspiratory muscle tone after the inspiratory phase of the ventilatory cycle has come to an end i.e., is there a *central control* of "passive" expiration? (3) *Delaying* re-activation of the inspiratory ramp, i.e., is there an inspiratory *'gate'*?

We have examined these questions in anaesthetised, paralysed, thoracotomised, and artificially ventilated cats in which the phrenic nerves and in some instances the vagus nerves were cut. We present a summary of the results here, on the basis of which we advance a theory about the minimum requirements of the basic respiratory central pattern generator.

---

1 Physiol. Institute, University of Heidelberg, D – 6900 Heidelberg, Fed. Rep. of Germany

Central Neurone Environment
(ed. by M.E. Schläfke, H.P. Koepchen and W.R. See)

The Theory

The central respiratory rhythm is an emergent property of the connectivity between neurones. The basic pattern generator includes only those neurones which are acitve when inspiratory muscles are active. These muscles are under the control of an inspiratory 'ramp oscillator', the activity of which produces the inspiratory phase of the ventilatory cycle and initiates inhibitory events leading to its own termination. This is coupled to a 'relaxation oscillator' which 'gates' the inspiratory ramp preventing its redevelopement during passive expiration. The time course of this 'relaxation' determines the point of transition from passive to active expiration and, via this control, the time at which the next inspiration begins. Each of these components is represented in the activity of identifiable bulbar neurones. The operation of these two oscillators is not dependent on the activation of the 'expiratory ramp oscillator'.

The Central Respiratory Cycle is Composed of Three Phases

The 3-phase composition of the central (neural) respiratory cycle is revealed in the activity of the diaphragm or efferent discharge of the phrenic nerve (Fig. 1). Each burst is composed of an initially augmenting (ramp) pattern of discharge which comes to an abrupt stop and a following declinig 'post-inspiratory' discharge, often separated from the ramp by a brief pause and extending well into the interval between consecutive ramps. The ramp component of discharge corresponds to the inspiratory phase of the ventilatory cycle; post-inspiratory discharge corresponds to the phase of passive expiration with reference to the behaviour of thoracic and abdominal expiratory muscles though it is accompanied by active control of expiratory airflow by the diaphragm and the larynx (Bartlett et al. 1973). We refer to this post-inspiratory phase of discharge in inspiratory muscles and neurones as *Stage I expiration*. Finally, the period of silence in the phrenic nerve corresponds to the phase of active expiration which we refer to as *Stage II expiration.*

These observations imply: (1) that termination of inspiration ('off-switching') is closely linked to the mechanisms which generate the ramp itself; the effectiveness of this 'off-switch' is clearly considerable since it is frequently accompanied by a distinct 'gap' in phrenic discharge; (2) that post-inspiratory discharge does not reflect a slow decay of activity in the mechanisms generating the inspiratory ramp but is due to the activation of a mechanism quite different from those underlying the inspiratory and expiratory ramp oscillators.

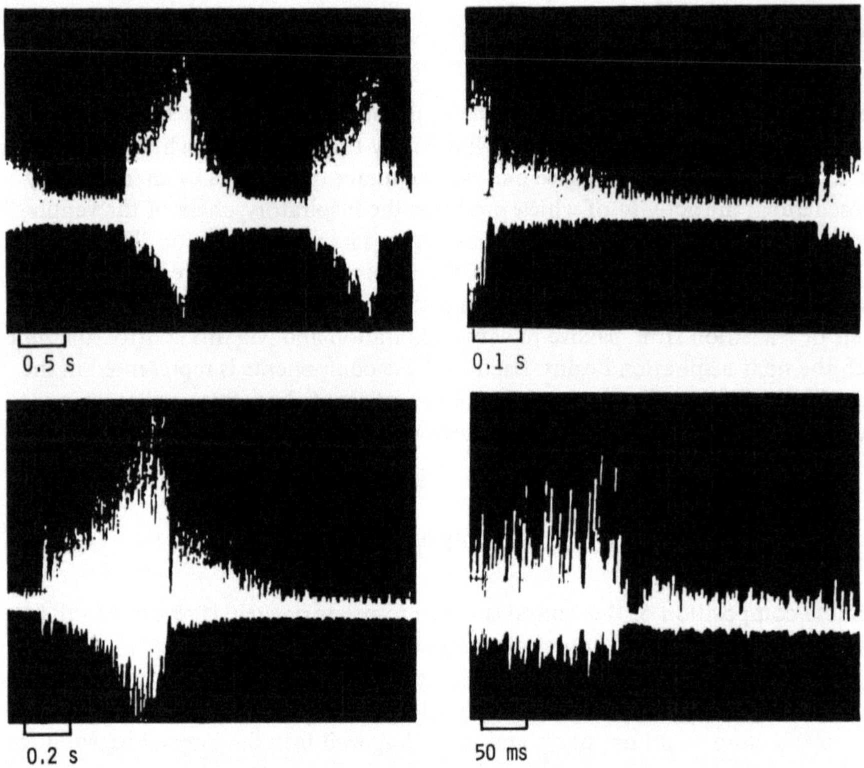

Fig. 1. Efferent discharge recorded from the phrenic nerve showing its division into an augmenting (ramp) component (inspiration), a post-inspiratory declining component (Stage I expiration) and a silent period (Stage II expiration)

Central Reprensentation of the 3-Phase Cycle

That a 3-phase neural cycle underlies the ventilatory rhythm is revealed by the patterns of activation of bulbar respiratory neurones, representative examples of the three major populations being shown in Fig. 2. Inspiratory bulbospinal neurones reveal a ramp-like synaptic depolarization during ramp discharge in the phrenic nerve (i.e., during inspiration). This is terminated in a rapid repolarization at the point of transition from inspiratory to post-inspiratory discharge in phrenic motoneurones. Post-inspiratory discharge (i.e., Stage I expiration) is accompanied by a very much slower repolarization or plateauing of membrane potential. Significant post-synaptic inhibition does not begin until phrenic nerve discharge has ceased (i.e., Stage II expiration) (Fig. 2a). This latter is consistent with the finding that augmenting discharge of bulbar expiratory fibres converging on inspiratory neurones does not begin before Stage I expiration comes to an end (Fig. 2b). Finally, there is a population of neurones which discharge action potentials only during Stage I expiration (Fig. 2c) and which are synaptically inhibited during both inspiration and Stage II expiration. We refer to these neurones as *'post-inspiratory' neurones* rather than 'early expiratory' neurones (Feldman and

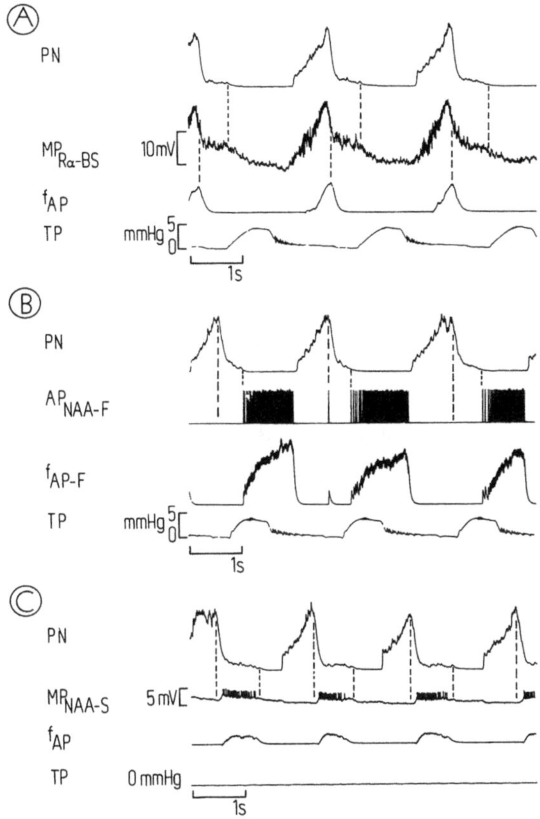

Fig. 2. The three-phase central respiratory cycle revealed in the activation pattern of **(A)** an inspiratory bulbospinal (BS) neurone of R-alpha type; **(B)** an expiratory non-bulbospinal fibre (NAA-F); **(C)** a post-inspiratory non-bulbospinal neurone (NAA-S). *PN* integrated phrenic nerve discharge; *MP* membrane potential; f_{AP} integrated discharge frequency; *TP* tracheal pressure. A non-inflation test is made during recording in C

Cohen 1978) because they exhibit a quite different pattern of output connectivity from those (Stage II) expiratory neurones the discharge of which is associated with the activation of spinal expiraroty motoneurones, and because of their precise temporal relationship to post-inspiratory discharge in phrenic motoneurones. These heurones seem to constitute a fairly large population, intermingled with inspiratory neurones mainly in the region of the nucleus retroambigualis.

We infer from these patterns of neuronal activity that the central respiratory cycle is composed of three phases.

Are all Three Neural Phases Essential for a Stable Rhythm?

The fact that post-inspiratory discharge of phrenic motoneurones may extend throughout the interval between successive inspiratory discharges (Fig. 3) would strongly suggest that rhythmic production of an inspiratory ramp is not dependent on the prior occurance of Stage II expiration. More certain evidence on this point is provided by the fact that bulbar inspiratory and bulbar post-inspiratory neurones are synaptically inhibited

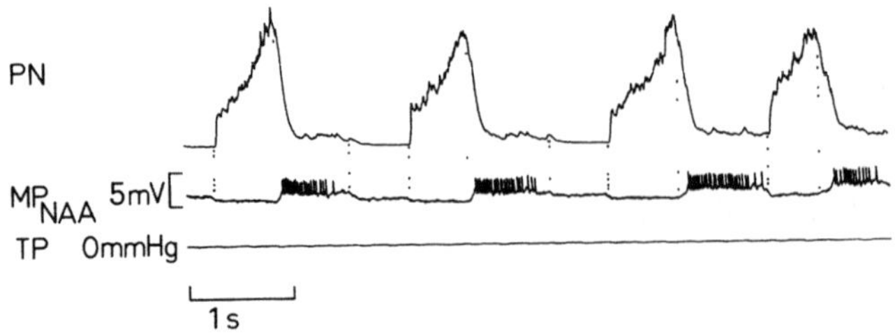

Fig. 3. The inspiratory ramp is not dependent on the occurrence of Stage II expiration. *PN* integrated phrenic nerve discharge showing its inspiratory, post-inspiratory and silent phases; MP_{NAA} membrane potential of a post-inspiratory neurone which is inhibited during inspiration and, when present, Stage II expiration; *TP* tracheal pressure

during Stage II expiration. The absence of such inhibition constitutes strong evidence that Stage II expiration may fail. This is illustrated in the case of a post-inspiratory neurone in Fig. 3 which shows that Stage II expiration can fail and that this failure does not prevent redevelopment of a normal inspiratory ramp. On this basis we conclude that a stable, if more rapid, rhythm may be generated using only two phases.

The antagonistic relationship between inspiratory and post-inspiratory phases of central respiration is revealing for the light it throws on the questions posed at the beginning of this paper and which are discussed below.

The Inspiratory "Off-Switch" Mechanism

This is a mechanism which is expected to be closely linked to inspiratory activity itself. Our results indicate that its synaptic correlate is represented in a usually transient but intense wave of inhibition which begins to appear late in inspriation (Fig. 4). The close relationship of this 'off-switch' inhibition to the preceding inspiratory activation becomes evident when late-discharging inspiratory neurones are examined (Fig. 5). All functional types of such late-discharging neurones reveal a pattern of synaptic potentials which indicate that they are activated in a ramp-like manner from the very beginning of inspiration. Late-discharging neurones are late because during early inspiration they receive a very powerful inhibition which very effectively shunts excitatory inputs and virtually 'clamps' the membrane potential at a relatively hyperpolarized level. Only when this inhibition comes to an end do these neurones depolarize rapidly and discharge at high frequency. The highly nonlinear trajectory of their membrane potential imposed by such early inspiratory inhibition and consequent intense late inspiratory discharge makes these neurones very likely canditates for the source of 'off-switch' inhibition.

The ramp-like synaptic depolarization of inspiratory neurones during inspiration stands in marked contrast to the powerful inhibition of post-inspiratory neurones. What is the probable significance of these neurones?

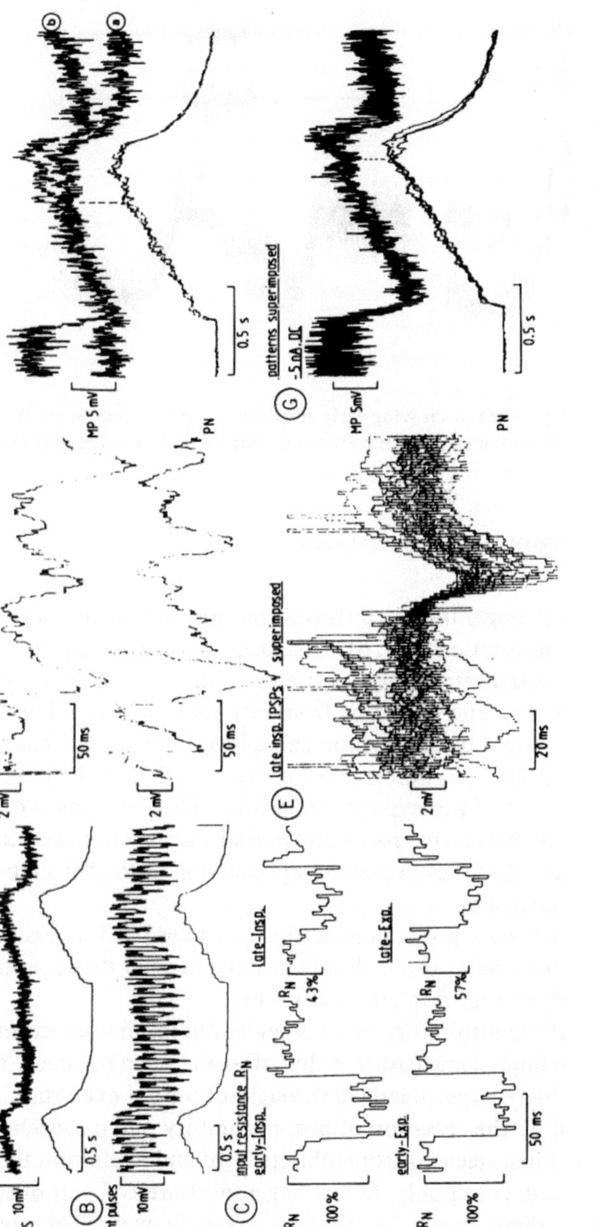

Fig. 4. Termination of the inspiratory ramp phase. **(A)** Membrane potential in an inspiratory bulbospinal neurone (MP_{BS}) and integrated phrenic nerve discharge (PN). **(B)** Voltage responses to negative current pulses showing reduction of neurone input resistance at end-inspiration and during Stage II expiration. **(C)** Relative change in input resistance between early and late inspiration and between Stage I and Stage II expiration. **(D)** Late inspiratory i.p.s.p.s shown in single sweeps. **(E)** Superimposition of late inspiratory i.p.s.p.s. **(F)** Pattern of membrane potential changes (*a*) control and (*b*) during injection of negative d.c. current showing the reversal of late inspiratory i.p.s.p.s. **(G)** Superimposed membrane potential patterns on successive sweeps showing constancy of late inspiratory inhibition

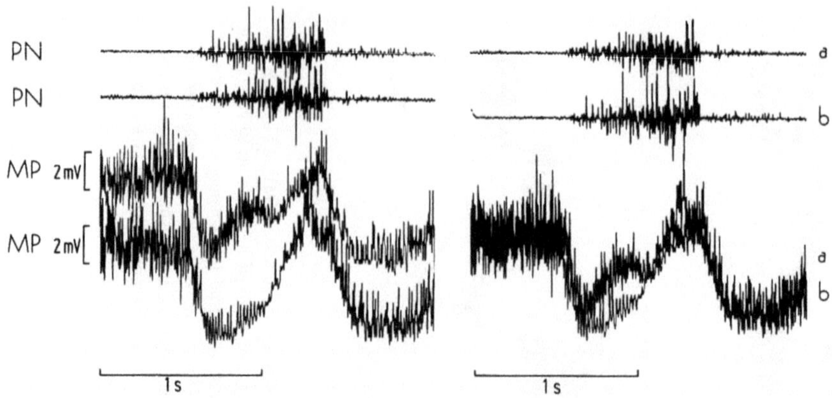

Fig. 5. A late-discharging inspiratory neurone showing early inspiratory i.p.s.p.s and Stage II expiratory i.p.s.p.s (trace *b*) which are reversed on injection of negative d.c. current and chloride ions (trace *a*)

Gating Features of Post-Inspiratory Neurones

Post-inspiratory neurones are strongly inhibited throughout inspiration with a maximal inhibition occcurring in early inspiration (Fig. 6). This means that although their discharge is accompanied by increased inspiratory muscle tone they cannot be regarded as conventional inspiratory neurones. The transition from inspiration to Stage I expiration is marked in these neurones by a rapid depolarization and a high frequency discharge. The transition from State I to Stage II expiration is signalled by cessation of their declining pattern of discharge and the onset of postsynaptic inhibition. Their inhibition during the augmenting discharge of expiratory neurones means that they are not classical expiratory neurones. We conclude that they represent a population distinct from both classical types of respiratory neurones.

Since none of these neurones have projections to the spinal cord and we assume that some of them are propriobulbar, their output should be detectable in the synaptic activity of other types of higher order respiratory neurones.

Post-inspiratory *'gating'* of the inspiratory ramp is evident in inspiratory interneurones as shown in Fig. 7. The important feature is that after the ramp has been 'off-switched', the neurone is strongly hyperpolarized throughout Stage I expiration. During Stage II expiration (indicated by the cessation of post-inspiratory phrenic discharge) inhibition is less intense but when released from inhibition at end-expiration the neurone is abruptly depolarized, commonly discharging a short-interval pair of spikes (Fig. 7a), before the synaptic ramp begins, i.e., these neurones show signs of rebound excitation, which may be important in starting the next inspiration.

Fig. 8. Membrane potential *($MP_{R\beta}$)* and membrane noise (high-pass filtered) (MNA) recorded in a R-beta neurone (**A**) shortly after penetration, after injection of chloride ions (**B** and **C**) showing reversal of late inspiratory, post-inspiratory and Stage II expiratory i.p.s.p.s towards the new equilibrium potentials established after chloride injection (**C** and **D**). (**E**) Voltage responses to injected current pulses showing reduction in input resistance during late inspiration, during Stage I expiration and during Stage II expiration ►

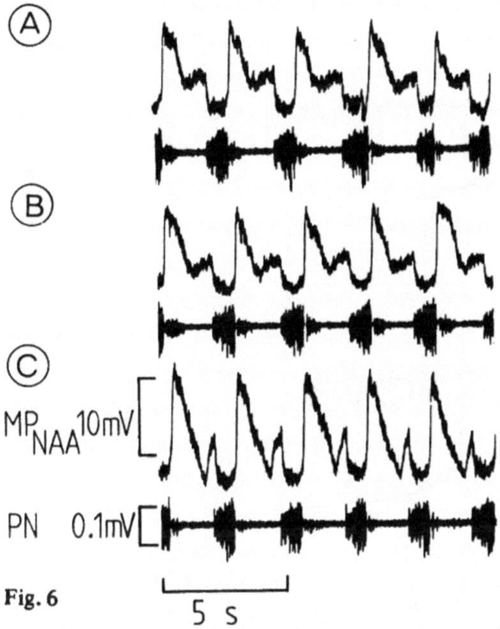

Fig. 6

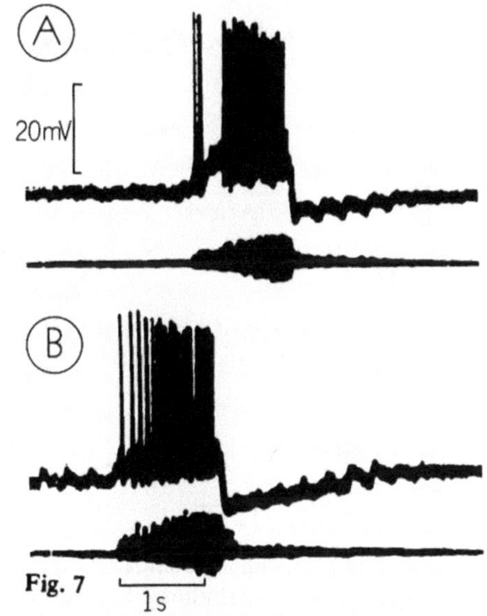

Fig. 7

Fig. 6. Membrane potential pattern in a post-inspiratory neurone (MP_{NAA}) and phrenic nerve discharge (PN). This neurone is strongly inhibited during inspiration and during Stage II expiration and disinhibited just prior to the onset of inspiration

Fig. 7. Post-inspiratory gating of inspiratory higher order interneurones. Membrane potential pattern of an R-alpha type inspiratory neurone showing augmenting discharge during inspiration and inhibition during Stage I expiration

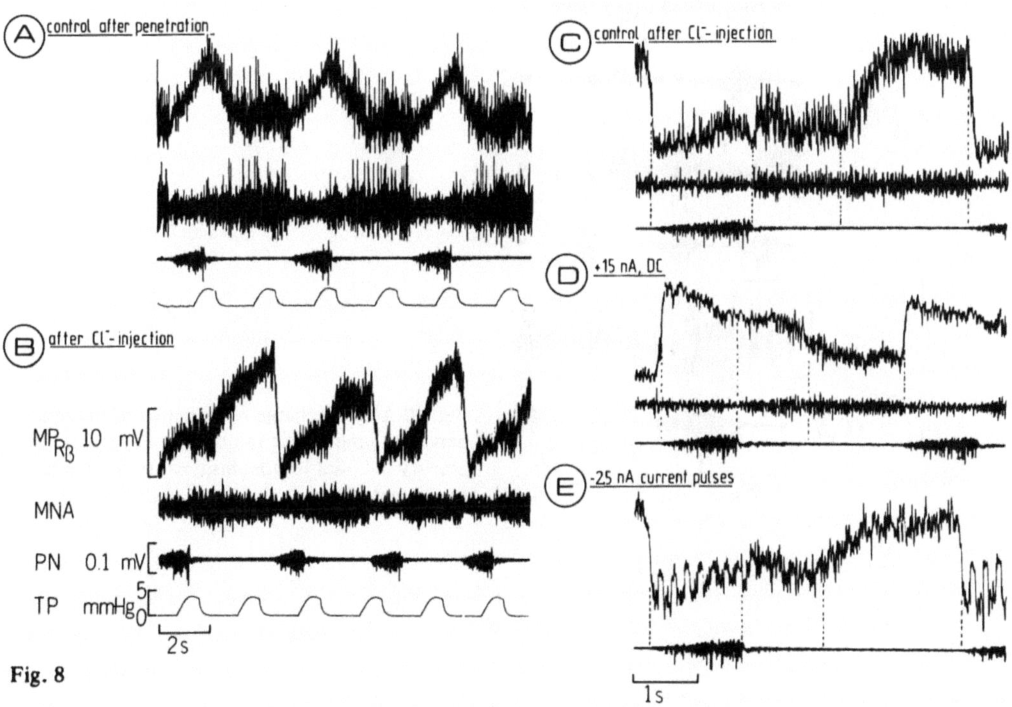

Fig. 8

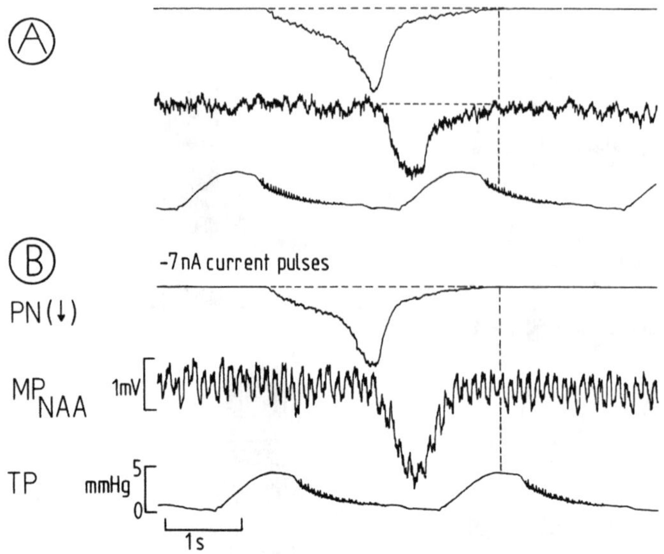

Fig. 9. (A) Membrane potential pattern recorded in a neurone of the medullary reticular formation (MP_{NAA}) showing rhythmic hyperpolarization locked to post-inspiratory discharge in the phrenic nerve (PN). **(B)** Voltage response to injected current pulses showing reduction in neurone input resistance during this hyperpolarizing wave. TP tracheal pressure

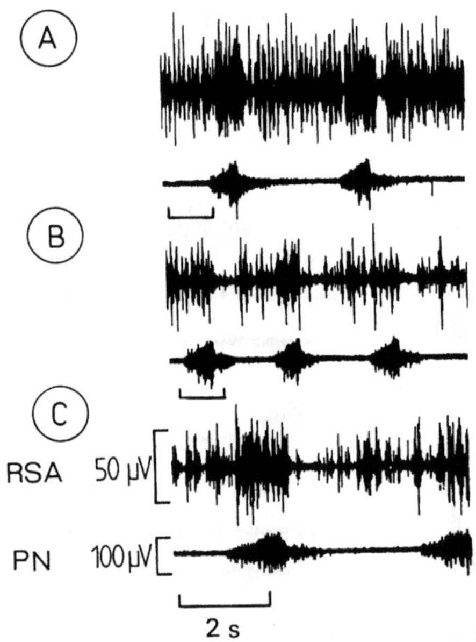

Fig. 10 A-C. Discharge recorded from the renal nerve showing phasic inhibition of sympathetic activity coupled to post-inspiratory discharge in the phrenic nerve

R-beta neurones show both inhibition and excitation during Stage I expiration and a delayed onset of intense inhibition during Stage II expiration. The three phases are particularly apparent following reversal of i.p.s.p.s by chloride injection (Fig. 8).

These results suggest that post-inspiratory bulbar neurones may form part of an inhibitory system one of whose functions is to suppress or "gate" the inspiratory ramp oscillator during Stage I expiration. That such an inhibitory system is not restricted in its action to classically defined respiratory neurones is shown in Fig. 9 which is a recording from a neurone forming part of the medullary reticular formation. This otherwise tonically active neurone – which might be called a 'phase-spanning' neurone – is rhythmically inhibited, the onset and time course of this inhibition corresponding closely to the onset and time course of post-inspiratory activity. Such inhibition of reticular neurones may serve a more general function of eliminating excitatory inputs to respiratory neurones and possibly also to cardiovascular neurones, as seen for example in the renal nerve (Fig. 10).

Post-Inspiratory Neurones and the Transition to Stage II Expiration

The gating during post-inspiratory activity of high order bulbar inspiratory neurones may be presumed to prevent re-development of the inspiratory ramp during the period of passive (Stage I) expiration. The further suppression of the ramp when Stage I expiration comes to an end is most certainly a consequence of the inhibition of inspiratory neurones during Stage II expiration and when the latter is absent the onset of inspiration is advanced.

What then is the relationship between these two stages of expiration? During inspiration expiratory neurones are inhibited. During Stage I expiration, however, they are disinhibited and not yet excited (Fig. 11). When Stage I expiration is prolonged facilitation of expiratory neurones active during Stage II expiration is delayed. The simplest interpretation of this effect is that bulbar expiratory neurones are disfacilitated by post-inspiratory neurones, quite possibly via their inhibitory action against reticular neurones. Only when Stage I expiration comes to an end are expiratory neurones synaptically de-

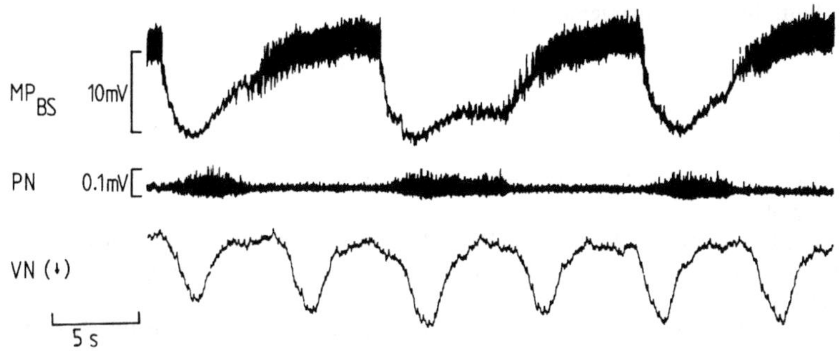

Fig. 11. Post-inspiratory disfacilitation of Stage II expiratory neurones. The membrane potential of a bulbospinal Stage II expiratory (MP_{BS}) neurone reveals disfacilitation when the post-inspiratory discharge in the phrenic nerve (PN) is prolonged. (VN) Integrated activity of the intact vagal nerve. An increase in activity is shown as a downward deflection of the trace

polarized to discharge, i.e., the transition from passive to active expiration is accomplished. Any marked delay in the onset of Stage II expiration delays the onset of the follo' ing inspiration. This this extent then the time course of Stage I expiration and post-inspiratory neurone discharge represents a powerful determinant of when the next inspiration will begin.

In conclusion we propose:

1. that the central respiratory cycle is composed of three phases;
2. that all three phases are represented by rather large populations of neurones within the lower brainstem;
3. that two populations of these neurones discharge during a period when the diaphragm is active, i.e., during the phase of inspiration and controlled passive expiration;
4. that at the level of higher order interneurones, inspiratory and post-inspiratory neurones, are antagonistic in their actions;
5. that the same two types of neurones form a part of the basic pattern generator as outlined at the beginning of this paper whereas neurones active in the third phase (Stage II expiration) are not essential to the function of this basic pattern generator;
6. that the mechanisms employed by this pattern generator include:

   6.1. a ramp oscillator which provides a common drive to inspiratory neurones and determines the inspiratory phase of the ventilatory cycle;
   6.2. early inspiratory inhibition;
   6.3. late inspiratory inhibition which *'off-switches'* inspiration;
   6.4. a relaxation oscillator represented by the activity of post-inspiratory neurones which determines the duration of Stage I expiration and which:

      6.4.1. *'gates'* both the inspiratory and expiratory ramp oscillators during this stage of passive expiration;
      6.4.2. *controls* inspiratory and possibly expiratory laryngeal muscle tone durin passive expiration;
      6.4.3. *'times'* the onset of the following inspiration via shifts in the Stage I to Stage II transition.

Supported by the Deutsche Forschungsgemeinschaft.

References

Bartlett Jr D, Remmers JE, Gautier H (1973) Laryngeal regulation of respiratory airflow. Respir Physiol 18:194–204

Euler von C, Hayward JN, Marttila I, Wyman RJ (1973) Respiratory neurones of the ventrolateral nucleus of the solitary tract of cat: vagal input, spinal connetions and morphological identification. Brain Res 61:1–22

Feldman JL, Cohen MI (1978) Relation between expiratory duration and rostral medullary expiratory neuronal discharge. Brain Res 141:172–178

Mitchell RA, Herbert DA (1974) Synchronized high frequency synaptic potentials in medullary respiratory neurones. Brain Res 75:350–355

Richter DW, Camerer H, Meesmann M, Röhrig N (1979) Studies on the interconnection between bulbar respiratory neurones of cats. Pflügers Arch 380:245–257

Functional Relations Between Dorsal Medullary Inspiratory Neurons and Phrenic Motoneurons

M.I. COHEN[1], J.L. FELDMAN[1,2], and D.F. DONNELLY[1]

In a recent review on neurogenesis of respiratory rhythm (Cohen 1979), I reached the conclusion that very little is known on the subject. Therefore, in the present paper I shall confine myself to a limited aspect of the system, namely, the transmission of rhythmic respiratory output from medullary inspiratory (I) neurons to phrenic (phr) motoneurons.

Many investigators have recorded the activity of brain-stem respiratory neurons, defined as neurons whose activity oscillates together with an indicator of respiratory output, such as tidal volume or phr nerve discharge. These respiratory-modulated neurons are concentrated in three major regions (Fig. 1): (a) the region of the "pneumotaxic

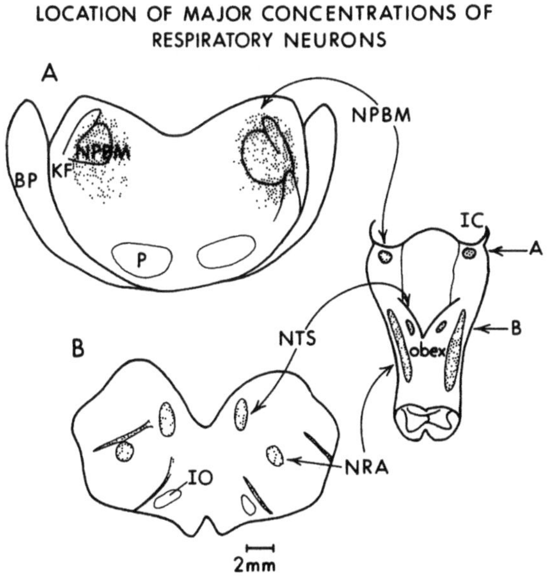

Fig. 1. Locations of major concentrations of respiratory neurons (stippling) in the brain-stem of the cat. Right: dorsal view of the brain-stem (cerebellum removed) with levels of sections A and B indicated. A, section through rostral pons. B, section just rostral to the obex. Abbreviations: BP, brachium pontis; IC, inferior colliculus; IO, inferior olive; KF, Kölliker-Fuse nucleus; NPBM, nucleus parabrachialis medialis; NRA, nucleus retroambigualis; NTS, nucleus of tractus solitarius; P, pyramidal tract

[1]Department of Physiology, Albert Einstein College of Medicine, Bronx, NY 10461, USA
[2]Present address: Department of Physiology, Northwestern University Medical School, Chicago, IL 60611, USA

Central Neurone Environment
(ed. by M.E. Schläfke, H.P. Koepchen and W.R. See)

center" in the rostral pons (Bertrand and Hugelin 1971, Cohen and Wang 1959); (b) the region of the dorsal respiratory group (DRG) in the medulla, corresponding to the ventro lateral nucleus of tractus solitarius (vlNTS), where predominantly I neurons are found (Berger 1977, Bianchi 1971, von Euler et al. 1973a); and (c) the region of the ventral respiratory group (VRG) in the medulla, which consists of a bilateral continuous rostrocaudal column extending from the lower pons to the first cervical segment (Bianchi 1971, Cohen 1979, Merrill 1970).

Since much evidence indicates that there is an intimate relation between DRG I neurons and phr motoneurons, I shall further limit the discussion to consideration of such relations. These have been analyzed by three types of experimental approach: (a) stimulation of descending axons in the spinal cord to evoke antidromic responses; (b) lung inflation tests to compare responses by members of the two populations; (c) cross-correlation between discharges of medullary and spinal neurons.

Antidromic Studies

A number of investigators have looked for spinal projections of DRG I neurons by stimulating descending axons in the spinal cord (usually at C2 to C3 levels) and recording antidromic spike responses (Berger 1977, Bianchi 1971, von Euler et al. 1973b, Lipski et al. 1979). These studies have reported that the great majority (80–90%) of these neurons send axons to the cervical spinal cord, mainly contralaterally.

The results of antidromic studies, as well as the observation that DRG I neurons and phr motoneurons have similar discharge patterns, led to the hypothesis that the medullar neurons are "upper motoneurons" that excite phr motoneurons. However, these antidromic studies do not give information about the ultimate destinations of the descending axons, nor about the effects of their activity on their target neurons. Therefore, analysis by other experimental methods must be pursued.

Lung Inflation Studies

Since pulmonary stretch receptor (PSR) afferents markedly affect the timing and pattern of I neuron discharge, it is of interest to compare their effects on DRG I neuron and on phr motoneuron activity. In our laboratory, these effects have been studied in decerebrate, paralyzed cats by use of the cycle triggered pump (CTP) system (Feldman and Gautier 1976, Feldman et al. 1976). In this system, the mass phr nerve discharge is used as an indicator of respiratory output, and timing pulses marking the onset of the I and the expiratory (E) phases are derived from the phr potentials. By means of a syster of logic circuits and solenoid valves, lung inflation is delivered during the period of phr discharge, thus avoiding a phase shift between central I activity and lung inflation.

In order to evaluate the effects of inflation, inflation is withheld for one cycle. As shown in the example of Fig. 2, this elimination of phasic PSR input results in lengthening of the I phase and of the discharge burst of a medullary I neuron. To obtain moı

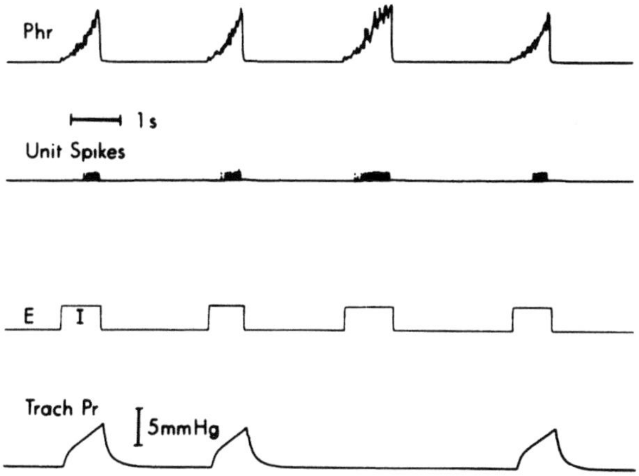

Fig. 2. Inkwriter traces of phr responses and responses of a DRG I neuron to withholding inflation by the CTP for one test I phase. *Phr* integrated phr discharge (time constant 0.1 s). *Unit spikes* standard pulses derived from spikes. *E* and *I* logic levels marking expiratory (E) and inspiratory (I) phases. *Trach Pr* airway pressure measured by transducer attached to tracheal cannula

precise information on the changes of discharge pattern produced by this maneuver, cycle triggered histograms (CTH) of I unit and of phr activity were obtained by an averaging computer, with the pulse marking the onset of I being used as a time reference.

The withholding inflation test allowed differentiation between three types of I neurons of the DRG (Cohen and Feldman 1977): (1) I-alpha neurons (Fig. 3A). These are not directly affected by inflation because withholding of inflation produces no change in the slope of frequency augmentation during most of the I phase, but rather only a lengthening of burst duration. This response is similar to the most common phr response (Fig. 3). (2) I-beta neurons (Fig. 3B). These are excited by PSR inputs, as shown by the reduction of their discharge frequency (reduction of slope of the CTH) when inflation is withheld. Thus these neurons receive excitatory inputs from the PSRs as well as from the central I pattern generator. The PSR inputs are probably monosynaptic, as indicated by the latency of I-beta neurons' responses to vagal electrical stimulation (Richter et al. 1979b). (3) Late I-beta neurons (Fig. 3C). This recently discovered subpopulation (Cohen and Feldman 1977, Baker and Remmers 1980, Marino et al. 1981) consists of neurons that start to fire a short time before the end of the I phase. When inflation is withheld the time of onset of discharge is delayed and the time to reach peak frequency is increased. The early-onset I-beta neurons and the late-onset I-beta neurons are distinct populations, as the distribution of times of discharge onset during I is multimodal (J.L. Feldman and M.I. Cohen, unpublished observations).

Since reduction of I-beta neurons' discharges is associated with lengthening of I, it might be suggested that these neurons are the mediators of the Breuer-Hering I-inhibitory reflex, as hypothesized by the original discoverers of the I-beta neurons (von Baumgarten and Kanzow 1958). A difficulty with this hypothesis is that many of the I-beta neurons start firing early during the I phase, but I-alpha neuron and phr activity do not change markedly till the end of I. In contrast, the late-onset I-beta neurons have

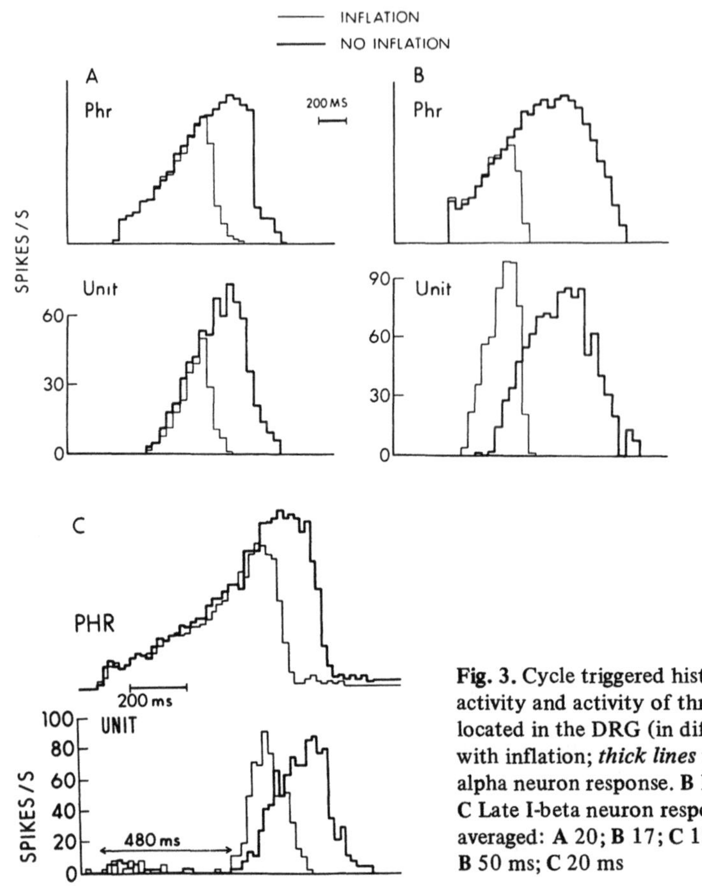

Fig. 3. Cycle triggered histograms (CTH) of PHR activity and activity of three I neurons (**A, B, C**) located in the DRG (in different cats). *Thin lines* with inflation; *thick lines* without inflation. **A** I-alpha neuron response. **B** I-beta neuron response. **C** Late I-beta neuron response. Number of cycles averaged: **A** 20; **B** 17; **C** 18. Bin durations: **A** and **B** 50 ms; **C** 20 ms

discharge properties that are consistent with their playing a role in the termination of I. Therefore, the following hypothesis on I termination has been proposed (Fig. 4): As the I phase progresses, both the I-alpha and the early-onset I-beta neurons gradually increase their firing rate, the discharges of the latter group being additionally increased by PSR inputs. The late-onset I-beta neurons are tentatively identified as part of an I-E switching system. These neurons receive excitatory inputs from the early-onset I-beta neurons; and when excitation has grown sufficiently large during late I, the late-onset neurons start firing, exert inhibitory actions on neurons of the central I pattern generator, and all I activity is shut off.

There are other possible hypotheses about the functions of the I-beta neurons. Large lung inflations produce facilitation of phr activity and presumably facilitation of activity of all types of I neuron (Pack et al. 1981). Further, under some ill-defined anesthetic conditions, even low (and static) lung inflations produce facilitation of phr and external intercostal activity (DiMarco et al. 1980). Under these conditions the distinction between I-alpha and I-beta neurons breaks down, since "alphas become betas" (C. von Euler, this Symposium). It is therefore conceivable that the I-beta neurons might ultimately be excitatory rather than inhibitory to phr motoneurons, but that under conditions of

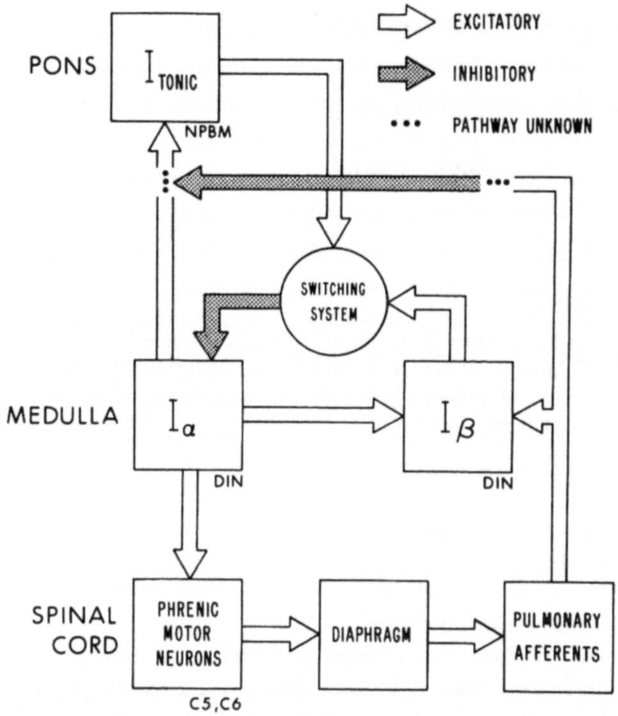

Fig. 4. Schema of possible interactions between neurons of different types in relation to I-E phase-switching. Excitatory connections: *white*, Inhibitory connections: *stippled*. NPBM nucleus parabrachialis medialis; NTS nucleus of tractus solitarius; C5, C6 fifth and sixth cervical segments

deep anesthesia or decerebration the excitatory actions would be subthreshold. This possibility is consistent with the observation that some of the I-beta neurons have descending axons in the spinal cord (Richter et al. 1979a).

In order to deal with these uncertainties, a quantitative comparison was made between the changes of DRG I neuron activity and of mass phr activity in response to the withholding inflation test. The method of analysis is shown schematically in Fig. 5. The discharges in I phases with inflation (control: C) and in I phases without inflation (test: T) were compared in two ways: (1) Activity at the end of the C phase (time of peak phr activity) was compared with activity at the corresponding time in the T phase. This evaluated the effect of inflation by measuring the change of activity at a fixed time from the start of I. (2) Activity at the end of the C phase was compared with activity at the end of the T phase. This evaluated the overall effects of PSR afferent input during the I phase.

With the use of this method, the relative changes of phr vs medullary activity were evaluated from a sample of 69 DRG I neurons (J.L. Feldman and M.I. Cohen, unpublished observations). The resulting graph showed that, while withholding inflation produced relatively small changes in phr activity (integrated peak amplitude) at times in the T phase corresponding to the end of the C phase (comparison 1 above), there was a large scatter of changes in I neuron activity. Many neurons showed little change of

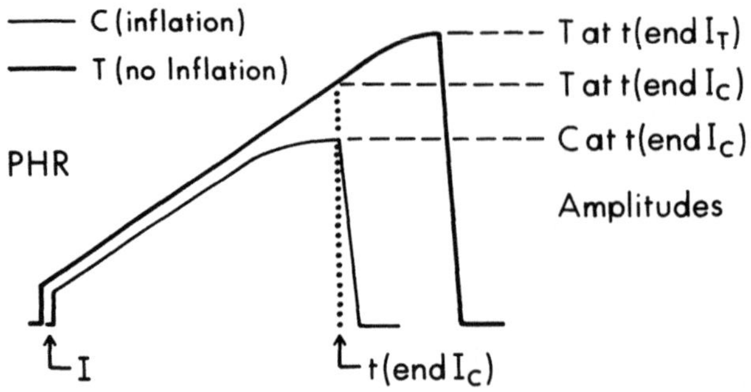

Fig. 5. Schema of method for quantitation of effects of withholding inflation. Traces represent integrated phr activity. C control, with inflation; T test, without inflation. *Vertical dotted line* reference time (time of peak phr discharge in C phase). *Horizontal lines* activity levels (top to bottom) at end of T phase; at reference time in T phase; at end of C phase. Similar traces of unit activity are not shown

activity. Others showed a large increase of activity (indicating inhibition by inflation). Neurons of these types are designated as I-alpha. Other neurons, designated as I-beta, showed large decreases of activity. Thus the analysis confirmed the impression that the DRG I neurons comprise a heterogeneous population.

The comparison of peak activity in the C and T phases (comparison 2 above) showed that peak phr and I-alpha neuron activity increased when inflation was withheld. This would be expected if the rate of frequency growth increased or remained the same, since the lengthening of I would allow discharge to reach a higher level at the end of I. However, it was of interest that in almost all cases the relative increase of peak phr activity was larger than the relative increase of peak unit activity. This observation might be explained by the hypothesis that the increase of excitatory input to the phr motoneuron pool produced differential actions on individual phr motoneurons.

In order to study such differential activation, recordings from individual phr motor fibers (N = 29) were taken and their responses to the inflation withholding test were observed and compared to the responses of contralateral mass phr activity (Donnelly and Cohen 1981, Cohen and Donnelly 1981). The picture that emerged from these observations was quite different from that seen for DRG I neuron responses. For the population as a whole, there was an approximately linear relation between change of mass phr activity and change of individual phr fiber activity. There were no responses comparable to those of the medullary I-beta neurons. Interestingly, although phr fiber discharge patterns were of two types, early-onset and late-onset, the responses of the late-onset neurons were opposite in direction to those of the late-onset medullary I-beta neurons: inflation withholding caused a large increase, rather than a decrease, in firing frequency of late-onset phr neurons.

In confirmation of the results of other workers (Hilaire and Monteau 1979, St. John and Bartlett 1981), we found that the distribution of firing onset times of phr motoneurons is not unimodal, i.e. there are two subpopulations, consisting of early- and late-onset neurons. [However, a conflicting result was found by Iscoe et al. (1976).] The two populations might correspond, respectively, to the B and A types of phr motoneuron described by Berger (1979).

Another finding of interest is that withholding inflation produced a larger increase of peak firing frequency in late-onset than in early-onset phr motoneurons. This finding implies that the major portion of the increase in mass phr peak discharge is due to the increase in discharge frequency of the late-onset neurons. This difference in responses between the early- and late-onset neurons might be due to differences in the excitatory inputs arriving from the medulla, and/or to differences of excitability related to cell size. At present we cannot distinguish between these possibilities.

These results furnish no support for the hypothesis that I-beta neurons excite phr motoneurons. In fact, the opposite responses of late-onset I-beta neurons and late-onset phr neurons suggest that the former group may exert inhibitory actions on the latter group, either directly or by graded inhibition of DRG I-alpha neurons which in turn are excitatory to late-onset phr neurons. Nevertheless, since low-threshold PSR afferents may be excitatory to phr motoneurons (DiMarco et al. 1980, Iscoe et al. 1979), we cannot exclude the possibility that this excitation is transmitted via early-onset I-beta neurons, either directly or by excitation of I-alpha neurons.

Cross-Correlation Studies

In order to obtain information on connections between medullary and spinal respiratory neurons, several investigators have used the method of cross-correlation (or the similar method of spike-triggered averaging). The cross-correlation histogram (CCH) of two neuronal signals (1 and 2) shows the probability of occurrence of signal 2 at different times before or after the occurrence of signal 1. From various features of the CCH, the functional connections between the two neurons may be deduced (Perkel et al. 1967, Kirkwood 1979). Thus, the occurrence of a sharp short-latency peak in the CCH strongly suggests that one neuron is monosynaptically driving the other.

This method has been used to analyze the short time scale relations between DRG I neurons and phr motoneurons by cross-correlating mass phr activity with the spike activity of a medullary neuron (Cohen et al. 1974, Hilaire and Monteau 1976). In the majority of the I neuron-contralateral phr CCHs, short-latency peaks were found, indicating that these I neurons were monosynaptically exciting contralateral phr motoneurons whose axons were in the nerve being recorded. In contrast, the I neuron-ipsilateral phr CCHs had either no peaks, or only small dispersed peaks. A typical example is shown in Fig. 6. These studies were done in vagotomized cats, so that it was impossible to determine whether the I neurons that monosynaptically excited contralateral phr neurons were of the I-alpha or I-beta type.

In another study (J.L. Feldman and M.I. Cohen, unpublished observations), unit-phr CCHs were derived for DRG I units that had been identified as I-alpha or I-beta. The interpretation of these CCHs was complicated by the existence of high-frequency oscillations (HFO). These HFOs, having periods of 8–17 ms, are found in all types of I discharge in decerebrate cats (Cohen 1973, 1976). In most of the CCHs the prominent HFOs tended to obscure any sharp peaks (indicative of monosynaptic excitation) that might have been present. In general, CCHs of I-alpha and I-beta neurons were indistinguishable, the common high-frequency rhythm being present in all types of I discharge. Thus, intrinsic limits of interpretation have so far prevented the use of the cross-correla-

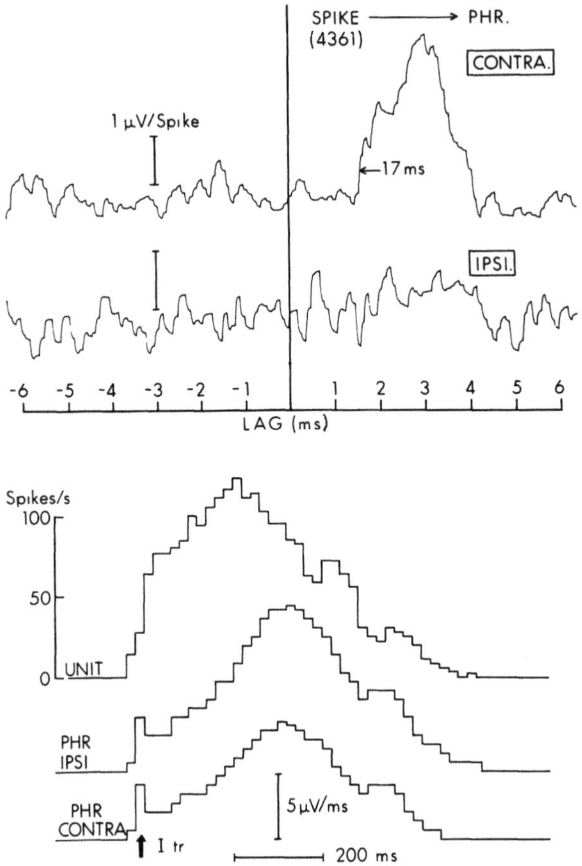

Fig. 6. Time relations between a DRG I neuron's activity and bilateral phr activity. *Top* CCHs (0.1 ms bins) of standard spike-derived pulses (N = 4361 spikes in 89 respiratory cycles) vs. mass phr discharge (contralateral *above*, ipsilateral *below*). Spikes occur at zero delay (*center line*); trace to right shows average phr activity following spikes. Latency of onset of contralateral peak (1.7 ms) is indicated by *arrow*. *Bottom:* CTHs (20-ms bins) of bilateral mass phr and of neuron's activity in time relation to start of I phase

tion method to differentiate connections between phr motoneurons and the several type of DRG I neuron.

Another possible approach to the problem is the use of cross-correlation between medullary I neuron spike activity and intracellular potentials of phr motoneurons. This method would increase the signal-to-noise ratio for the computations and would reveal subthreshold relations (e.g. EPSPs locked to spikes of a medullary unit).

Conclusion

There is no doubt that a predominant action of DRG I neurons is excitation of phr motc neurons. The main question that remains to be answered is: do all phr motoneurons receive similar inputs from the medulla, or are there preferential connections between di ferent types of medullary I neuron and different types of phr motoneuron?

Acknowledgements. This research was supported by USPHS Grants HL-20800 and HL-27300.

References

Baker JP, Remmers JE (1980) Temporal correlation of graded reversible inspiratory inhibition with discharge patterns of late inspiratory neurons located in the dorsal respiratory group in cats. Brain Res 200:331–340

Baumgarten von R, Kanzow E (1958) The interaction of two types of inspiratory neurons in the region of the tractus solitarius of the cat. Arch Ital Biol 96:361–373

Berger AJ (1977) Dorsal respiratory group neurons in the cat: Spinal projections, responses to lung inflation and superior laryngeal nerve stimulation. Brain Res 135:231–254

Berger AJ (1979) Phrenic motoneurons in the cat: subpopulations and nature of the respiratory drive potentials. J Neurophysiol 42:76–90

Bertrand F, Hugelin A (1971) Respiratory synchronizing function of nucleus parabrachialis medialis: pneumotaxic mechanisms. J Neurophysiol 34:189–207

Bianchi AL (1971) Localisation et étude des neurones respiratoires bulbaires. Mise en jeu antidromique par stimulation spinale ou vagale. J Physiol (Paris) 63:5–40

Cohen MI (1973) Synchronization of discharge, spontaneous and evoked, between inspiratory neurons. Acta Neurobiol Exp 33:189–218

Cohen MI (1976) Synaptic relations between inspiratory neurons. In: Duron B (ed) Respiratory centres and afferent systems. Colloq Inst Natl Santé Rech Méd 59:19–29

Cohen MI (1979) Neurogenesis of respiratory rhythm in the mammal. Physiol Rev 59:1105–1173

Cohen MI, Donnelly DF (1981) Inhomogeneity of phrenic motoneuron responses to lung inflation. Neurosci Lett Suppl 7:S441

Cohen MI, Feldman JL (1977) Models of respiratory phase-switching. Fed Proc Fed Am Soc Exp Biol 36:2367–2374

Cohen MI, Wang SC (1959) Respiratory neuronal activity in pons of cat. J Neurophysiol 22:33–50

Cohen MI, Piercey MF, Gootman PM, Wolotsky P (1974) Synaptic connections between medullary inspiratory neurons and phrenic motoneurons as revealed by cross-correlation. Brain Res 81: 319–324

DiMarco AF, Euler von C, Romaniuk JR, Yamamoto Y (1980) Low threshold facilitation of inspiration by lung volume increments. Acta Physiol Scand 109:343–344

Donnelly DF, Cohen MI (1981) Responses of early and late onset phrenic neurons to lung inflation. Fed Proc Fed Am Soc Exp Biol 40:429

Euler von C, Hayward JN, Marttila I, Wyman RJ (1973a) Respiratory neurones of the ventrolateral nucleus of the solitary tract of cat: vagal input, spinal connections and morphological identification. Brain Res 61:1–22

Euler von C, Hayward JN, Martilla I, Wyman RJ (1973b) The spinal connections of the inspiratory neurones of the ventrolateral nucleus of the cat's tractus solitarius. Brain Res 61:23–33

Feldman JL, Gautier H (1976) Interaction of pulmonary afferents and pneumotaxic center in control of respiratory pattern in cats. J Neurophysiol 39:31–44

Feldman JL, Cohen MI, Wolotsky P (1976) Powerful inhibition of pontine respiratory neurons by pulmonary afferent activity. Brain Res 104:341–346

Hilaire G, Monteau R (1976) Connexions entre les neurones inspiratoires bulbaires et les motoneurones phréniques et intercostaux. J Physiol (Paris) 72:987–1000

Hilaire G, Monteau R (1979) Facteurs déterminant l'ordre de recrutement des motoneurones phréniques. J Physiol (Paris) 75:765–781

Iscoe S, Dankoff J, Migicovsky R, Polosa C (1976) Recruitment and discharge frequency of phrenic motoneurones during inspiration. Respir Physiol 26:113–128

Iscoe S, Feldman JL, Cohen MI (1979) Properties of inspiratory termination by superior laryngeal and vagal stimulation. Respir Physiol 36:353–366

Kirkwood PA (1979) On the use and interpretation of cross-correlation measurements in the mammalian central nervous system. J Neurosci Meth 1:107–132

Lipski J, Trzebski A, Kubin L (1979) Excitability changes of dorsal inspiratory neurons during lung inflations as studied by measurement of antidromic invasion latencies. Brain Res 161:25–38

Marino PL, Davies RO, Pack AI (1981) The responses of I_β cells to increases in the rate of lung inflation. Brain Res 219:289–306

Merrill EG (1970) The lateral respiratory neurones of the medulla: their associations with nucleus ambiguus, nucleus retroambigualis, the spinal accessory nucleus and the spinal cord. Brain Res 24:11–28

Pack AI, de Laney RG, Fishman AP (1981) Augmentation of phrenic neural activity by increased rates of lung inflation. J Appl Physiol 50:149–161

Perkel DH, Gerstein GL, Moore GP (1967) Neuronal spike trains and stochastic processes. II. Simultaneous spike trains. Biophys J 7:419–440

Richter DW, Camerer H, Meesmann M, Röhrig N (1979a) Studies on the synaptic interconnection between bulbar respiratory neurones of cats. Pflügers Arch 380:245–257

Richter DW, Camerer H, Röhrig N (1979b) Monosynaptic transmission from lung stretch receptor afferents to R_β-neurones. In: Euler von C, Lagercrantz H (eds) Central nervous control mechanisms in breathing. Pergamon Press, New York, p 267–270

St. John WM, Bartlett D (1981) Comparison of phrenic motoneuron activity during eupnea and gasping. J Appl Physiol 50:994–998

A New Aspect of the Functional Organization of Respiratory Neurons in the Brainstem with Respect to the Rhythmogenesis of Respiration: Different Stabilities of Reticular Respiratory Neurons in the Rabbit

T. HUKUHARA, Jr., K. GOTO, K. TAKANO, Y. KIGUCHI, Y. HATTANMARU, and N. KIMURA[1]

In viewing the present state of our knowledge concerning the functional organization of the central respiratory mechanisms (Cohen 1979, Hukuhara Jr. 1973, Wyss 1964), a systematic survey of that part of the reticular neuron population which discharges spontaneously and which is correlated with phrenic nerve activity would be still indispensable as a first step, and an essential prerequisite for further investigations of the neuronal organization of the central respiratory mechanisms (Cohen 1979, Hukuhara Jr. 1980, Hukuhara Jr. et al. 1981), concurrently with studies on the neuron groups in the high density foci in the lower brain stem which may be the more typical ones (Cohen 1979, von Euler and Trippenbach 1976, Wyss 1964).

The spontaneous burst activity of respiratory neurons varies more or less in the cat and rabbit under anesthetized as well as paralyzed experimental conditions. Some reticular respiratory neurons exhibit fairly stable spontaneous discharge patterns for several hours, but on the contrary, other respiratory neurons show unstable discharge with time during simultaneous recording of two neuronal discharges (Hukuhara Jr. 1973, Hukuhara Jr. et al. 1979, 1981).

In the course of our systematic exploration experiments in rabbits, pontine and bulbar respiratory unit activities were identified by means of a new quantitative evaluation for the periodicity of the units on the basis of auto- and cross-correlation analyses (Hukuhara Jr. et al. 1981) besides the conventional criteria for identification of respiratory units (Cohen 1979, Fallert and Wassermeyer 1977, Gromysz 1971, Hukuhara et al. 1969, Woldring and Dirken 1951, Wyss 1964). Furthermore the temporal variation of spontaneous respiratory unitary discharges was analyzed quantitatively to examine whether different degrees of variation of the units exist and whether the degree of variation may relate to the discharge type and the localization of respiratory neurons in the brain stem.

[1]Department of Pharmacology II, Jikei University School of Medicine, 3–25–8 Nishishinbashi, Minato-ku, Tokyo 105, Japan

Central Neurone Environment
(ed. by M.E. Schläfke, H.P. Koepchen and W.R. See)

Methods

Under ether anesthesia, rabbits were vagotomized, paralyzed, and ventilated artificially. The end-tidal O_2 (16.8 ± 1.0%, N = 40) and CO_2 levels (3.80 ± 0.84%, N = 40) were monitored continuously together with arterial blood pressure. All the recorded unit spike potentials were confirmed to be from the soma-dendritic part of a neuron (Hukuhara Jr. et al. 1969). Locations of the micro-electrode tip were obtained by the use of stereotaxic co-ordinates. Respiratory unit activities have been identified from auto-correlograms of units and cross-correlograms between unit activity and phrenic nerve discharge on the basis of spike incidence distribution within the respiratory cycle using a digital correlator (Hukuhara Jr. et al. 1981) besides the conventional criteria for identi fication. As a measure of the stability of unit activities the coefficient of variation (CV: the ratio of the standard deviation vs the mean as a percentage) of the variables of a volley was calculated for individual neurons which exhibited intermittent bursting discharges for ten respiratory cycles.

Results

Identification of Respiratory Units

As shown in the auto-correlogram (Fig. 1, 3 in A and B) of respiratory unit activities and the cross-correlogram (Fig. 1, 5 in A and B), the periodicity of the respiratory units was in good agreement with that of the phrenic nerve discharge (Fig. 1, 4 in A and B). The discharge type of respiratory units was determined on the basis of the relation to the respiratory phases, defined as the period of phrenic activity and inactivity, by means of cross-correlograms (Fig. 1, 5 in A and B) between unit activity and the phrenic discharge.

Three main populations of reticular respiratory neurons, inspiratory (Fig. 1, A), expiratory, and a third, broader group were defined. Units in the third group did not show clearly defined locking to a phase transition of the respiratory cycle, but nevertheless showed respiratory periodicity. The third group consisted of: (a) inspiratory-expiratory phase spanning; (b) expiratory-inspiratory phase-spanning; and (c) four categories of reticular neurons which discharge tonically and exhibit respiratory modulation in the following phases of the respiratory cycle: inspiratory (Fig. 1, B), expiratory, inspiratory-expiratory, and expiratory-inspiratory phases.

In addition, there were a few reticular neurons discharging with a relative co-ordination of 1:4 with the respiratory phases, although the co-ordination in their activity patterns was unsteady and transient.

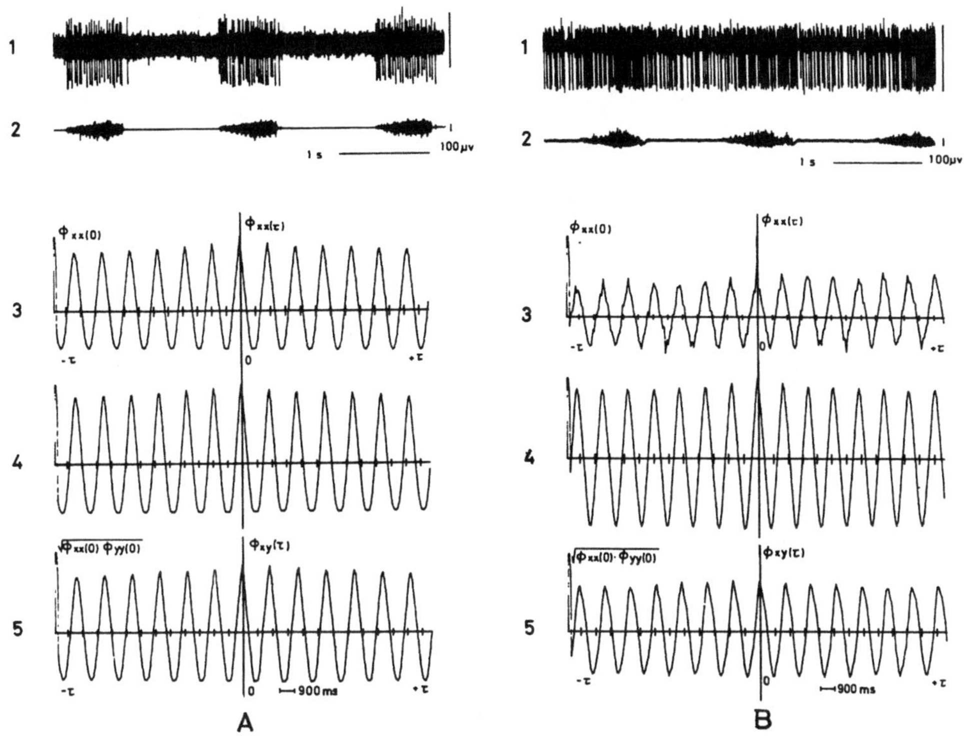

Fig. 1. Identification of respiratory unitary discharge. *1* = Respiratory unit; *2* = Phrenic discharge; *3* = Auto-correlogram of 1; *4* = Auto-correlogram of 2; *5* = Cross-correlogram between 1 and 2. A An inspiratory unit; B An inspiratory frequency modulated type. Paralyzed and vagotomized rabbit

Spatial Distribution of Pontine and Bulbar Respiratory Neurons Obtained by Systematic Exploration Experiments in the Rabbit

Using correlation analysis for identification, respiratory unitary discharges were systematically searched for within the entire area of the bulbar and pontine reticular formations from the caudal border of the inferior colliculi to the level of C_1 and their neighbouring structures involving the known high-density regions in the cat. Of the 151 respiratory units identified, 29 were found in the pons and the remainder in the bulbar structures.

Respiratory reticular units (137 units of 151) of three main populations were scattered and intermingled throughout most regions of the pontine and bulbar reticular formation. However, neurons were found more frequently in the lateral than in the medial region of the bulbar reticular formation and in the pontine reticular formation (Fig. 2, I and II). In contrast to the findings described in previous reports (Cohen 1979, Fallert and Wassermeyer 1976, Gromysz 1971, Woldring and Dirken 1951, Wyss 1964), no circumscribed regions where respiratory units were grouped according

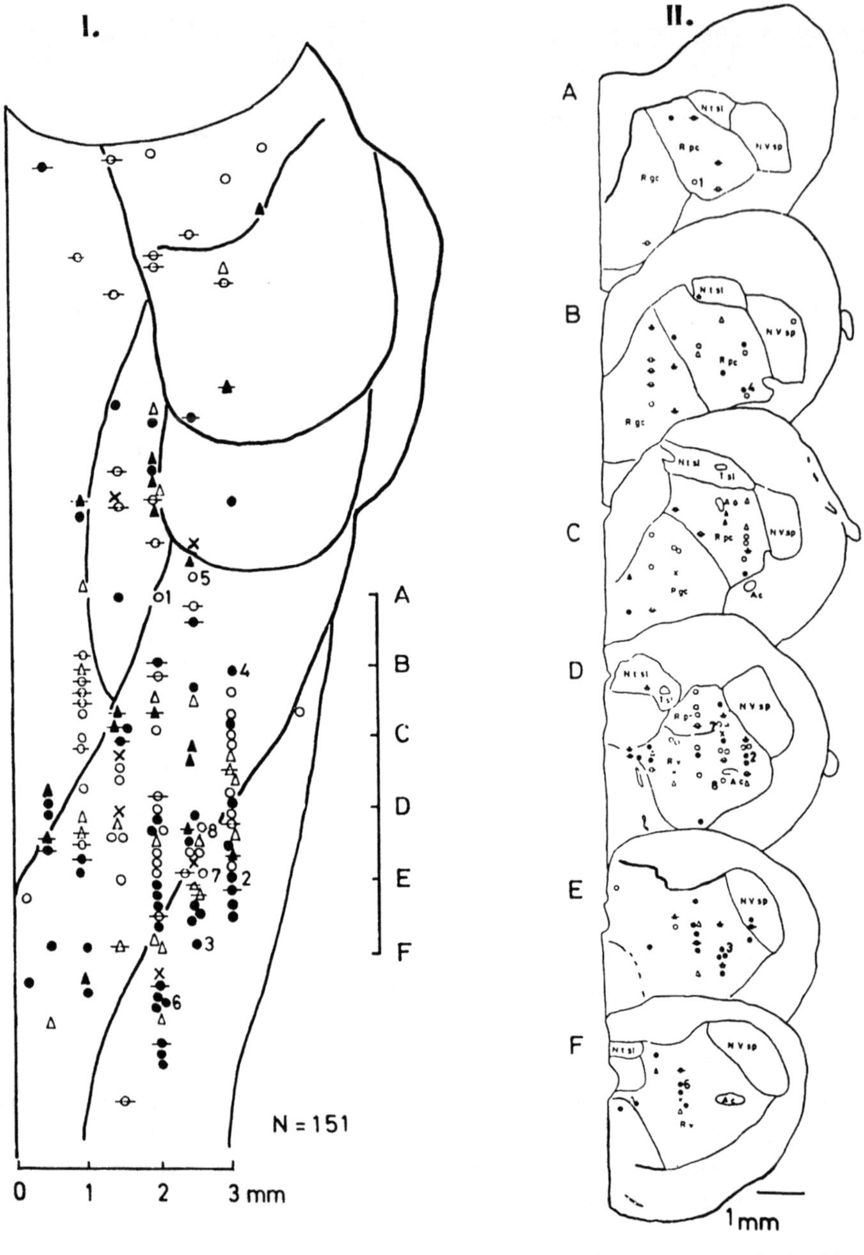

Fig. 2. Location of respiratory neurons in the brain stem. The positions were obtained by systematic exploration experiments over the pontine and bulbar reticular formation as well as the neighbouring structures in 42 rabbits paralyzed, vagotomized, and ventilated artificially. I. Dorsal projection of the brain stem showing distribution of 151 respiratory neurons. II. Six frontal sections of the medulla taken at obex level (E), at 1 mm intervals rostrally (D, C, B, A), and caudally (F). Each of the following symbols indicates respectively the position where different types of respiratory unitary discharge were recorded. ○ = inspiratory (i); ● = expiratory (e); △ = i-e phase-spanning (ph-s); ▲ = e-i ph-s; ⊖ = i-modulated (m); -●- = e-m; -△- = i-e ph-s-m; -▲- = e-i ph-s-m. For explanation of the numbers of 1–8 in I and II see the legend for Fig. 4

to discharge type were found, although inspiratory and expiratory units were found predominantly in the bulbar reticular substance (Fig. 2, I and II).

Different Stability of Spontaneous Discharge Pattern Between Respiratory Neurons

It has been shown that the spontaneous activity of respiratory neurons varies considerably in the cat (Hukuhara Jr. 1973, 1980, Hukuhara Jr. et al. 1969). In the rabbit the stability of the spontaneous discharge of respiratory neurons also varies somewhat with time (Fig. 3). When two respiratory units were simultaneously recorded, different degrees of stability of spontaneous discharge between an unstable unit and another, stably discharging unit were noted. In the present experiments, we attempted to study quantitatively the stability of the spontaneous unit activity of respiratory neurons identified by means of auto- and cross-correlation analysis in aiming to examine whether there exists any difference in the stability of spontaneous unitary discharges among respiratory neurons in the brain stem and if so whether the degree of variation relates to the discharge type and the location of respiratory neurons in the brain stem.

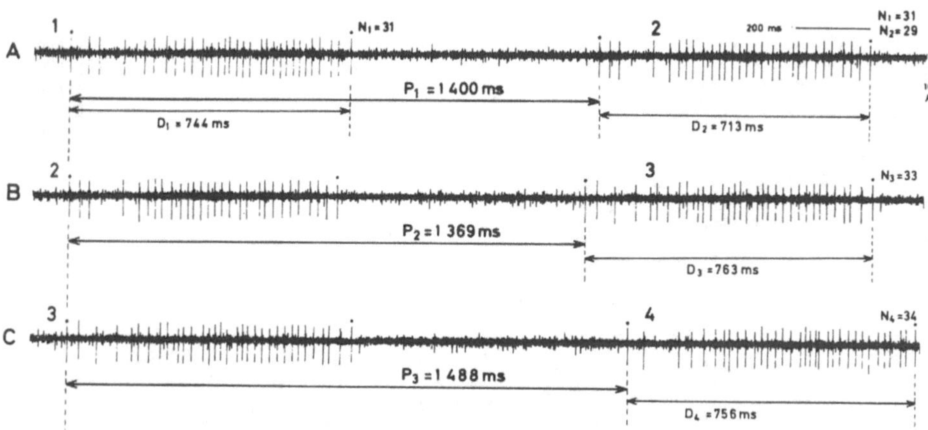

Fig. 3. Variability of the unitary discharge of a bulbar expiratory neuron in the course of three respiratory cycles in a paralyzed, vagotomized rabbit. The first and last spike in each volley numbered from 1 to 4 are each indicated by a *dot*. For comparison of the variables for the volleys, each first spike of the 1st and 2nd volley is placed on a *vertical dashed line* in the left part of the figure; thus, the volley 2 in **A** and **B** is identical, and so on. P = period of volley; D = duration of volley; N = number of spikes in volley. Note the remarkable fluctuation in the three variables

Coefficient of Variation of Variables of Unit Burst Activity

As a quantitative measure of the stability of respiratory unitary discharges the coefficient of variation (CV) as a percentage of four variables of the volley (period of volley, duration of volley, number of spikes per volley (Fig. 3), and mean frequency in volley) was calculated in the course of seven to ten respiratory cycles for individual neurons which exhibited intermittent bursting discharges. CV for each variable of the volley varied over a wide range. Mean values of CV with standard deviation and range in percentage for 87 neurons were as follows; period of volley, 5.9 ± 3.2, 1.3–17.2; duration of each volley, 15.4 ± 13.7, 1.8–89.4; number of spikes per volley, 18.5 ± 12.5, 3.2–70.5; mean frequency in each volley, 17.7 ± 12.0, 2.3–75.5. Thus, CV for the period was the smallest among four variables. When two respiratory unitary discharges belonging to the same group of discharge type were recorded simultaneously in an experimental animal, different degrees of the stability of spontaneous discharges between an unstable unit having a larger CV for volley period and another, stably discharging unit showing a smaller CV, were noted under the same experimental conditions.

Respiratory units of different discharge types having a larger CV for all four variables were found in both the bulbar and the pontine structures. However, the units of different discharge patterns having smaller CVs were located only in the medulla, not in the pons.

Although the CVs for the period of volley in each of the three main groups: inspiratory, expiratory, and phase-spannings populations covered also a considerably wide

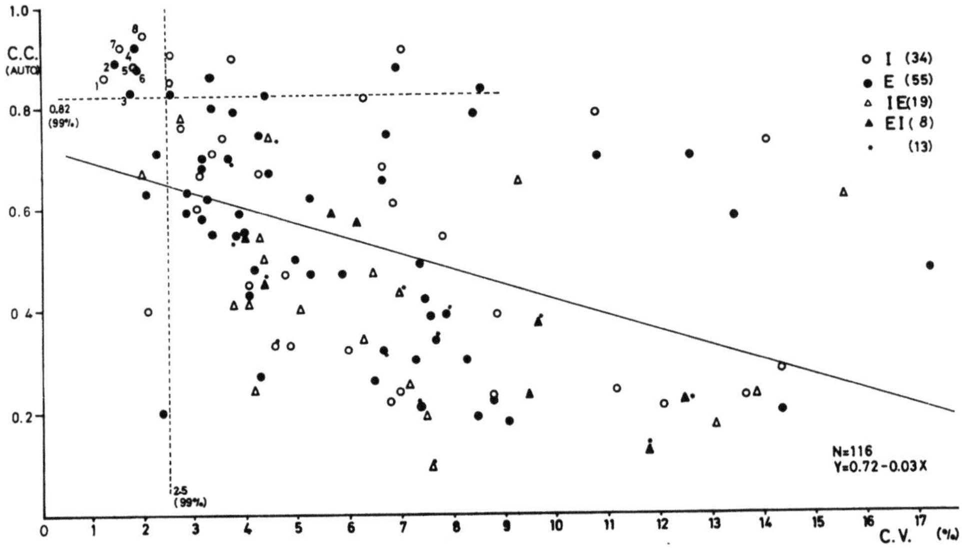

Fig. 4. Correlation diagram between the correlation coefficient (C.C.) of respiratory unitary discharges (*ordinate*) and the coefficient of variation (C.V.) for the volley period of respiratory unit activity in the course of consecutive 10 respiratory cycles (*abscissa*). Correlation coefficient between two variables is –0.47 ($P < 0.001$, Y = –0.03 X + 0.72). Each symbol represents a respiratory unit: ○ = inspiratory (n = 34); ● = expiratory (n = 55); △ = inspiratory-expiratory phase-spanning (n = 19); ▲ = expiratory-inspiratory phase-spanning (n = 8); * = pontine unit (n = 13). The eight respiratory neurons (*1–8* in the correlation diagram) discharging stably with a small coefficient of variation and a large correlation coefficient are plotted in the right half of the dorsal projection of the brain stem (I in Fig. 2) and frontal sections of the medulla (II in Fig. 2) with the numbers 1–8

range, very small CV values of less than 2.1% were found only in the inspiratory and expiratory groups, not in the group of phase-spanning units (Fig. 4).

Correlation Coefficient (CC) of Auto-Correlation

The correlation coefficient of auto-correlation of respiratory unitary discharges was computed in order to quantify the stability of spontaneous unit activity of individual respiratory neurons as another variable for examining the degree of the variation of spontaneous unitary discharges. Correlation coefficients of auto-correlation also exhibited a considerable range of variation from 0.06–0.94. Although respiratory units of different discharge types having a smaller CC were located both in the medulla and the pons, those yielding a larger CC than 0.79 were found in the medulla, not in the pons. Though CC in each of the four groups: inspiratory, expiratory, phase-spannings, and frequency-modulated populations, covered a wide range, larger CC values than 0.79 were found only in the bulbar inspiratory and bulbar expiratory population.

Correlation Between CC and CV

Figure 4 shows the relation between the CC of auto-correlation and CV for the period of the volley for a total of 116 respiratory units (bulbar 103, pontine 13). There was a significant correlation ($r = -0.47$, $P < 0.001$, $Y = 0.72 - 0.03\,X$) between the two variables. The coefficient of variation for the period of the phrenic burst discharges in 85 rabbits ranged from 0.9 to 9.1% and the mean with standard deviation of CV was 3.0 ± 1.6% with the 99% confidence interval from 2.5% to 3.5%. The correlation coefficients of auto-correlation of the phrenic nerve activity in these 85 experiments ranged

Table 1. Correlation coefficient of auto-correlation (C.C.), coefficient of variation (C.V.), and anatomical location of bulbar respiratory neurons displaying stable burst discharges in the rabbit. C.V. for each parameter was calculated for 10 respiratory cycles. P, period of volley; N/D, mean frequency of spikes within volley in Hz; $\bar{N}$, mean value of number of spikes per volley; R.pc, Nucleus reticularis parvocellularis; R.v, Subnucleus reticularis ventralis medullae oblongatal

Neuron	Type	C.C.	C.V. (%)			Localization
			P	N/D	$\bar{N}$	
1	I	0.86	1.3	7.3	23.0	R.pc
2	E	0.89	1.5	3.3	32.2	R.pc
3	E	0.83	1.8	4.9	78.0	R.pc
4	E	0.92	1.9	3.5	80.0	R.pc
5	I	0.88	1.9	7.8	124.0	R.pc
6	E	0.88	1.9	7.7	75.0	R.v
7	I	0.92	1.6	2.0	111.1	R.v
8	I	0.94	2.1	5.7	141.3	R.pc

from 0.51 to 0.96 giving the mean and the standard deviation of 0.80 ± 0.09 with the 99% confidence interval from 0.78 to 0.82. Four bulbar inspiratory and four bulbar expiratory neurons (Table 1) were spontaneously discharging stably with a CV for the period in a range from 1.3% to 2.1% smaller than the lower limit of the 99% confidence interval of the phrenic CV for the volley period of 2.5% indicated by the vertical broken line in Fig. 4, while their CCs of auto-correlation ranging from 0.83 to 0.94 were larger than the upper limit of the 99% confidence interval of the phrenic CC of 0.82 indicated by the horizontal broken line in Fig. 4. These inspiratory and expiratory neurons discharging stably were located in the Nucleus reticularis parvocellularis or Subnucleus reticularis ventralis medullae oblongatae (Fig. 2, I and II).

Thus, this evidence indicates that the degree of stability of spontaneous unitary discharges of reticular respiratory neurons is related to the type of the discharge pattern and the location of the neurons in the brain stem.

Effects of Thiamylal on Spontaneous Discharge of Respiratory Neurons in Relation to their Characteristics of Variables of Stability

When two respiratory unitary discharges belonging to the same population were recorded simultaneously, intravenous injection of thiamylal in small doses of 2, 4, or 8 mg/kg caused concomitantly different types of changes of discharge patterns of each neuron. It has been examined whether the type of change of individual respiratory units relates to the degree of the stability of discharges as measured by its CC of auto-correlation and CV for the volley period. Thiamylal injection (4 mg/kg IV) caused a decrease in the mean frequency in each volley, prolongation of burst duration and of the period of the

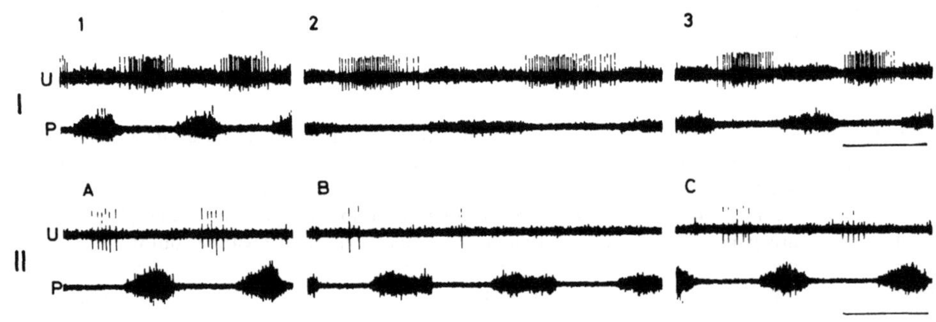

Fig. 5. Effect of thiamylal on respiratory unitary discharges. U = expiratory unit; P = phrenic discharge. Time calibration, 1 s indicated by the *horizontal bar* at the right. Each *vertical bar* at the right is 100 μV for each tracing respectively. I: 1 = before; 2 = 60 s after thiamylal (4 mg/kg IV); 3 = 300 s after the drug. Correlation coefficient (C.C.) and coefficient of variation (C.V.) were 0.859 and 3.4% respectively. F_{ECO_2} was 3.0%, 2.8% and 3.0% in 1, 2 and 3 respectively. II: A = before; B = 21 s after thiamylal (2 mg/kg IV); *C* = 300 s after the drug. C.C., 0.783; C.V., 8.5%. F_{ECO_2}: 3.0%, 2.9%, and 3.0% in A, B, and C respectively. Note that the drug in a small dose of 2 mg/kg IV caused a cessation of spontaneous discharge the expiratory unit (U) in II, while the drug with 4 mg/kg IV elicited only an inhibitory effect of the spontaneous firing of the expiratory unit in I

Table 2. Parameters of spontaneous unitary discharge of bulbar respiratory neurons and effect of thiamylal on their spontaneity in the rabbit. C.C., correlation coefficient of auto-correlation; C.V., coefficient of variation for period of volley calculated for 10 respiratory cycles; $\bar{N}$, mean value of number of spikes per volley; N/D, mean frequency of spikes within volley. The plus sign in the column of spontaneity means the presence of spontaneous unitary discharge of respiratory neurons and the minus sign represents a cessation of spontaneous unit firing of respiratory neurons after intravenous injection of thiamylal in doses of 2, 4 or 8 mg/kg. The plus sign means the existence of burst discharge of the phrenic nerve activity and the minus sign represents an absence of significant periodic changes confirmed by examination of the autocorrelogram of the phrenic nerve activity. Note the plus in unit 1–4 showing a larger C.C. and a smaller C.V. and the minus in unit 5–10 having a smaller C.C. and a larger C.V.

Unit No.	Type	C.C.	C.V. (%)	$\bar{N}$	N/D	Spontaneity of unitary discharge	Phrenic burst discharge
1	I	0.91	2.6	71.0	151.3	+	+
2	I	0.91	7.1	68.0	89.6	+	+
3	E	0.83	2.6	24.5	55.7	+	+
4	E	0.86	3.4	25.0	37.8	+	+
5	E	0.59	2.9	4.8	12.0	–	+
6	E	0.75	4.3	15.0	32.6	–	–
7	E	0.74	6.8	13.5	22.7	–	+
8	E	0.39	7.6	4.7	18.8	–	+
9	E	0.70	12.7	5.6	16.0	–	+
10	E	0.79	8.5	18.2	26.6	–	+

volley in some neurons (Fig. 5, I), on the contrary, in other neurons the drug (2 mg/kg IV) caused a cessation of spontaneous firing for several minutes (Fig. 5, II) accompanied with a decrease in the total power of the phrenic nerve activity (Hukuhara Jr. et al., unpublished observations), prolongation of burst duration and of the period of the volley (Fig. 5, I and II). Changes of spontaneous unitary discharges of 124 respiratory neurons induced by thiamylal (2–8 mg/kg) injection were studied. Respiratory units (39) which discharged unstably, having a large CV for the volley period and a small CC of auto-correlation, ceased to fire after the drug injection, while respiratory units (85) having a small CV for the volley period and a large CC of auto-correlation exhibited a change into an inhibited pattern, but continued to discharge. As shown in Table 2 in which the results of some typical experiments are summarized, no particular relation between the spontaneity of firing of respiratory units after drug administration and the number of spikes per volley or the mean frequency in each volley in control was found.

Discussion and Conclusions

It has been shown that the neural mechanism generating the respiratory rhythms exists in the brain stem involving pontine and medullary structures (Hukuhara Jr. 1973, 1974, Hukuhara Jr. et al. 1978). Periodic activity similar to eupnoea originating in the medullary neural mechanisms is dominant with respect to spontaneity (Hukuhara Jr. 1973, 1976, Hukuhara Jr. et al. 1978) over other rhythmic activities which have been de-

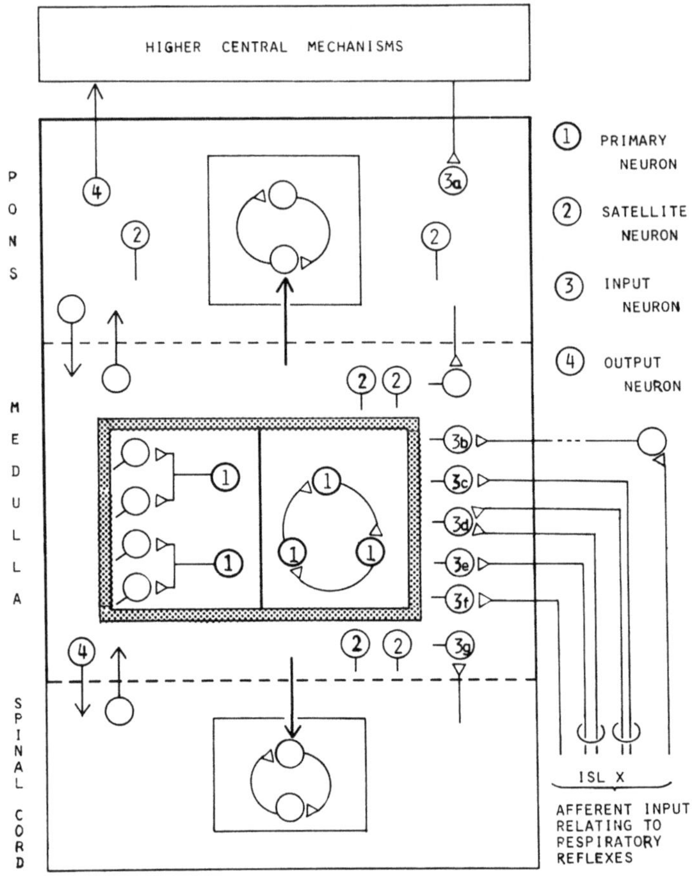

Fig. 6. Functional organization of respiratory neurons in the central respiratory mechanisms in the brain stem. For explanation see text

scribed in other neural mechanism in the brain stem and the spinal cord (Cohen 1979, von Euler and Trippenbach 1976, Hugelin 1978, Wyss 1964). It has been proposed that the neuronal network for the central respiratory mechanisms consists of 4 neuronal functional subsystems: primary, satellite, input and output respiratory neuron systems which are characterized in their specified relations to the rhythmogenesis and the possible different roles in the regulatory function of the central respiratory mechanisms within the neuronal network (Hukuhara Jr. 1973, 1976, 1980) (Fig. 6). Not only in the paralyzed but also in the anesthetized condition in cats and rabbits a variability of spontaneous discharge of respiratory neurons to a greater or lesser degree was observed (Hukuhara Jr. 1973, Hukuhara et al. 1979, 1981). In the present experiments it has been shown that the stably discharging inspiratory and expiratory neurons having a small CV for the volley period and a large CC of auto-correlation were found in the bulbar reticular formation, but not in corresponding areas in the medulla described as the high density regions in cats (Cohen 1979, von Euler and Trippenbach 1976, Hugelin

1978) and in pontine structures (Cohen 1979, Hugelin 1978). This fact is not contradictory to and in accordance with the results of previous experiments obtained by brain stem transection techniques in combination with correlation analysis (Hukuhara Jr. 1973, 1974, 1976, Hukuhara Jr. et al. 1978).

Injection of thiamylal in small doses caused different types of changes of spontaneous unitary discharge patterns in good correlation to the degree of the stability of spontaneous unitary discharges of individual respiratory neurons. Respiratory neurons showing a low degree of stability of their unitary discharges ceased to fire during the period when spontaneous burst discharges of the phrenic nerve activity still persisted after drug administration. Thiamylal in a small dose blocks the synaptic transmission at the presynaptic site in the monosynaptic reflex pathway in the spinal cord (Løyning et al. 1964). In the rabbits used in the present experiments the vagus nerves were cut and in most cases the carotid sinus nerves were also severed bilaterally. Therefore, before thiamylal at the spontaneous rhythmic discharge of the unstably firing respiratory neurons was thought to be maintained by multiple inputs mediated by the synaptic mechanism and fed from various sources including the neural mechanism generating the respiratory rhythms (Hukuhara Jr. 1973, 1974, Hukuhara Jr. et al. 1978), the central chemosensitive mechanism in the ventral medulla, other respiration related structures in the neuraxis (Cohen 1979, Hugelin 1978), and other possible sources.

Thus, we would like to suggest that the respiratory neurons, inspiratory and expiratory, which are discharging stably with a large correlation coefficient of auto-correlation and a small coefficient of variation for the volley period, and are found in the bulbar reticular formation, are playing an essential role as the primary respiratory neurons in generating the respiratory rhythm in the bulbar mechanism.

Acknowledgements. We would like to thank Miss M. Mii for typing the manuscript. This work was supported partly by the Ministry of Education, Science, and Culture, Japan, Grant-in-Aid for Special Project Research No. 537005 and Grant-in-Aid for Environmental Science No. 40391 and No. 56030083.

References

Cohen MI (1979) Neurogenesis of respiratory rhythm in the mammal. Physiol Rev 59:1105–1173

Euler von C, Trippenbach T (1976) On the respiratory phase-switching mechanisms. In: Duron B (ed) Respiratory centers and afferent systems. Inst Natl Sante Rech Med , vol 59. INSERM, Paris, p 11

Fallert M, Wassermeyer B (1977) Spatial distribution of various types of bulbar respiratory neurons in the rabbit. Exp Brain Res 30:339–351

Gromysz H (1971) Studies on the activity of the brain stem respiratory neurons of the rabbit. Ph D Thesis Med Res Center (Warsaw) (in Polish) (cited from Gromysz H, Karczewski WA (1976) Acta Neurobiol Exp 36:581–592)

Hugelin A (1978) Regional effects of nembutal anesthesia on brain stem respiratory neurones. In: Fitzgerald RS, Gautier H, Lahiri S (eds) The regulation of respiration during sleep and anesthesia. Plenum Press, New York, p 5

Hukuhara T Jr (1973) Neuronal organization of the central respiratory mechanisms in the brain stem of the cat. Acta Neurobiol Exp 33:219–244

Hukuhara T Jr (1974) Functional organization of brain stem respiratory neurons and rhythmogenesis. In: Umbach W, Koepchen HP (eds) Central-rhythmic and regulation. Hippokrates, Stuttgart, p 35

Hukuhara T Jr (1976) Functional organization of brain stem respiratory neurons and its modulation induced by afferences. In: Duron B (ed) Respiratory centers and afferent systems. Inst Natl Sante Rech Med, vol 59. INSERM, Paris, p 41

Hukuhara T Jr (1980) Spontaneous activity pattern and anatomical distribution of brain stem reticular neurons showing correlation with phrenic and renal sympathetic nerve activities. In: Koepchen HP, Hilton, SM, Trzebski A (eds) Central interaction between respiratory and cardiovascular control systems. Springer, Berlin Heidelberg New York, p 21

Hukuhara T Jr, Saji Y, Kumadaki N, Kojima H, Tamaki H, Takeda R, Sakai F (1969) Die Lokalisation von atemsynchron entladenden Neuronen in der retikulären Formation des Hirnstammes der Katze unter verschiedenen experimentellen Bedingungen . Naunyn-Schmiedebergs Arch Pharmakol Exp Pathol 263:462–484

Hukuhara T Jr, Kageyama S, Kiguchi Y, Goto K, Nishikawa Y, Takano K (1978) Bulbar neural mechanisms originating intrinsic respiratory rhythms in the central respiratory mechanisms. In: Ito M (ed) Integrative control function of the brain, vol 1. Kodansha, Tokyo/Elsevier, Amsterdam, p 254

Hukuhara T Jr, Goto K, Kiguchi Y, Takano K (1979) Unterschiedliche Stabilität respiratorischer Einzelneuronenaktivität im Hirnstamm der Katze. Jikeikai Med J 26:245–261

Hukuhara T Jr, Goto K, Takano K, Hattammaru Y (1981) Localization and functional organization of bulbar and pontine respiratory neurons in the rabbit. In: Hutás I, Debreczeni A (eds) Adv Physiol Sci vol 10, Respiration. Pergamon Press, London/Akadémiai Kiado, Budapest, p 579

Løyning Y, Oshima T, Yokota T (1964) Site of action of thiamylal sodium on the monosynaptic spinal reflex pathway in cats. J Neurophysiol 27:408–428

Woldring S, Dirken MNJ (1951) Site and extension of bulbar respiratory center. J Neurophysiol 14:227–241

Wyss OAM (1964) Die Nervöse Steuerung der Atmung. Ergeb Physiol 54:1–479

The Influence of Chemosensory, Laryngeal, and Vagal Afferents on Respiratory Phase-Switching Mechanisms and the Generation on In- and Expiratory Efferent Activities

W. MAREK, N.R. PRABHAKAR, and A. MIKULSKI[1]

The ponto-medullary respiratory neurons receive afferent inputs from a variety of sources. The arterial gas tensions are sensed by the arterial chemoreceptors (Heymans and Heymans 1927) and the pH of the extracellular fluid (ECF) surrounding the nervous tissue of the ventral medulla by the central chemosensitive structures (for reviews see Loeschke 1974, Schläfke 1981). Another important input to the respiratory neurons comes from the slowly and the rapidly adapting stretch receptors of the trachea and lungs (Paintal 1973). The activity arriving from the peripheral chemoreceptors is tonic and irregular (Biscoe et al. 1970) whereas the information from the airways is phase locked to the inspiration, i.e. depending on the lung volume (Paintal 1973). From the tonic chemosensory and the volume-locked vagal input, the respiratory neurons generate the alternating activity in the inspiratory and expiratory muscles, which is the basis for rhythmic breathing.

To obtain further information about the central mechanisms responsible for the generation of respiration and for the respiratory phase-switching, chemosensory, laryngeal, and vagal afferents were stimulated either with a continous-sequence of impulses or with short trains (0.2–0.4s) of electrical pulses in different phases of the respiratory cycle. The experiments were carried out on chloralose-urethane anaesthetized, spontaneously breathing cats. The tidal volume was recorded with a spirometer along with the arterial blood-pressure, the P_{ECO_2} and the electrical activities of in- and expiratory nerves and muscles.

Chemosensory Afferent Stimulations

The respiratory responses to brief electrical stimulations of the afferents from the arterial chemoreceptors and the central chemosensitive structures are of the same pattern, but differ in their amplitude. This suggests, that all these chemosensory afferents project to the same pool of respiratory neurons (Marek et al. 1981). Inspiratory stimulations, with a latency of 15–25 ms, are followed by an increase in the inspiratory muscle activity (diaphragm as well as inspiratory intercostal muscles), causing an increased inspiration. The amplitude of the response depends on the timing of the stimulation (Eldridge 1972, Black et al. 1973, Marek et al. 1981). The later the stimu-

1 Institut für Physiologie, Ruhr Universität, Bochum, D–4630 Bochum, Fed. Rep. of Germany

Central Neurone Environment
(ed. by M.E. Schläfke, H.P. Koepchen and W.R. See)

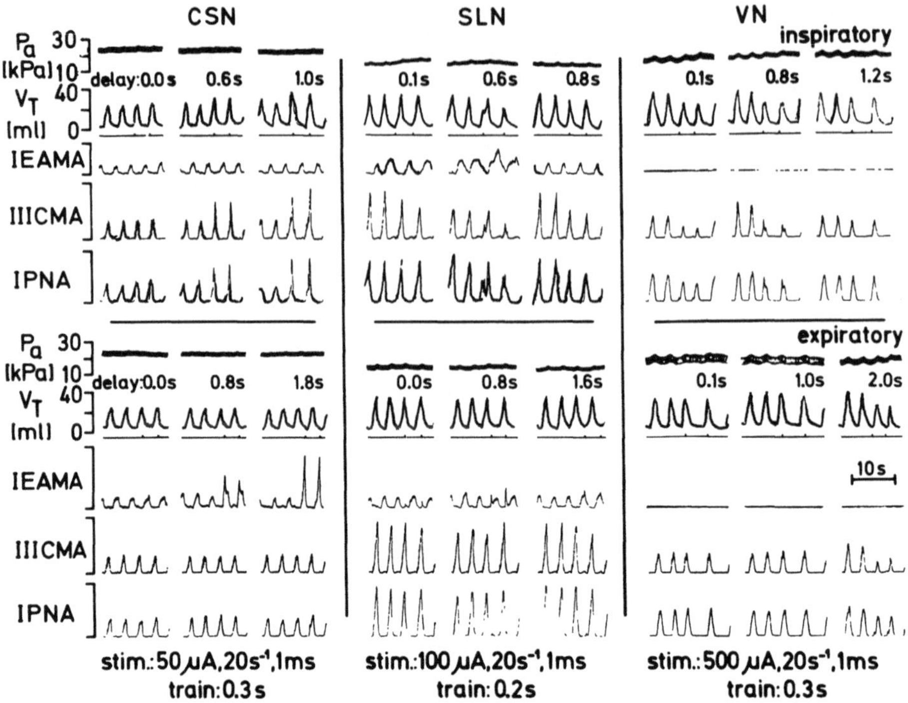

Fig. 1. Respiratory responses to brief electrical stimulations of chemosensory (carotid sinus nerve, CSN), left side of the picture, laryngeal (superior laryngeal nerve, SLN) middle, and vagal (vagus nerve, VN) afferents, right side, in early, middle and late inspiration (*upper part*) and expiration (*lower part*). The delay from the onset of inspiration or expiration to the onset of the stimulation is given in the figure. From top to bottom: P_a arterial blood pressure; V_T tidal volume recorded by spirometer; the stimulation period is marked under the spirometric signal; IEAMA integrated expiratory abdominal muscle activity; IIICMA integrated inspiratory intercostal muscle activity, IPNA integrated phrenic nerve activity, time constants for integration are all 0.1 s

lation starts in the inspiratory cycle, the bigger is the increase in the tidal volume and in the activity of the respiratory muscles. Early inspiratory stimulations result in a shortening of the inspiration while late stimulations cause a prolongation. The greatest increase of tidal volume is obtained when the stimulation starts at 70–90% of the normal inspiratory duration. Together with the increase in tidal volume, the next expiration is prolonged without further activation of the expiratory muscles. The greatest increase of minute ventilation occurs during midinspiratory stimulation. Late inspiratory stimulations, even those starting at 100% of T_I, or on the inspiratory peak of the spirometric signal, regularly result in an inspiratory response, i.e. a prolongation of the inspiration and an increase of tidal volume. Stimulations, however, starting less than 0.1 s after the peak of inspiration already lead to expiratory responses.

Stimulations in expiration usually cause an activation of expiratory muscles with a latency of 80–120 ms and an increase of the depth of expiration below the normal endexpiratory gas volume level (FRC). The change in the depth of expiration, however

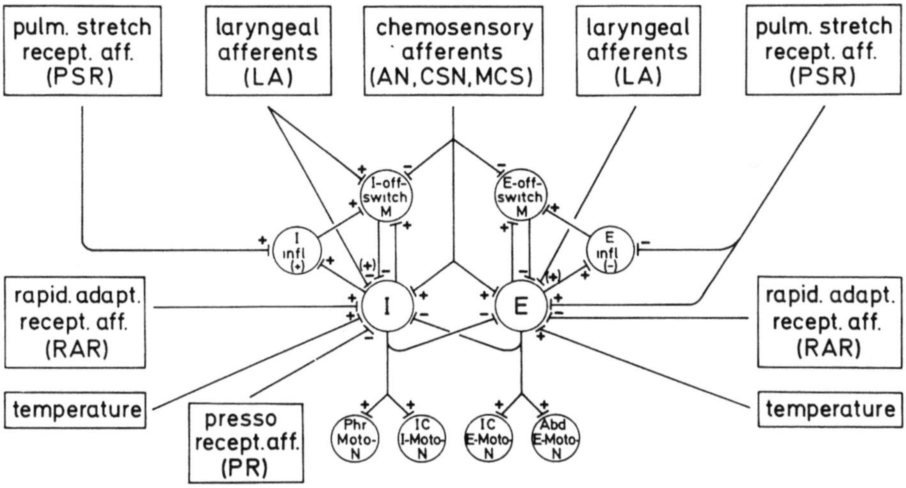

Fig. 2. Model from the functional point of view for the organization of the central respiratory control system, being responsible for the generation of the alternating activity in the in- and expiratory muscles and the respiratory phase switching, developped from input-output relations during the stimulation of afferent nerves. I and E are the central generators for the inspiratory and expiratory activity receiving afferent excitatory (+) and inhibitory (–) inputs from the chemosensory organs, AN aortic nerve; CSN carotid sinus nerve; MCS medullary chemosensor, the airways and lungs. I infl. (+), a population of inspiratory neurons receiving excitatory input from I and the lung stretch receptors (PSR). E infl. (–), a population of expiratory neurons, receiving excitatory input from E and inhibitory input from the lung stretch receptors (PSR). I-off-switch M, and E-off-switch M, are the inspiratory and expiratory off-switch mechanisms, terminating the I- and E-activities when alternately activated. For further explanation see text

is only about 20% of the maximal increase in the inspiratory gas volume (see Fig. 1, left side, lower part). The amplitude of the responses to expiratory stimulations depends on the timing of the stimulation as was also observed in inspiration. Early expiratory stimulations cause a relatively small increase of the depth of the expiration and a shortening of the expiratory phase. The later in the expiratory phase the stimulation starts, the larger is the increase of the depth and duration of expiration. Stimulations, starting in the last 85–100% of the expiration either result in a maximal increase in the depth and duration of expiration or have no effect on the expiratory half-cycle. In such cases, where the expiratory mechanisms are not excited, the stimulation elicits a forced inspiration and a shortening of the inspiratory phase. This type of response is also typical for early inspiratory stimulations.

The obtained results have shown, that the amplitude of the respiratory responses to stimulation of the peripheral as well as the central chemosensory afferents depends on the timing of the stimulation in the in- and expiratory half-cycle. This indicates cyclic alterations of the sensitivity of the generators of the inspiratory (I) and expiratory (E) output (see Fig. 2) within the respective half-cycles of respiration.

Laryngeal Afferent Stimulations

Brief inspiratory stimulations of laryngeal afferents with a stimulus strength just above the threshold result in a phase locked diminution of the tidal volume, a shortening of the inspiratory duration and a prolongation of the expiration without an activation of the expiratory muscles (see Fig. 1, middle). With a short latency of only 5–10 ms the phrenic nerve activity and the activity of inspiratory intercostal muscles are inhibited (Iscoe et al. 1979, Marek et al. 1981). Early inspiratory stimulations only result in a transient diminution of the inspiratory activity and a decrease in the inspiratory gas flow. The tidal volume however is nearly unchanged, depending on a prolongation of the inspiratory activity in the phrenic nerve and the inspiratory intercostal muscles. Stimulations starting in the middle of inspiration result in two types of respones: (1) a termination of the inspiration or (2) a transient inhibition of the inspiration with a slower and prolonged inflation. Stimulations starting in the second half of inspiration, however, after about 50 ms always lead to a termination of the inspiration. The later the stimulation starts in the inspiratory half-cycle, the more effective is the inhibition of inspiration and the prolongation of expiration.

Expiratory stimulations with a stimulus strength just above the threshold result in a phase dependent prolongation of the expiration without further activation of the expiratory muscles (see Fig. 1, middle, lower part). The next inspiration is prolonged and the slopes of the integrated phrenic nerve and inspiratory intercostal muscle activities are decreased. On the other hand, suprathreshold stimulations elicit respiratory protective reflexes (Korpas and Tomori 1979). In the case of the expiration reflex, the expiratory muscles get excited, with a latency of 30–40 ms, for a period of about 50 ms. The stimulation is most effective in early expiration, but the responses can be elicited in late inspiration too. The result is a forced active expiration to the level of FRC. The inspiration reflex most commonly is elicited at the end of expiration and in the early inspiration, causing a short and rapid inflation. These reflexes may be effective in cleaning the upper airways from inhaled small particles. The responses to stimulation of the superior laryngeal nerve afferents are phase-locked. This indicates, that the central respiratory mechanisms (I, E, see Fig. 2) cyclically change their sensitivity to inhibitory inputs in the respiratory cycle. The amplitude of the respiratory protective reflexes also depends on the respiratory phase.

Vagal Afferent Stimulations

Brief inspiratory stimulations of vagal afferents elicit a phase locked gradual inhibition of the inspiratory activity. The inspiratory gas flow gets diminished and the tidal volume is reduced. Stimulations in the second half of inspiration result in a termination of inspiration within less than 200 ms and a prolongation of the expiratory period. This response is known as the classic Breuer-Hering-Inflation inhibitory reflex. Brief expiratory stimulations result in a prolongation of expiration, typical for the Breuer-Hering-expiratory facilitating reflex. The later the stimulation begins in the expiratory half-cycle, the more the expiration gets prolonged (see Fig. 1, right side, lower part). Stimulations starting in the last 70–100% of the expiration result either in a prolongation of expiration or in no change . In the following inspiration, the inspiratory muscle activity transiently gets diminished. Brief expiratory stimulations may also result in an activation of the expiratory muscles. This causes an increase in the depth of expiration below the level of the normal endexpiratory volume (FRC). During vagal stimulation in experiments with stimulus frequencies modulated by the oesophageal pressure, (1) a diminution of the inspiratory muscle activity, (2) a shortening of inspiration and (3) a reduction of the inspired gas volume are observed. The inspiratory inhibition depends on the stimulation frequency, frequencies at the end of inspiration between 150–200 $imp \cdot s^{-1}$ being most effective. Endexpiratory stimulation frequencies between 30–100 $imp \cdot s^{-1}$ were most effective in an excitation of expiratory muscles. As a result of the activation of the expiratory muscles there is an increase in the depth and a prolongation of the expiration.

Depending on the stimulation parameters, vagal afferent stimulation causes, besides the classical Breuer-Hering-reflexes, also a transient inhibition of the inspiratory activity.

Model for the Functional Organization of the Central Respiratory Control Mechanisms

Based on the characteristics of the central respiratory neurons, Von Euler (1977) and Cohen and Feldman (1977) have proposed models for the genesis of the respiratory rhythm and the inspiratory phase-switching mechanisms. From our results based on the input-output relations, a model is developed for the functional organization of the central respiratory control system. Our model of the central neuronal mechanisms responsible for the generation of efferent in - and expiratory activity, from the functional point of view, should contain two generators, one for the inspiratory (I) and a second for the expiratory (E) activity (Fig. 2). Each of them is able to inhibit the other during its active period. I and E contain the different populations of inspiratory and expiratory neurons. The neurons for the generation of inspiratory as well as for expiratory efferent activity receive excitatory afferent input from the arterial chemoreceptors and the medullary chemosensor. Stimulation of these afferents results in an increase of the slopes of the integrated activities of the inspiratory and expiratory muscles. The increase of the amplitude of in- and expiratory muscle activities further indicates an elevation of the threshold for the activation of the respiratory phase-switching mechanisms (I-off-switch and E-off-switch M), or a direct inhibition of the "off-switch" mechanism by the chemosensory afferents.

Stimulation of the laryngeal afferents (LA) results in an immediate termination of inspiration . The inhibition of the phrenic nerve and the nerves to the inspiratory intercostal muscles begins with a latency of 5–10 ms, and the termination of inspiration is irreversible after less than 50 ms. This may indicate a direct activation of the inspiratory "off-switch"-mechanism by laryngeal afferents. The observed transient inhibition in early inspiratory stimulations and the prolongation of expiration without activation of the expiratory muscles shows an inhibitory input to the generator of the inspiratory activity (I) itself. On the other hand, in the cases when respiratory protective reflexes are elicited, I and E directly get excited in the inspiration and expiration reflex respectively.

Stimulation of the slowly adapting lung stretch receptor afferents (PSR) by lung inflations or by electrical stimulations results in a shortening of inspiration. The termination of inspiration is due to an activation of a sub-group of inspiratory (I_β or Infl.+) neurons, activating the inspiratory "off-switch" mechanism. In the present study, expiratory stimulation of the slowly adapting stretch receptor afferents (PSR) resulted in an activation of the expiratory muscles and an increase in the depth of expiration below the functional residual volume. This would mean that the slowly adapting stretch receptor afferents (PSR) have an excitatory input to the generator of expiratory activity (E) and an inhibitory input to a subgroup of expiratory neurons (E Infl -) The discharge of these expiratory neurons (E Infl -) is postponed by lung inflations. The result is a prolongation of expiration, conceivably because of the later activation of the expiratory "off-switch" mechanism. The discharge pattern of the E Infl (-) neurons is equivalent to that of the inspiratory I_β or Infl (+) neurons.

The so called rapidly adapting (RAR) or "irritant" receptor fibres, being activated by quick and large lung inflations and deflations have an inspiratory facilitatory and an expiratory inhibitory influence. Further inhibitory input to the generator of inspiration (I), is regularly observed by aortic and carotid sinus presso-receptor afferents (PR). Moderate hyperthermia leads to an increase in the slope of the integrated respiratory muscle activity, without changing the amplitude. In this situation, the activity of the I and E generators reaches the "off-swich" threshold earlier. The result is an increase in the respiratory frequency.

It can be summarized from the results of the present study, that the generators of inspiration (I) and expiration (E) can be activated independently. In other words inspiratory stimulations of the chemosensory afferents only result in activity changes of the inspiratory muscles, whereas expiratory stimulations cause the excitation of the expiratory output. Only the shortening of inspiration and the expiratory prolongation observed during inspiratory stimulations of the laryngeal and vagal afferents are due to an inhibition of the inspiratory (I) mechanism. The expiratory prolongations occuring in expiratory stimulations may be due to (1) an activation of the expiratory generator leading to an activation of the expiratory muscles and (2) to an activation of the E-generator together with an inhibition of the I-generator, which would postpone the onset of the next inspiration.

References

Biscoe TJ, Purves MJ, Sampson SR (1970) The frequency of nerve impulses in single carotid body chemoreceptor afferent fibres recorded in vivo with intact circulation. J Physiol (London) 208:121–131

Black AMS, Goodman N, Nail BS, Rao PS, Torrance RW (1973) The significance of timing of chemoreceptor impulses for their effect upon respiration. Acta Neurobiol Exp 33:139–148

Cohen MI, Feldman JL (1977) Models of respiratory phase-switching. Fed Proc Fed Am Soc Exp Biol 36:2367–2374

Eldridge FL (1972) The importance of timing on the respiratory effects of intermittent carotid sinus nerve stimulation. J Physiol (London) 222:297–318

Euler von C (1977) The functional organisation of the respiratory phase-switching mechanisms. Fed Proc Fed Am Soc Exp Biol 36:2375–2380

Heymans JF, Heymans C (1972) Sur les modifications directes et sur la régulation réflexe de l'activité du centre respiratoire de la tête isolée du chien. Arch Int Pharmacodyn 33:272–370

Iscoe S, Feldman JL, Cohen MI (1979) Properties of inspiratory termination by superior laryngeal and vagal stimulation. Respir Physiol 36:353–366

Korpas J, Tomori Z (1979) Cough and other respiratory reflexes. Karger, Basel

Loeschcke HH (1974) Central nervous chemoreceptors. In: Widdicombe JG (ed) Respiration physiology. Butterworths, London, 2:167–196

Marek W, Mikulski A, Prabhakar NR (1981) Atemzyklische Empfindlichkeitsschwankungen des Reglers Atmung bei Stimulation von chemosensiblen und laryngealen Afferenzen. Atemwege/Lungenerkrankungen 7/3:145–147

Paintal AS (1973) Vagal sensory receptors and their reflex effects. Physiol Rev 53:159–227

Schlaefke ME (1981) Central chemosensitivity: A Respiratory Drive. Rev. Physiol Biochem Pharmacol 90:171–244

Central and Peripheral Chemosensitivity in Early Essential Hypertension in Man

A. TRZEBSKI, M. TAFIL, M. ZOLTOWSKI and J. PRZYBYLSKI[1]

One of the mechanisms, not yet systematically explored in essential hypertension, is a possible enhancement of peripheral and/or central chemosensitivity of the respiratory drive. This possibility appears interesting in view of the close central interaction between the respiratory and sympathetic control systems (Koepchen et al. 1980). Significant alveolar hyperventilation and respiratory alkalosis was observed in young spontaneously hypertensive rats (SHR), and the hypothesis was proposed that the arterial chemoreflex is hyperactive in early hypertension (Przybylski 1978, 1981). Another question which might also be relevant to hypertension is whether central chemosensitivity to CO_2 is increased, on account of a stimulating effect on sympathetic activity and neurogenic vascular tone (Trzebski et al. 1974, Lioy et al. 1978), which may be independent of the central respiratory drive (Trzebski and Kubin 1981).

The present study was undertaken to test if peripheral and/or central chemosensitivity is augmented in early essential hypertension in young men. For this purpose responses of ventilation, blood pressure, and heart rate to isocapnic hypoxia – a primary stimulus for arterial chemoreceptors also in man (Lugliani et al. 1971, 1973) and to hyperoxic hypercapnia, a stimulus acting mainly upon the central chemosensitive areas (Lloyd 1966), were measured and compared in young age-matched hypertensive and normotensive subjects.

Subjects and Methods

The investigation was carried out in 20 hypertensive and 20 normotensive age-matched male volunteers, 20–21 years of age except for two hypertensive subjects 24 and 26 year old respectively (group mean 20.6 years) and two normotensive subjects 23 and 24 years repectively (group mean 20.3 years). In the control normotensive group, mean diastolic pressure (DBP) was 10.17 kPa ± 0.86 SD. (76.5 mmHg ± 6.51), range: 7.98 kPa (60 mmHg) – 11.3 kPa (85 mmHg), mean systolic pressure (SBP) was 15.36 kPa ± 1.54 SD (115.5 mmHg ± 11.57), range: 12.63 kPa (95 mmHg) – 17.29 kPa (130 mmHg). In the hypertensive group mean DBP was 13.06 kPa ± 0.34 SD (98.25 mmHg ± 2.59),

1 Department of Physiology, Warsaw Medical Academy, Krakowskie Przedmiescie 26/28, 00–927; and Institute of Tuberculosis, Warsaw, Poland

Central Neurone Environment
(ed. by M.E. Schläfke, H.P. Koepchen and W.R. See)

range: 12.63 kPa (95 mmHg) – 13.3 kPa (100 mmHg) mean SBP was 18.8 kPa ± 0.66 SD (141.25 mmHg ± 4.96). In no case had the hypertension been treated during at least one month preceding the study. Progressive hyperoxic hypercapnia and isocapnic hypoxia were induced using the rebreathing method (Read and Leigh 1967). End-tidal CO_2 and O_2 were sampled at the mouth by a mass spectrometer (Centronic Residual Gas Analyzer MGA 007) and the gas returned to the 6-litre rebreathing bag. Inspiratory minute ventilation $\dot{V}_I$ was recorded by a Godart pneumotachograph, accumulated up to 20 litres, and then started again. Ventilation records corrected to BTPS were sampled over 10 s-periods which corresponded to each successive increase of P_{A,CO_2} of 0.32 kPa (2.5 mmHg) and to each successive decrease of P_{A,O_2} of 0.58 kPa (4.5 mmHg). For studying the ventilatory response to CO_2 the rebreathing bag was filled with a 7% CO_2 in 93% O_2 gas mixture.

Rapid equilibration was thus established on rebreathing between P_{CO_2} in the bag, lungs, and pulmonary capillary blood and the P_{CO_2} of the whole equilibrated system began to rise at a steady state. Despite vigorous rebreathing end-tidal P_{O_2} never fell below 43.13 kPa (324 mmHg) so hyperoxia was maintained in all of the CO_2 studies.

For studying the ventilatory response to progressing isocapnic hypoxia, the rebreathing bag was filled with a 27% O_2, 6% CO_2, and 67% N_2 gas mixture. 20 s after the start of rebreathing the CO_2 scrubbing circuit was turned on. By adjusting a rheostat a variable flow of bag gas was drawn by the pump through the CO_2-absorbing bypass. Thus, over the total period of hypoxia, P_{CO_2} in the whole equilibrated system was maintained constant within ± 0.2 kPa (1.5 mmHg) of the mean level which was 1.04 kPa ± 0.416 SEM (7.8 mmHg ± 3.13) below initial $P_{\bar{V},CO_2}$ mean value 5.72 kPa ± 0.43 SEM (43 mmHg ± 3.23). For assessing the force of inspiratory muscles and respiratory center output the technique of airway occlusion pressure was adopted (Whitelaw et al. 1975). Instantaneous heart rate was recorded as the length of the cardiac cylce by a circuit provided with an analog output which was linearly proportional to the intervals between successive R-R waves of the ECG. All measurements were continuously recorded on a 6-channel recorder (Watanabe WTR R81). Arterial blood pressure measurements were taken by a mercury sphygmomanometer at 15 s intervals during each study. The magnitude of respiratory sinus arrhythmia (RSA) was calculated as the maximal difference between the length of the cardiac cycle in expiration and inspiration or its reciprocal value in beats min^{-1}.

$\dot{V}_I - P_{A,CO_2}$ linear function was calculated using the least squares linear regression method. The CO_2 response curve was expressed in terms of its slope S_{CO_2} (litres $min^{-1} \cdot kPa^{-1}$) and intercept on the P_{CO_2} axis, B_{CO_2} (kPa) according to the procedure proposed by Lloyd et al. (1958).

Hyperbolic $\dot{V}_I - P_{A,O_2}$ curves were calculated by the empirical equation $\dot{V}_I = \dot{V}_{I_0} + A\,(P_{A,O_2} - 4.26)$, where $\dot{V}_{I_0}$ is the asymptote for ventilation obtained by extrapolation and a constant 4.26 representing in kPa the P_{A,O_2} (32 mmHg) at which the slope of the $\dot{V}_I - P_{A,O_2}$ curve usually approaches infinity. Thus parameter A determines the shape of the curve such that the higher the value for A, the greater the hypoxic ventilatory drive (Weil et al. 1970). To determine the approximated hyperbolic curve of best fit a nonlinear hyperbolic regression formula $y = a\,x^b$ and the analysis of variance was used for both hyperbolic ventilatory and pressure response to hypoxia. Correlations of the hypoxic and hypercapnic responses (Rebuck et al. 1973) and of the responses in normotensive and in hypertensive subjects were calculated by linear regression of least squares. Pulse interval durations (R-R intervalss of the ECG) during expiration and inspiration were plotted against decreasing P_{A,O_2} values in progressing hypoxia and the resulting curve was approximated by a logarithmic regression of best fit according to the formula y = a + b ln x and the analysis of variance. The standard Student's t-test was used to determine statistical significance.

Results

Ventilatory Response to Isocapnic Hypoxia in Normotensive and Hypertensive Subjects

Figures 1 and 2 present examples of the response to hypoxia in a normotensive and in a hypertensive subject. Starting from a P_{A,O_2} of 14.04 kPa the difference of $\dot{V}_I$ between normotensive and hypertensive subjects became significant abd increased with progressing hypoxia (Fig. 3). Parameter A denoting the hyperbolic curve shape in the normotensive group was 28.78 ± 2.68 SEM (range: 6.98–44.25) and in the hypertensive group 116.1 ± 10.52 SEM (range: 71.63–234.77). This difference was highly significant (t = 9.119, $P < 0.001$).

Normotensive subjects were able to rebreathe longer and to produce a lower level of hypoxia up to P_{A,O_2} 6.11 kPa ± 0.24 SEM. Hypertensive subjects tolerated hypoxia poorly and felt a stronger discomfort compelling them to take out the mouthpiece earlier than did normotensives. Thus the whole period of the rebreathing procedure was significantly shorter in the hypertensive group – 4 min. 36 s ± 39 s SEM – than in the normotensive group – 5 min. 42 s ± 18.8 s SEM ($P < 0.001$).

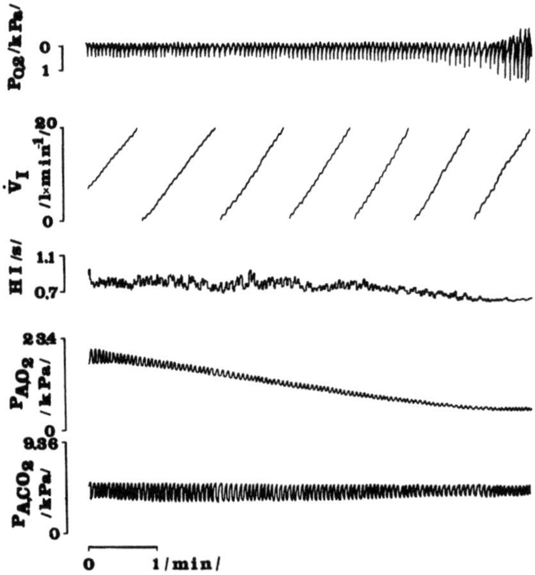

Fig. 1. Response of a normotensive 21-year-old subject to isocapnic hypoxia induced by rebreathing. From the top: (1) airway occlusion pressure, (2) minute ventilation, (3) pulse intervals linearily proportional to R-R waves of ECG, (4) oxygen tension at the mouthpiece; the level of downward deflections corresponds to end-tidal values, (5) CO_2 tension at the mouthpiece the level or upward deflections corresponds to end-tidal values. Initial systolic blood pressure was 15.96 kPa and diastolic blood pressure was 10.64 kPa. At the end of the rebreathing period respective values were 16.29 kPa and 11.97 kPa. Conversion: 1 kPa = 7.519 mmHg

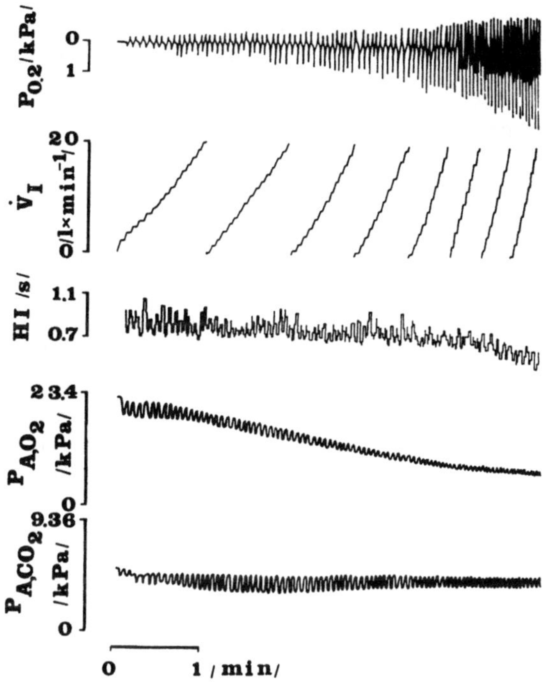

Fig. 2. Response of a hypertensive 21-year-old subject to isocapnic hypoxia. Records the same as in Fig. 1. Initial systolic blood pressure was 19.95 kPa and diastolic blood pressure 13.33 kPa. Respective values at the end of rebreathing were 23.94 kPa and 15.96 kPa. Respiratory sinus arrhythmia was preserved over the whole period of rebreathing. Conversion: 1 kPa = 7.519 mmHg

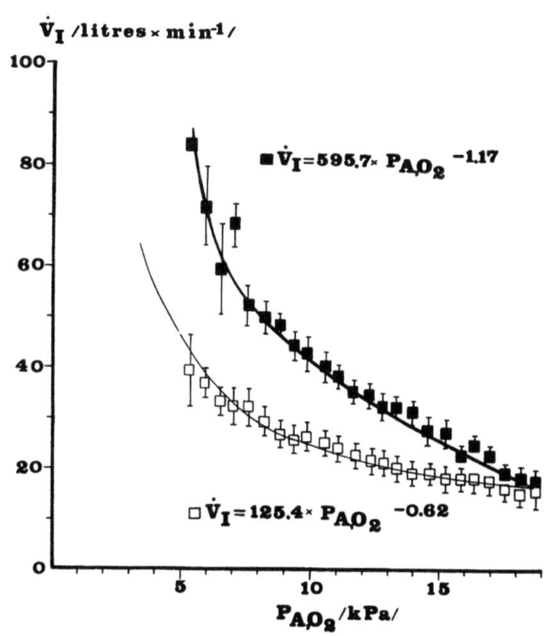

Fig. 3. Mean ventilatory response to hypoxia in the hypertensive subjects (*black squares*) and in the normotensive subjects (*white squares*). *Vertical bars* correspond to SEM. Curves of best fit were approximated according to equations. Conversion: 1 kPa = 7.519 mmHg

In hypertensive subjects the increase of the occlusion pressure $\Delta P_{0.2}$ measured at a P_{A,O_2} of 7.11 kPa ± 0.23 SEM obtained at the end of rebreathing reached 2.05 kPa ± 0.14 SEM. In the normotensive group at the same level of hypoxia the $\Delta P_{0.2}$ value was 1.2 kPa ± 0.08 SEM, the difference being highly significant ($P < 0.001$).

Arterial Blood Pressure Response to Isocapnic Hypoxia

With progressing hypoxia, blood pressure rose in a curvilinear fashion resembling the hyperbolic ventilatory response. The rise of the systolic blood pressure (ΔSBP) and diastolic blood pressure (Δ DBP) plotted against declining P_{A,O_2} values was significantly steeper in the hypertensive group (Fig. 4). At a P_{A,O_2} value of 10.53 kPa both ΔSBP and ΔDBP became significantly greater in the hypertensive group ($P < 0.01$). In the latter part of the rebreathing period at a P_{A,O_2} of 7.11 kPa ΔSBP was 2.95 kPa ± 0.318 SEM (22.1 mmHg ± 2.4 SEM) and ΔDBP was 2.088 kPa ± 0.233 SEM (15.7 mmHg ± 1.75) in the hypertensive group. In the normotensive group at the same P_{A,O_2} level ΔSBP was only 1.33 kPa ± 0.169 SEM (10 mmHg ± 1.3) and ΔDBP 0.95 kPa ± 0.165 SEM (7.1 mmHg ± 1.25). These differences of blood pressure elevations between the two groups were highly significant ($P < 0.001$).

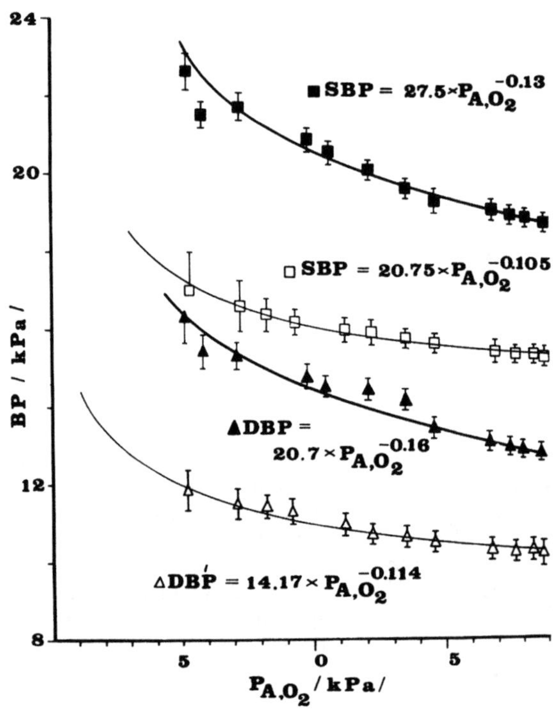

Fig. 4. Responses of the mean systolic blood pressure (SBP) and diastolic blood pressures (DBP) to isocapnic hypoxia in the hypertensive and normotensive group. Mean values in the hypertensive group are marked by *black squares* and *triangles*. Mean values in the normotensive group are marked by *white squares* and *triangles*. *Vertical bars* correspond to SEM. Lowest P_{A,O_2} values were reached by three hypertensive subjects only, their SEM values are higher. Approximated functions were calculated according to equations of best fit

Heart Rate and Respiratory Sinus Arrhythmia (RSA) During Progressing Hypoxia

At the start of rebreathing a respiratory sinus arrhythmia (RSA) was observed in both groups of subjects (Fig. 1 and 2). RSA was of significantly lesser magnitude in normotensive subjects during progressing hypoxia (Fig. 5A). A significant difference of the magnitude of RSA between the two groups began at a P_{A,O_2} of 9.36 kPa ($P < 0.01$) and remained until the end of rebreathing. In all normotensive subjects RSA disappeared entirely before rebreathing had been completed. In hypertensive subjects RSA was preserved until the end of rebreathing.

The difference of the magnitude of RSA during hypoxia in the two groups of subjects appeared much more evident after RSA values had been plotted against minute ventilation (Fig. 5B).

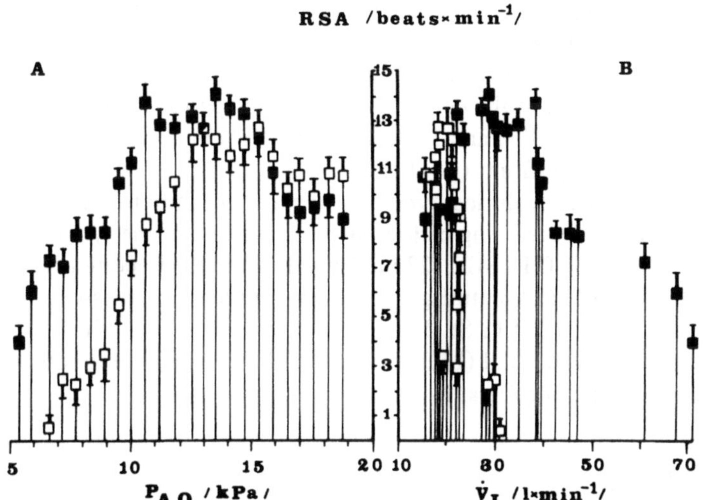

Fig. 5. **A** Changing magnitude of mean respiratory sinus arrhythmia (in beats min^{-1}) plotted against end-tidal O_2 tension during progressing hypoxia in the hypertensive group (*black squares*) and in the normotensive group (*white squares*). **B** Changing magnitude of mean respiratory sinus arrhythmia (in beats min^{-1}) plotted against minute ventilation during progressing hypoxia in the hypertensive group (*black squares*) and in the normotensive group (*white squares*). *Vertical bars* correspond to 0.5 SEM

The mean heart rate (HR) calculated over a 10 s period was initially 76.5 beats min^{-1} ± 0.82 SEM in the hypertensive group and 80.3 beats min^{-1} in the normotensive group, the difference being not significant. At the end of rebreathing HR in the hypertensive group rose to 107.7 beats min^{-1} ± 0.75 SEM and in the normotensive group to 100.4 beats min^{-1} ± 0.74 SEM. This difference in the response of the mean HR to hypoxia was significant ($P < 0.01$).

Ventilatory and Circulatory Responses to CO_2-Stimulus in Normotensive and Hypertensive Subjects

The tolerance to hyperoxic hypercapnia was about the same in the two groups. Both normotensive and hypertensive subjects continued rebreathing until the P_{A,CO_2} increased 2.03 kPa ± 0.15 SEM in normotensive subjects and 1.78 kPa ± 0.084 SEM in the hypertensive group above initial levels. The difference was not significant ($P > 0.2$). The slope S_{CO_2} of the mean regression line (Fig. 6) of ventilatory respones to CO_2 calculated for each subject was slightly higher in the hypertensive group: 17.25 ± 1.89 SEM (range : 5.29–40.2) than in the normotensive group 14.64 ± 1.53 SEM (range: 5.93–26.33, Fig. 6). Due however, to a great individual variability of the S_{CO_2} values in both groups this difference was not significant ($P > 0.02$). The value of the intercept of the CO_2 response line, B_{CO_2}, was 4.76 kPa ± 0.2 SEM (range: 2.92–5.87) in the hypertensive group and 5.139 kPa ± 0.23 SEM (3.45–6.51) in the normotensive group. However, this difference of mean B_{CO_2} values, which could suggest a lower threshold of the CO_2 response in hypertensive subjects, was insiginificant ($P > 0.1$) because of a very great individual variablility of the CO_2 responses.

Airway occlusion pressure $P_{0.2}$ at the end of the rebreathing period in hypertensive subjects was 1.56 kPa ± 0.12 SEM and 1.34 kPa ± 0.09 SEM in normotensive subjects. This difference was not significant ($P > 0.2$).

Blood pressure rose significantly during hyperoxic hypercapnia in both groups of subjects and at the end of the rebreathing period the mean increase of the systolic blood pressure (ΔSBP) was 3.06 kPa ± 0.18 SEM (23 mmHg – 1.35) in the hypertensive group and 3.26 kPa ± 0.26 SEM (24.5 mmHg ± 1.95 SEM) in the normotensive group. ΔDBP was respectively 2.23 kPa ± 0.18 SEM (16.7 mmHg ± 1.35 SEM) in the hypertensive group and 2.83 kPa ± 0.23 SEM (21.3 mmHg ± 1.72 SEM) in the normotensive group. The differences were not significant.

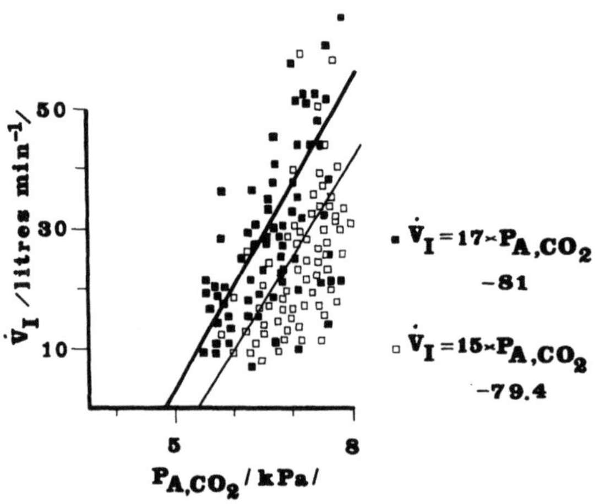

Fig. 6. Individual ventilatory responses to hyperoxic hypercapnia plotted against increasing end-tidal CO_2 tension in the hypertensive subjects (*black squares*) and in the normotensive subjects (*white squares*). Due to heavy overlapping some of the individual values were omitted in the figure. *Thick line* represents a mean linear regression line of ventilatory responses to CO_2 in the hypertensive subjects and *thin line* is a mean linear regression of ventilatory responses to CO_2 in the normotensive subjects according to equations of best fit. Slight differences of the intercept values B_{CO_2} and of the slope S_{CO_2} value between the hypertensive and normotensive subjects were not significant ($P < 0.1$ and $P < 0.2$ respectively)

Mean heart rate increased at the end of the rebreathing period 8.8 beats min^{-1} $\pm$ 0.74 SEM in hypertensive subjects and 7.7 beats min^{-1} $\pm$ 1.16 SEM in normotensive subjects, the differences being not significant.

Correlation of the Ventilatory Response to Hypoxia and Hypercapnia in Normotensive and Hypertensive Subjects

Correlations between hypoxic and hypercapnic responses were calculated by plotting reciprocal values of the hypoxic stimulus taken as parameter A against ventilatory responses to CO_2 (Fig. 7). A positive and significant correlation was found in normotensive subjects ($r = 0.56$, $t = 2.89$, $P < 0.01$). Conversely, in the hypertensive group there was no correlation of ventilatory responses to hypoxia and to hypercapnia ($r = 0.07$, not significant).

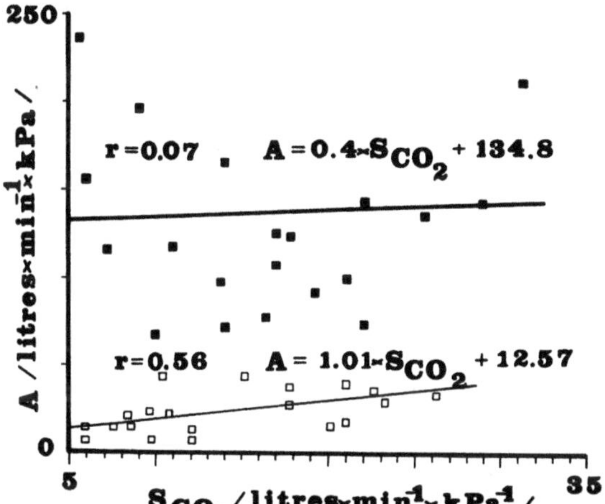

Fig. 7. Correlation of the ventilatory responses to hypoxia and to hypercapnia. Ventilatory response to hypoxia, taking parameter A, the reciprocal value of the stimulus, plotted against ventilatory response to CO_2 in the hypertensive subjects (*black squares*) and in the normotensive subjects (*white squares*). Great dispersion of mean values around the regression line and no significant correlation of two ventilatory responses ($r = 0.07$) in the hypertensive group. Positive and significant correlation in the normotensive group ($r = 0.56$, $P < 0.01$). The difference between two coefficients is significant ($t = 7.72$, $P < 0.001$)

Discussion

Increased ventilatory and pressor responses to acute hypoxia in young hypertensive subjects indicates that the arterial chemoreceptor drive is augmented. A significant difference of ventilation between normotensive and hypertensive subjects has been observed even at normal P_{A,O_2} values suggesting a higher chemoreceptor drive in hypertensive subjects even in normoxia. Parameter A by denoting the shape of the $\dot{V}_I - P_{O_2}$

function curve is a rough index of the hypoxic ventilatory drive (Weil et al. 1970). In hypertensive subjects the hypoxic ventilatory drive was four times higher than in normotensives. This striking difference is not to be accounted for simply by an unspecific hyper activity of the respiratory and cardiovascular systems as ventilatory, blood pressure, and heart rate responses to a CO_2-stimulus did not differ significantly in the two groups. Slightly, yet statistically not significant greater values of S_{CO_2} and B_{CO_2} in the hypertensive group may possibly be due to some chemoreceptor-dependent component in the stimulatory action of hyperoxic hypercapnia.

A dissociation of the peripheral and central chemosensitivity, the former being predominant, in early human hypertension is consistent with the hypothesis that the arterial chemoreceptor reflex is specifically hyperactive in hypertension and does not support the supposition of an increased central chemosensitivity.

The mechanism by which the arterial blood pressure elevation is greater in hypertensive subjects in hypoxia requires further study as it may depend on several factors such as an increased neurogenic vasomotor response, cardiovascular structural adaptation (Folkow 1978) secondary effects of vigorous breathing, and increased venous return with an increase of the cardiac output (Richardson et al. 1966).

The increased magnitude of respiratory sinus arrhythmia (RSA) observed in hypertensive subjects during hypoxia corresponds to a greater hypoxic drive as acute systemic hypoxia facilitates RSA in man (Hirsch and Bishop 1980). Our previous study (Tafil and Trzebski 1980) showed that even a normal, normoxic chemoreceptor drive in man contributed to expiratory bradycardia and thus to RSA. This finding was consistent with the results of Drysdale and Petersen (1977) who produced dynamic step changes in alveolar gas tension and found that the primary effect of withdrawing arterial chemoreceptor stimulation in conscious man is tachycardia. Thus a greater expiratory bradycardia in hypertensive subjects during hypoxia may be explained by the increased responsiveness of the arterial chemoreceptor reflex. A reported contribution to RSA of the low-pressure receptors in man (Melcher 1976) does not appear to play a role in the mechanism of augmented RSA during hypoxia in hypertensive subjects because differences in RSA between the two groups of subjects were even more accentuated at the same level of ventilation (Fig. 5), that is in similar mechanical conditions of breathing.

The present study does not provide information on the mechanism, peripheral or central, which causes an enhancement of the chemoreceptor reflex drive in early hypertension in man. Some reports suggest a peripheral site of the altered chemoreceptor function in hypertension. A positive correlation was observed between the weight of the carotid bodies and left ventricular hypertrophy in hypertension (Edwards et al. 1971). Similar changes were observed in the carotid bodies of humans, exposed to prolonged hypoxia (Steele et al. 1972). It has been suggested (Przybylski 1981) that a reduction of the arterial supply to the carotid bodies, unnecesarily exciting the chemoreceptors, may occur in hypertension. This reduction may be due to local arteriosclerosis at the branching of small arteries supplying the carotid body or to an increased thickness of the carotid artery wall, a well known adaptative change in hypertension (Folkow 1978). A possible role of an increased chemoreceptor drive as a factor in the pathogenesis of early essential hypertension in man requires further study.

References

Drysdale DB, Petersen ES (1977) Arterial chemoreceptors, ventilation and heart rate in man. J Physiol (London) 273:109–120

Edwards C, Heath D, Harrris P (1971) The carotid body in emphysema and left ventricular hypertrophy. J Pathol 104:1–13

Folkow B (1978) Cardiovascular structural adaptation: its role in the initiation and maintenance of primary hypertension. Clin Sci Mol Med 5, Suppl 4:3–22

Hirsch IA, Bishop S (1980) Respiratory sinus arrhythmia (RSA) in man: altered inspired O_2 and CO_2. Proc Int Un Physiol Sci 14:(0297) 140

Koepchen HP, Hilton SM, Trzebski A (eds) (1980) Interaction between Respiratory and Cardiovascular Control System. Springer, Berlin Heidelberg New York

Lioy F, Hanna BD, Polosa C (1978) CO_2-dependent component of the neurogenic vascular tone in the cat. Pflügers Arch 374:187–191

Lloyd BB (1966) The interactions between hypoxia and other ventilatory stimuli. In: Hatcher JD, Jennings DB (eds) Proc Int Symp Cardiovasc Resp Effects Hypoxia. Karger, Basel , p 146

Lloyd BB, Jukes MGM, Cunningham DJC (1958) The relation between alveolar oxygen pressure and the respiratory response to carbon dioxide in man. J Exp Physiol 43:214–227

Lugliani R, Whipp BJ, Seard C (1971) Effect of bilateral carotid body resection on ventilatory control at rest and during exercise in man. Engl J Med 285:1105–1111

Lugliani F, Whipp BJ, Wasserman K (1973) A role of the carotid body in cardiovascular control in man. Chest 63:744–750

Melcher A (1976) Respiratory sinus arrhythmia in man. A study in heart rate regulating mechanism. Acta Physiol Scand Suppl 435:1–31

Przybylski J (1978) Alveolar hyperventilation in young spontaneously hypertensive rats. IRCS Med Sci Cardiovasc Syst 6:315

Przybylski J (1981) Do arterial chemoreceptors play a role in the pathogenesis of hypertension? Med Hypotheses 7:127–131

Read DJC, Leigh J (1967) Blood-brain tissue P_{CO_2} relationships and ventilation during rebreathing. J Appl Physiol 23:53–70

Rebuck AS, Kangalee M, Pengelly LD, Campbell EJM (1973) Correlation of ventilatory responses to hypoxia ane hypercapnia. J Appl Physiol 35:173–177

Richardson DW, Kontos AA, Shapiro W, Patterson J L Jr (1966) Role of hypocapnia in the circulatory responses to acute hypoxia in man. J Appl Physiol 21:22–26

Steele RH, Hinterberger H (1972) Catecholamines and 5-hydroxytryptamine in the carotid body in vascular, respiratory and other diseases. J Lab Clin Med 80:63–70

Tafil M, Trzebski A (1980) Contribution of the central respiratory drive and of the arterial chemoreceptor reflex to the mechanism of respiratory sinus arrhythmia in man. Proc Int Un Physiol Sci 14:3367–735

Trzebski A, Kubin L (1981) Is the central inspiratory activity responsible for P_{CO_2}-dependent drive of the sympathetic discharge? J Aut Nerv Syst 3:401–420

Trzebski A, Zielinski A, Majcherczyk S, Lipski J, Szulczyk P (1974) Effect of chemical stimulation and depression of the medullary superficial areas of the respiratory motoneurone discharges, sympathetic activity and efferent control of the carotid area receptors. In: Umbach W, Koepchen HP (eds) Central rhythmic and regulation. Hippokrates, Stuttgart

Weil JV, Byrne-Quinn E, Sodal JE, Freisen WD, Underhill D, Filley GF, Grover RF (1970) Hypoxic drive in normal man. J Clin Invest 49:1061–1072

Whitelaw WA, Derenne IP, Milic-Emili J (1975) Occlusion pressure as a measure of respiratory center output in conscious man. Respir Physiol 23:181–199

The Role of the Ventral Medulla in the Control of Sympathetic Activity by Systemic Arterial CO_2

C. POLOSA[1], F. LIOY[2], and B.D. HANNA[2]

Background activity of sympathetic preganglionic neurons (SPNs), displayed by approximately one third of them in CNS-intact or midcollicular decerebrate cats, is responsible for the neurogenic sympathetic tone of cardiovascular and other effector organs (Polosa et al. 1979). The origins of this background activity remain largely unknown. Whether CO_2 could be one of the causes of this activity, by acting either on chemoreceptors or directly on the SPN or on antecedent neurons has been the topic of our investigations. This paper reviews our results. The conclusion we have reached is that sympathetic activity levels are directly related to arterial pCO_2 levels in the range from 14–90 mm Hg. This relation depends on input from both peripheral and ventral medullary surface chemoreceptors.

The Relation Between CO_2 Levels in Arterial Blood and Sympathetic Activity or Neurogenic Vascular Tone

We have studied a sample of 60 spontaneously active SPNs, the axons of which project to the cervical sympathetic trunk (Polosa and Preiss 1977, Hanna et al. 1981) and examined the characteristics of their CO_2 - sensitivity. These neurons have cell bodies in the Th1 to Th5 segments of the spinal cord. Targets of their axons are superior cervical ganglion neurons innervating vascular smooth muscle in skin and muscle of the face and neck and in the oro-pharyngeal and nasal mucosa, the smooth muscle of the eye and the salivary glands. These neurons are under baroreceptor control (Gerber and Polosa 1978) and display a number of properties and reflex responses common to SPNs located in other parts of the spinal cord. The firing frequency of seventy per cent of these neurons showed a direct relation to the arterial CO_2 tension in a range from 14–90 mm Hg. The remaining 30% were neurons which were either insensitive to CO_2 or showed a decrease in firing rate as $paCO_2$ increased. The $paCO_2$-firing rate relation for the whole sample was sigmoid in shape (e.g., Fig. 1) and the mean firing rate increased by a factor of

1 Department of Physiology, McGill University, McIntyre Medical Sciences Building, 3655 Drummond Street, Montreal, Quebec H3G 1Y6, Canada
2 University of British Columbia, Vancouver, Canada

Central Neurone Environment
(ed. by M.E. Schläfke, H.P. Koepchen and W.R. See)

five between the extremes of $paCO_2$ given above (Hanna, et al. 1981). A CO_2-sensitivity of this magnitude is reminiscent of that of brainstem respiratory neurons (Cohen, 1968) and phrenic motoneurons (Gill 1963).

In other experiments we measured perfusion pressure in the innervated hindquarters, vascularly isolated and perfused at constant flow with normocapnic blood (Lioy et al. 1978). In this preparation, perfusion pressure is a measure of vascular resistance. The latter is a function, at least in part, of sympathetic nerve activity. When the pCO_2 of the inspired gas was changed we found a direct relation between pCO_2 of the arterial blood perfusing the body and hindlimb vascular resistance.(Fig. 2). By measuring the perfusion pressure after functional sympathetic denervation of the hindquarters by i. v. administration of the alpha-adrenergic blocking drug phentolamine, we could partition vascular resistance into neurogenic and non-neurogenic components (Fig.2). For a change in body $paCO_2$ from 14–60 mm Hg the neurogenic component of vascular resistance increased 2.5 times. We also estimated that 20–30 % of the neurogenic vasoconstrictor tone in normocapnia was generated by CO_2. These results demonstrate that the above described neurophysiological data are relevant to cardiovascular control, and that both sympathetic neuron activity and neurogenic vasoconstrictor tone are a continuous function of $paCO_2$.

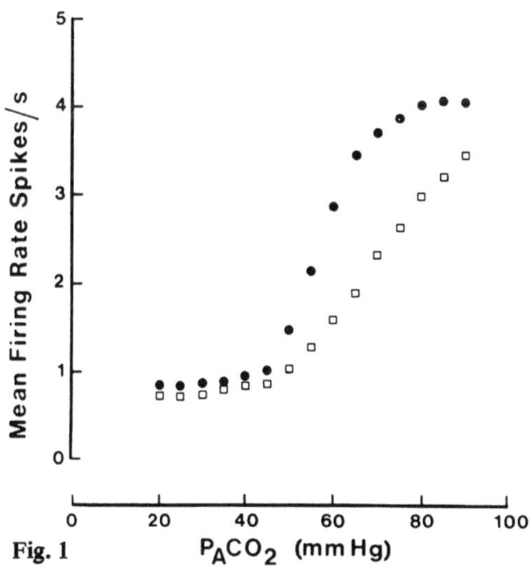

Fig. 1

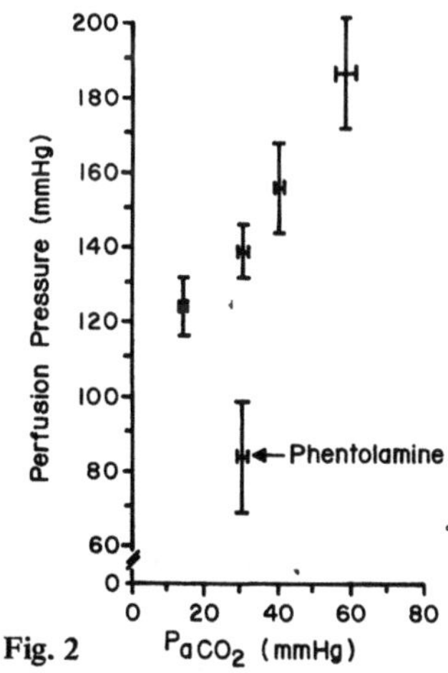

Fig. 2

Fig. 1. Depressant effect of bilateral carotid sinus nerve section on the average CO_2 response of sympathetic preganglionic neurons in normoxia (n = 7). *Dots* response before; *squares* response after nerve section. Range of $paCO_2$ values obtained by rebreathing. (Courtesy of Hanna et al. 1981)

Fig. 2. Effect of changes in $paCO_2$ on the perfusion pressure of the isolated, normocapnic hindlimbs of six cats. *Horizontal* and *vertical bars* indicate standard errors of the means. *Lower point* gives perfusion pressure after phentolamine injection into perfusion system. (Courtesy of Lioy et al. 1978)

Role of Peripheral Chemoreceptors in the CO_2-Response of Sympathetic Neurons

When isolated carotid sinus is perfused with hypercapnic blood of various pCO_2 the vascular resistance of the hindquarters increases in proportion to the pCO_2 of the perfusate (Pelletier 1972). Thus, the carotid sinus chemoreceptors are likely to mediate the CO_2 response of SPNs. The question is whether the peripheral chemoreceptors are the only source of these CO_2 responses. In the study of the innervated, perfused hindquarters of the cat (Lioy et al. 1978) we found that vascular resistance still increased significantly as a function of body $paCO_2$ after section of the aortic and carotid sinus nerves. In order to assess how much of the CO_2 response of SPNs is due to the peripheral chemoreceptors, we compared the CO_2 response curves of seven SPNs during rebreathing before and after carotid sinus nerve section in cats in which the cervical vagosympathetic trunk and the aortic nerves had been previously cut (Hanna et al. 1981). In the cat the carotid sinus nerve also contains baroreceptor afferents, which are known to exert an inhibitory action on the SPN proportional to the level of systemic arterial pressure (SAP). This may complicate the interpretation of the experimental results, because after baro-deafferentiation the mean firing rate of the SPN may be higher and therefore the $paCO_2$-sympathetic firing curve may be displaced. The gain (slope) of the curve may also be changed by the suppression of baroreceptor input. To minimize these sources of variability the animals were pretreated with hexamethonium or phentolamine, so as to lower SAP and consequently the mean level of baroreceptor activity. In addition, a SAP stabilizing device was used so that hypotension did not occur as a result of the direct vasodilator effect of CO_2. Proof that changes in baroreceptor afferent were not appreciably interfering with this experiment is in the finding that at low CO_2 values, i.e., in the region in which SPN firing is relatively CO_2-independent (Fig. 1), the mean firing rate before and after carotid sinus nerve section was the same. After peripheral chemoreceptor denervation the CO_2 response persisted, although there was a significant displacement of the CO_2 response curve to the right and a decrease in it's maximum slope (Fig. 1). Therefore, the CO_2 response of SPNs must be mediated also by structures different from the peripheral chemoreceptors.

Role of the Ventral Medulla in the CO_2 Response of Sympathetic Neurons

An important source of the sympathetic response to changes in systemic CO_2 is the ventral surface of the medulla. This has been shown by experiments of cold-block or superfusion of the ventral surface of the medulla. The experiments utilizing cold block (Hanna et al. 1979) were performed by exposing the ventral medulla and positioning thermodes on a bilateral area corresponding to the intermediate ares "S" described by Schlaefke and Loeschcke (1967). After denervation of the peripheral chemo- and baroreceptors, the vascular resistance of the isolated, innervated hindquarters was measured as perfusion pressure (PP) at constant flow at various $paCO_2$ levels. The perfused vascular bed, in this series of experiments, was not kept normocapnic but was exposed to the same $paCO_2$ as the rest of the animal. This was done to simplify the experimental procedure. The local vasodilator effect of CO_2, which would act in opposition to the CNS-mediated vasoconstrictor effect and could mask it, was assessed by repeating the experiments after denervation of the hindquarters. Within the $paCO_2$ range examined this local effect was found to be small or absent in this preparation and not to interfere with the results. Cooling area "S" caused a significant reduction in the slope of the CO_2 response curve from 1.1 to 0.7 mm Hg PP/mm Hg $paCO_2$ in a series of 12 cats. (Fig. 3). The y-intercept of the curves did not change and this is in agreement with the information obtained from the CO_2 response curves of SPNs (Hanna et al. 1981): in hypocapnia SPNs have low sensitivity to CO_2 and therefore removal of an input mediating CO_2 sensitivity should have no significant effect. The significant change in slope of the CO_2 response of the hindquarters vascular resistance caused by cold block of area "S" showed that the response of vasoconstrictor neurons to systemic CO_2 is mediated by the ventral medulla. Prior to these experiments a relation between ventral medulla and cardiovascular tone had been suggested, based on experiments showing that local anesthesia or cooling may lead to dramatic decreases in SAP (Loeschcke and Koepchen 1958, Schlaefke and See 1980) whereas electrical or chemical stimulation (Loeschcke et al. 1970, Schlaefke et al. 1970, Trzebski et al. 1974) could cause an increase. However, a role of the ventral medulla in the cardiovascular response to changes in systemic arterial pCO_2 had not been demonstrated. Similar results were obtained in another series of experiments in adrenalectomized cats, which showed that cooling area "S" reduced the slope of the vascular resistance vs CO_2 relation by a large amount. This experiment demonstrated that the effect of cooling was due primarily to a change in vasoconstrictor neuron activity and not to a change in adrenal medullary secretion. Thus, unless CO_2 controls the release of some vasopressor agent other than catecholamines, the effects removed by cooling area "S"

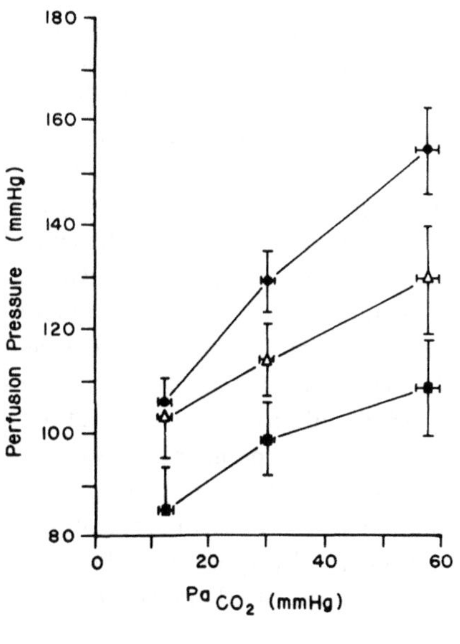

Fig. 3. Effect of cooling the ventral surface of the medulla on the relation between $paCO_2$ and hindlimb perfusion pressure. *Dots control* (medullary surface temperature 37° C). *Triangles* selective cooling of area S to 12° C. *Squares* cooling of the whole ventral suface to 12° C (Courtesy of Hanna et al. 1979)

must have been mediated by the sympathetic system. This effect of cooling on the CO_2 response must have been due to the inactivation of some mechanisms superficially located on the ventral surface sinde the pressor responses evoked by a stimulating electrode placed 2 mm or deeper below the cooling thermode were not modified by the cold block.

If the CO_2 response of vasoconstrictor neurons, in peripherally chemodenervated cats, i due to superficially located structures on the ventral aspect of the medulla, it should be possible to activate these structures by superfusing the ventral medulla with hypercapnic and/or acid fluid and obtain responses similar to those observed in systemic hypercapnia. Cardiovasular responses to variations in the pH of the fluid perfusing the ventral medulla had been described previously (Schlaefke et al. 1970) but were small. This is possibly due to the use of spontaneously breathing animals, in which the changes in intrathoracic pressure, systemic gas tensions (mostly CO_2) and lung volume, resulting from the changes in ventilation may have produced cardiovascular effects which would have masked those evoked by the central chemoreceptors. In a series of curarized and artificially ventilated cats (in which the activity of the respiratory centre was monitored by recording phrenic nerve activity) we perfused the exposed ventral medulla with mock CSF (Lioy et al. 1981) and recorded hindlimb vascular resistance and heart rate. Both parameters increased with acid/hypercapnic perfusion and decreased with alkaline/hypocapnic perfusion. The responses of vascular resistance and heart rate were parallel to those of the phrenic nerve.

Conclusions

Our studies have shown that SPN activity is dependent on the level of arterial pCO_2 and this dependence is mediated by both peripheral and ventral medullary chemoreceptors. There is the possibility that an additional source of CO_2 sensitivity exists in the spinal cord, as suggested by the experiments of Szulcyk and Trzebski (1976) who observed an increase in sympathetic activity in the vertebral and cardiac nerves on acid superfusion of the Th_1 spinal segment. The extent of the control exerted by CO_2 on SPNs can be assessed by noticing that an average SPN increases its firing rate by a factor of 5 when pCO_2 is changed from 14 to 90 mm Hg. The conclusion derived from our experiments, that ventral medulla chemoreceptor activity influences SPN activity and vascular resistance, is concurrent with that of Trzebski et al. (1974) and Szulczyk and Trzebski (1976) who observed changes in SAP, sympathetic activity and vascular resistance resulting from acid superfusion of the ventrolateral medulla. Our experiments, in addition, by defining the quantitative relation between $paCO_2$ and either SPN firing rate or vascular resistance, enable us to predict the magnitude of the effect on either variable, which may be expected as a result of experimental procedures eliminating the function of ventral medulla chemoreceptors. Our earlier experiments on vascular resistance (Lioy et al. 1978) showed that only between 20 and 30% of neurogenic vascular tone is CO_2-dependent in normocapnia. Our more recent studies on single units (Hanna et al. 1981) show that the SPN firing rate decreases by approximately one third when $paCO_2$ levels are reduced from normocapnia to a value of 20 mm Hg. Thus elimination of central chemoreceptors should not reduce vascular resistance or sympathetic activity by more than these amounts (i.e, 20–30%). This is, in fact, the extent of reduction of vascular resistance that we observed in normocapnia upon cooling area "S" (see Fig. 3). Some other experiments in which the ventral surface of the medulla was treated with procaine (Loeschcke and Koepchen 1958), or cold blocked (Schlaefke and See 1980) showed falls in SAP comparable to those that would be obtained if sympathetic activity was nearly completely suppressed. It is clear, therefore, that these experiments demonstrate the existence of mechanisms other than chemoreceptors which make a significant contribution, greater than that of the central chemoreceptors, to sympathetic tone and which are located close to the ventral surface of the medulla. Thus, investigations of the ventral surface of the medulla, which started with the classical experiments of Loeschcke and Koepchen (1958) have not only explained part of the CO_2 response of SPNs but also promise to contribute to our understanding of CO_2-independent mechanisms participating in the generation of neurogenic vasomotor tone.

Acknowledgements. The experimental work by the authors was supported by grants from the Medical Research Council, the Quebec and the British Columbia Heart Foundations.

References

Cohen MI (1968 Discharge patterns of brainstem respiratory neurons in relation to carbon dioxide tension. J Neurophysiol 31: 142–165

Gerber U, Polosa C (1978) Effects of pulmonary stretch receptors afferent stimulation on sympathetic neuron firing. Can J Physiol Pharmacol 56: 191–198

Gill PK (1963) The effects of end-tidal CO_2 on the discharge of individual phrenic motoneurones. J Physiol (London) 168: 239-257

Hanna BD, Lioy F, Polosa C (1979) The effect of cold blockade of the medullary chemoreceptors on the CO_2 modulation of vascular tone and heart rate. Can J Physiol Pharmacol 57: 461–468

Hanna BD, Lioy F, Polosa C (1981) Role of carotid and central chemoreceptors in the CO_2 response of sympathetic preganglionic neurons. J Autonom Nerv Syst 3: 421–435

Lioy F, Hanna BD, Polosa C (1978) CO_2 -dependent component of the neurogenic vascular tone in the cat. Pflügers Arch 374: 187–191

Lioy F, Hanna BD, Polosa C (1981) Cardiovascular control by medullary surface chemoreceptors. J Autonom Nerv Syst 3: 1–7

Loeschcke HH, Koepchen HP (1958) Versuch zur Lokalisation des Angriffsortes der Atmungs- und Kreislaufwirkung von Novocain Liquor cerebrospinalis. Pflügers Arch 266: 628–641

Loeschcke HH, Lattre J de, Schlaefke ME, Trouth CO (1970) Effects on respiration and circulation of electrically stimulating the ventral surface of the medulla oblongata. Respir Physiol 10: 184–197

Pelletier CL (1972) Circulatory responses to graded stimulation of the carotid chemoreceptors in the dog. Circ Res 31: 431–443

Polosa C, Preiss G (1977) The relation between end-tidal CO_2 and discharge patterns of sympathetic preganglionic neurons. Brain Res 122: 255–267

Polosa C, Mannard A, Laskey W (1979) Tonic activity in the autonomic nervous system: functions, properties, origins. In: Brooks C McC, Koizumi K, Sato A (eds) Integrative functions of the autonomic nervous system. Univ Press Tokyo, p 342

Schlaefke ME, Loeschcke HH (1967) Lokalisation eines an der Regulation von Atmung und Kreislauf beteiligten Gebietes an der ventralen Oberfläche der Medulla oblongata durch Kälteblockade. Pflügers Arch 297: 201–220

Schlaefke ME, See WR (1980) Ventral medullary surface stimulus response in relation to ventilatory and cardiovascular effects. In: Koepchen HP, Hilton SM, Trzebski A (eds) Central interaction between respiratory and cardiovascular control systems. Springer, Berlin Heidelberg New York, p 56

Schlaefke ME, See WR, Loeschcke HH (1970) Ventilatory response to alterations of H^+ ion concentration in small areas of the ventral medullary surface. Respir Physiol 10: 198–212

Szulczyk P, Trzebski A (1976) The local effect of pH changes in the cerebrospinal fluid on the ventrolateral areas of medulla oblongata and on spinal cord surface on the activity of cardiac and vertebral sympathetic nerves. Acta Physiol Pol 27: 9–17

Trzebski A, Zielinski A, Majcherczyk S, Lipski J, Szulczyk P (1974) Effect of chemical stimulation and depression of the medullary superficial areas on the respiratory motoneurones discharges, sympathetic activity and efferent control of carotid area receptors. In: Umbach W, Koepchen HP (eds) Central-rhythmic and regulation. Hippokrates, Stuttgart, p 170

Respiratory and Cardiovascular "Centres": Functional Entirety or Separate Structures?

H.P. KOEPCHEN[1]

In 1958 Loeschcke and Koepchen described decreases in ventilation and blood pressure caused by addition of procaine to perfusion of the brain ventricles with mock CSF, an action which could be localized roughly somewhere at the ventral base of the brain (Loeschcke and Koepchen, 1958c). In many later communications analysing further the meaning and the finer localization of these phenomena by various methods of blockade, stimulation, variation of H^+ concentration or synaptic drugs, actions on respiration as well as circulation in various combinations were described (e.g., Loeschcke and Katsaros 1959: Loeschcke and Koepchen 1958a, Loeschcke et al. 1970; Mitchell et al. 1963; Schläfke and Loeschcke 1967; Schwanghart et al. 1974). This raises the question whether the observed common actions on circulation and respiration are due to a common superficial structure which delivers drive to both central control substrates, whether there are two separate superficial structures, each of which acts on one control system separately, or whether the common actions are caused by an obligatory intracentral coupling between the centres where the action on one automatically is reflected in simultaneous changes of the other.

Obviously the action of the ventral medullary structures couples centrally respiratory and cardiovascular control. This is one case of central coupling of both systems. Before further analyzing this special phenomenon, it seems to be necessary to undertake a basic consideration upon the coupling or interaction between the "centres" because frequent errors, apparent questions and apparent answers can be avoided by a theoretical re-examination which reminds of classical findings and includes semantic and logical questions. On the basis of these general considerations, we will come back to the action of central chemosensitivity on respiration *and* circulation.

Cardiovascular-respiratory (abbreviated in the following as "cardiorespiratory") coupling is of functional importance because circulation and respiration lie in series with respect to the transport of O_2 and CO_2 where the transport rhythm of respiration originates in the central nervous system, whereas that of circulation is produced in the periphery. Moreover, central nervous cardiorespiratory coupling is of general interest as a paradigm of somatomotor-autonomic coordination and the problem of coaction and interaction of nervous centres.

The discovery in the 19th century that the maintenance of respiratory rhythm and cardiovascular tone and the function of homeostatic respiratory and cardiovascular

1 Institute of Physiology, The Free University of Berlin, Arnimallee 22, 1000 Berlin 33, FRG

Central Neurone Environment
(ed. by M.E. Schläfke, H.P. Koepchen and W.R. See)

reflexes depends on the integrity of the lower brainstem has led to the postulate of the existence of a "respiratory centre" and a "cardiovascular centre" (or its subdivisions "cardiac centre" and "vasomotor centre") at the pontomedullary level. These terms can be found in nearly all physiological textbooks, also in their recent editions. The recognition that these are differentiated and extended structures is frequently expressed by adding quotation marks to the term "centre" or by speaking of "the central control system" for respiration or circulation, respectively. This does not change the presumption of the existence of two different control systems which is implied also in the questions how these systems are coordinated or coupled or interact with each other This presumption is favoured by the usually separate research on both control systems by different laboratories. Thus it corresponds also today with the actual state of affairs to start the general considerations from the assumption of two different or specific control systems or centres. There are two basic possibilities for central coupling: First, action of common afferences on both systems; secondly, actions originating from one control system and influencing the other by intracentral connections. A third possibility will appear in the course of our considerations.

For each of the central control systems stabilizing feedback circuits have been detected with a double – chemical and mechanical – feedback in the case of respiratory control and a preferentially mechanical feedback in the circulatory control. A special case of coupling by common afferents is the mutual connection of both feedback control circuits through the action of the afferent limb of one control circuit on the control centre of the other one. Well-known examples of this kind of coupling are the action of arterial baroreceptors on respiration and of peripheral chemoreceptors or pulmonary stretch receptors on circulation (Anrep et al. 1936, Daly et al. 1964, Heymans and Neil 1958). Another important group of common afferences come from working muscles driving respiration and circulation during muscular exercise (Kao et al. 1967).

In principle three kinds of coupling by common afferences and their combinations are conceivable: First, separate action on both of the control systems; second, a action on one control system which transmits the effect to the other by intracentral connections; third, action on a substrate superposed on both control systems.

Before the mechanisms can be analyzed the known manifestations of cardiorespiratory coupling shall be briefly reviewed. The most prominent phenomenon is the presence of respiratory rhythm in cardiovascular innervation discovered by Hering in 1869 (Hering 1869) in the form of respiratory blood pressure waves in the curarized dog. Adrian et al. (1932) first recorded respiratory rhythm in sympathetic nerves which was amply confirmed in many subsequent recordings. Recently, Preiss and Polosa have found inspiration-related respiratory rhythm in about half of the preganglionic sympathetic fibres projecting into the cervical sympathetic trunk (Preiss and Polosa 1977). With refined methods it can be shown that peripheral vasomotion can follow such fluctuations even in the range of normal respiratory frequencies (Koepchen et al. 1968). Another well-known phenomenon partly caused by intracentral mechanisms is respiratory cardiac arrhythmia (Koepchen and Thurau 1959).

The classical concept for the explanation of this group of phenomena is expressed by the term "irradiation". In this context "irradiation" means that impulses from the respiratory centre originally destined for the innervation of respiratory muscles flow over to the cardiovascular centre. Simple irradiation would cause pure superposisition

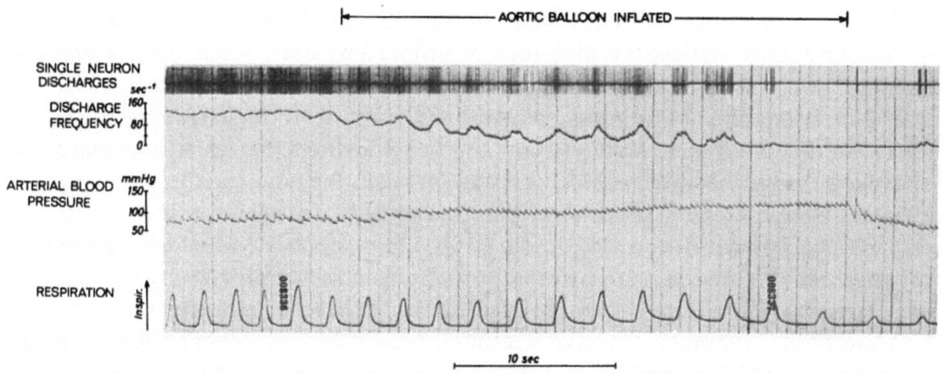

Fig. 1. Neuronal unit activity in the medulla oblongata of an anaesthetized dog. The recorded neurone changes its discharge pattern during gradual experimental increase of arterial blood pressure induced by slow inflation of a balloon in the aorta descendens. Tonic pattern with initially slight respiratory modulation becomes distinctly "respiratory" with increasing blood pressure and finally activity nearly ceases in response to enhanced baroreceptor stimulation parallel with depression of respiration. Thus this neurone shares the properties of respiratory as well as cardiovascular neurones according to the usual labelling, where the degree of respiration-related rhythmicity depends on the actual level of arterial blood pressure. (From Koepchen et al. 1975)

of respiratory rhythm on cardiovascular innervation. This can occur, but frequently more complicated forms of coupling are reflected in the records of peripheral parameters. Sliding coordination, mutual entrainment of rhythms, superposition of several different rhythms, integer multiple relations, spontaneous appearance and disappearance of coordination can be observed in animal experiments as well as in human subjects (Golenhofen and Hildebrandt 1958). An extensive analysis in dogs under diffusion respiration – i.e., without respiratory movements – (Koepchen 1962) revealed that coordination between respiratory and cardiovascular innervation obeys the rules of "relative coordination", detected first in motor rhythms of fish and lateron found to be a general phenomenon in the interaction between motor neurone pools (v. Holst 1939). The consequenses for our picture of coupling between the presumed two centres are the following: Their interactions are not unidirectional but bidirectional. Not fixed but sliding, not obligatory but facultative and comprise not only one but several rhythms.

The elements of the central control system are single neurones the activity of which was extensively studied in the past decades. Thus for an understanding of central coupling we have to continue our considerations at the neuronal level. Nevertheless, the phenomena found in the mutual coordinations of peripheral parameters should be respected and utilized for the functional interpretation of single neurone recordings.

If we maintain the presupposition of two seperate control systems (which is done implicitly or explicitly in nearly all of the work in this field), we have to identify the elements of both control systems at the neuronal level, i.e., to define cardiovascular or respiratory neurones (Koepchen et al. 1975) The only meaningful definition in a functional sense is that according to the target organ which is influenced by the concerned neurone, irrespective of the number of neurones interposed on the way to the target organ. This definition can be realized for the output neurones from the C.N.S., but serious difficulties arise if we go backward into the central network.

The criteria in common use for the labelling of brainstem neurones as "respiratory" or "cardiovascular" reflect this difficulty. Neurones have been called "respiratory neurones" when their discharge pattern was somehow related to the phrenic rhythm. They have been called "cardiovascular neurones" when their discharge pattern was related to that of efferent cardiovascular and/or sympathetic nerves, to spontaneous or evoked changes in systemic blood pressure or when they did react to baroreceptor afferents. Simple considerations show that these criteria are necessary but not sufficient. On the basis of the classical findings it has to be postulated that e.g., respiratory neurones likewise react to baroreceptor afferents and cardiovascular neurones have respiratory discharge pattern. A large number of brainstem neurones at the same time meet the mentioned criteria of a respiratory as well as of a cardiovascular neurone (Hukuhara Jr. 1980, Koepchen et al. 1975, Langhorst et al. 1981). One characteristic example is shown in Fig. 1. It follows that the usual criteria are inappropiate for a qualitative discrimination between the single neuronal elements of both control systems.

In order to overcome this dilemma we have to realize the possible interrelations in the central network and to reexamine critically our nomenclature of control systems, neurones, irradiation, specificity and non-specificity. It will be shown that most of the difficulties do not arise from the complex central interactions but from our presupposition of the existence of two separate control systems.

When we start the specific output neurone, with defined target organ and function – as e.g., an intercostal motor neurone or a preganglionic vasoconstrictor neurone – and go backward into the network to the superposed neurones, all neurones projecting on a cardiovascular (vasomotor, cardiac) output neurone are likewise cardiovascular (vasomotor, cadiac) neurones and those projecting on a respiratiory output neurone are respiratory neurones according to the functional definition given above. But this way of thinking at first the artificial discrimination between a proper centre and precentral input and postcentral output stations becomes dispensable.

Common events of central origin in both peripheral systems must be caused by neurones projecting on both outputs. If we strongly maintain our definition such a neurone is a respiratory as well as a cardiovascular neurone, that means a cardiovascular-respiratory or "cardiorespiratory" neurone. It produces coupling of the systems by divergence. Coupling by divergence and therewith cardiorespiratory neurones can occur at any level of the network including the spinal output neurones. Therewith another basis of coupling appears, that is partial identity of the central control systems. At the same time, the term "interaction" becomes dispensable. (The possibility of nonsynaptic effects does not principally change the consequences of these considerations.) The difference between this coupling by divergence and the classical concept of "irradiation" is the following: "Irradiation" means that one of the projections of the neurone mediates the specific function, the other one a side-effect only. In the case of respiratory-cardiovascular coupling respiration-related rhythmicity usually is taken as sign for an original destination of this activity to innervate the respiratory output and consequently the interpretation results that the same pattern in cardiovascular outputs is a side-effect, i.e., "irradiation". This, however, would mean that we give up our original definition according to the peripheral function and introduce a teleological criterion assuming that cardiovascular innervation does not need rhythmicity (which is by no means sure). Therefore we should discard the Term "irradiation" and simply state that neurones project on both systems.

How is the existence of cardiorespiratory neurones, representing identity of the control systems, compatible with the experience that under certain conditions respiratory and cardiovascular parameters can vary quite independently? It should be stressed that cardiorespiratory neurones and coupling by divergence does not mean uniformity of a common control network. We have to introduce a new point of view: Inhomogeneity. At a single neurone level inhomogeneity means that the branching neurites of the neurone are unevenly distributed to the dependent neurones and outputs. Thus e.g., an 80% respiratory and 20% cardiovascular neurone is conceivable as well as other relations. The extreme would be a 100% respiratory or 100% cardiovascular neurone, that means truly specific neurones. It is probable that intermediate cases of the spectrum will occur more frequently than the extremes. In other words: Specificity of the central neurones is *not a qualitative but a quantitative question*. Specifity is "weighed". "Weighed specificity" is equivalent with weighed non-specificity. Therewith the apparent contradiction between specificity and non-specificity can be reduced to a quantifiable position on a continuous scale.

The experience that coupling phenomena are not fixed (Koepchen 1962) indicates that the relative weights can change. Changing weights at the neuronal level are easily imaginable when facilitatory or inhibitory additional inputs act on downstream dependent neurone pools. If we assume – as a hypothetical example– that in case of a 70% respiratory and 30% cardiovascular neurone the cardiovascular output is suppressed by interference from baroreceptor afferents, this neurone will become 100% respiratory in its functional efficiency without changing its properties. Thus variably weighed specificity results.

Proceeding to the afferent side of the common network we likewise find coupling by divergence – in this case by divergence of afferent inputs. Prominent and well-known examples are the actions of baroreceptor as well as chemoreceptor afferents on both neuronal outputs mentioned above under the auspice of mutual interaction between both feedback control systems. These and analogous phenomena have comprised under the term "irradiation of autonomic reflexes" (Schweitzer 1937). This term implies that each afferent produces its specific reflex, whereas the other actions are non-specific side effects. The same objections can be raised against this teleological nomenclature as in the case of branching of neurones within the network towards different outputs. An analogous argumentation leads to the statement that specificity or non-specificity of an afferent input is likewise a quantitative and not a qualitative property and depends on the relative distribution of the afferents in the network. Branching and therewith coupling by afferent inputs can take place at any level of the network including the output neurones.

Divergence by branching of afferents or within the network at the same time means the occurence of convergence. Again it is an old experience that the same respiratory or cardiovascular output reacts to several different inputs as e.g., from baroreceptors, chemoreceptors, muscle receptors, J-receptors, etc. Extensive somatosensory-vegatative convergence has been described in single brainstem neurones (Koepchen et al. 1975). The effect of convergence of chemoreceptor and pulmonary stretch receptor afferents could be analyzed in a quantitative manner in single respiratory neurones (Frank et al. 1976). These observations exclude the possibility to classify single neurones in the central network according to their afferent input, because also the re-

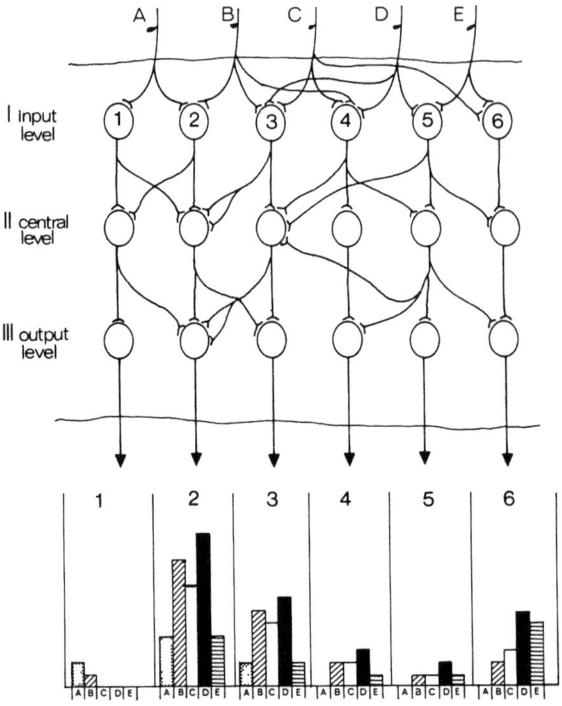

Fig. 2. Model of a simple network with 5 different inputs (A-E) and 6 different outputs (*1* –*6*), illustrating that slight inhomogenity in intracentral diverging and converging connections leads to a marked uneven distribution of excitation between the outputs. The model contains excitatory connections only and is based on the assumption that all inputs deliver the same amount of excitation and that pure additative interaction occurs at the synapses. The columns below the network indicate the relative amount of activity reaching the respective output from the different sources *A* –*E*

lative contribution of the different sources of excitation or inhibition is a quantitative and not a qualitative problem. Like divergence, convergence can occur at any level of the network including the output neurones or even the target organ.

After all, it is adequate to imagine the cardiovascular-respiratory control substrate as an inhomogeneous network with converging and diverging connections and numerous different inputs and outputs. Figure 2 shows a very simple model of such a network. It does not intend to simulate the central cardiorespiratory control, but to demonstrate that already a light inhomogeneity of the divergent and convergent connections leads to quite different weights in the contribution of different sources of excitation to different outputs even under the extremely simplifying presumption of homogeneous inputs. The network exhibits foci of activity and preferential pathways of excitation. Minor variatons are sufficient to effect considerable changes in the position on the quantitative scale between full specific and full non-specific behaviour of the network. Our model has in common with the real network that it lies somewhere between these extremes. For a better approach to reality it would have to be completed by inhibitory connections, recurrent pathways, reverberating circuits, etc. In addition, afferents from other parts of the C.N.S. ("intracentral afferents") have to be included.

It is a great advantage for the experimental analysis of the cardiorespiratory control network that it contains rhythms. Rhythms imposed by afferent inputs like cardiac rhythms as well as endogenous rhythms can be utilized as markers. I will confine the present considerations to the most prominent endogenous rhythm in the concerned

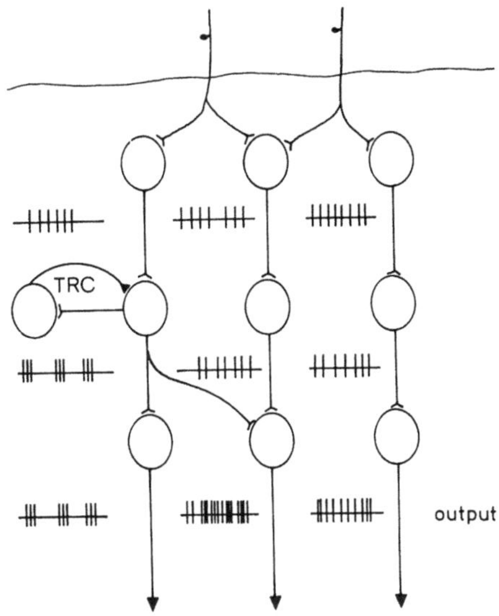

Fig. 3. Fraction of a network with a tone-rhythm-converter (*TRC*). The inserted patterns symbolize tonic input and rhythmic output of the TRC and admixture of rhythmic pattern to tonic pattern of the superposing type by convergence in the middle output neurone. The scheme in addition illustrates the possibilities that activity from an afferent source – as e.g., chemosensitivity – reach an output neurone either by way of the TRC or bypassing the TRC

region – the so-called respiratory rhythm. According to our above considerations the terms "respiratory rhythm" is not quite correct because it appears also in many other outputs and therewith does produce not only respiration. Therefore in the following this rhythm will be called "basic central respiratory-related rhythm" (BCR).

BCR is produced by a special part of the network which receives tonic input and delivers rhythmic output. Therefore it can be characterized as a tone-rhythm-converter (TRC). (Fig. 3). TRC is not identical with respiratory neurones, but only with part of the inspiratory neurones and all circuits which contribute to the "off-switch" function (Bradley et al. 1975), pontine, ventrolateral NTS, and others). The output of the TRC feeds rhythm into the network in an inhomogeneous manner. It gets tonic at a low level of activity if the input to the TRC is reduced below a certain level, it gets tonic at a high level if the off-switch feedback is impaired.

Projection of TRC output on other neurones within the network or on output neurones which in addition receive tonic inputs produces mixed tonic-rhythmic pattern at the output of these "mixing neurones" ("tonic" in this context means continuous discharge or rhythmic components with other than respiration-related rhythm). Thus various degrees of BCR admixture can appear in different outputs. Preferential pathways exist from the TRC output to phrenic and other respiratory motor neurones. Expiratory neurones do not belong to the TRC but are primarily tonic neurones which become secondarily rhythmic by "negative admixture" through inhibitory inputs during inspiratory phase.

In different sympathetic outputs the degree of admixture of BCR is quite different, e.g., strong in the cervical sympathetic and weak in the renal sympathetic under the same experimental condition (Koepchen et al. 1981a).

In the admixture of BCR we can distinguish two basic types: In the first one there is simple superposition of BCR on other patterns in an additive, hypo-additive or hyper-additive manner. The second one can appear if the mixing neurone itself is a member of another network with oscillatory properties and therefore shows excitability cycles with other time constants. In this case phase lags, multiple integer relations and other kinds of relative coordination with the BCR result as they are known from recordings of peripheral cardiorespiratory parameters as well as from recordings of sympathetic nerves (Barman and Gebber 1976, Cohen and Gootman 1979, Koepchen et al. 1962). I will call the first type "superposing admixture" and the second one "entraining admixture". All degrees of combination between these pure types are conceivable and will occur in the majority of cases.

The distinction between both types is of more than academic importance. In the superposing type of admixture, the connection with the TRC output is a source of net drive for the mixed output ("net drive" is defined as impulses per time irrespective of the actual pattern). In the entraining type the excitation of the mixing neurones can be compensated by negative phases of the oscillatory behaviour, so that no increase in net drive but only changes in pattern result. Only in the first type the projection of the TRC on more than the respiratory motor outputs could play an essential role in cardiorespiratory coupling during adaptive states as exertion, hypoxia or hypercapnia.

Adrian et al. (1932) looking for the functional significance of the discovered respiratory-related rhythm in sympathetic nerves suggested that "the connection of the highly chemosensitive respiratory centre with the cardiovascular centre will enable the latter to react promptly to changes which are too small to affect its own less chemosensitive cells". Replacing the classical nomenclature by the terms gained by our above considerations, Adrian's interpretation implies two assumptions: (1) excitation from chemosensitive sources reaches the sympathetic output by way of the TRC; (2) admixture is of the superposing type.

There is an experimental way to test these assumptions: Limited lesions or cooling in the region of the ventrolateral NTS somehow weakens the off-switch mechanism and induces an apneustic respiratory pattern similar as in the case of pontine lesions (Koepchen et al. 1974, Koepchen et al 1981a). During this state the net TRC output is markedly enhanced without any change in chemosensory input the network. All parts of the network and outputs dependent on the TRC output must be activated likewise to a corresponding amount. If this activation is much less than that observed during comparable net increase of phrenic activity caused by excitation of chemosensitivity, we can conclude that at least part of the chemosensory drive reaches the corresponding output on a way bypassing the TRC. Such an experiment is shown in Fig. 4. The following conclusions result from this and similar experiments: (1) The contribution of the TRC output is different in different sympathetic nerves; (2) Even in the distinctly BRC-related cervical sympathetic output, this contribution adds a relatively small amount of net drive; the admixture is partly of the entraining type; (3) Since it is known that a similar enhancement of net phrenic output by elevation of chemical drive is accompanied by considerably greater relative increase of cervical sympathetic activity (Hanna et al. 1981), part of the chemosensory cervical sympathetic drive must bypass the TRC; (4) The apneustic net increase of TRC output does not spread to other, not recorded sympathetic constrictor efferents to an essential degree since blood pressure does not change.

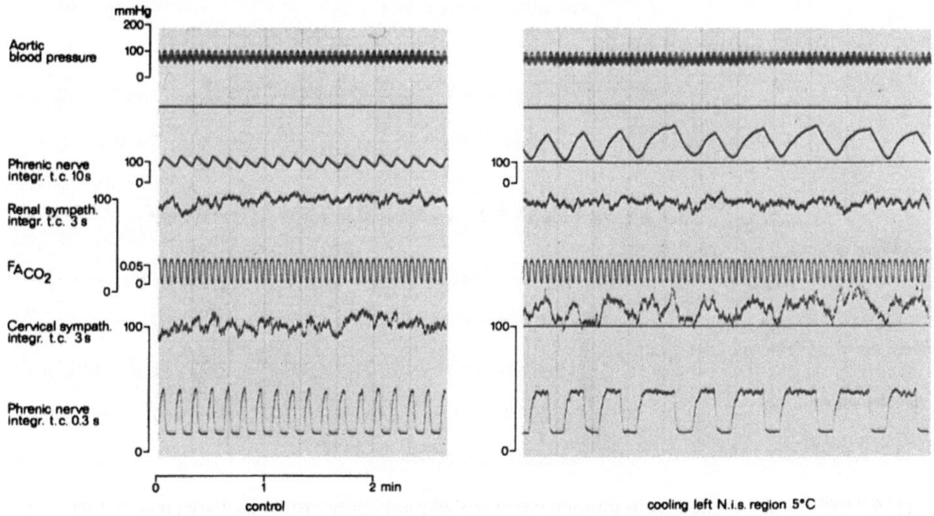

Fig. 4. Effect of local unilateral cooling in the region of nucleus infrasolitarius (N.i.s. region, equivalent to ventrolateral NTS region) on phrenic nerve (PNA), renal (RSA) and cervical sympathetic nerve (CSA) activities in an anaesthetized, paralyzed, artificially ventilated cat. Vagi and carotid sinus nerves are cut. *Left* before cooling, *right* during cooling. *t.c.* time constant of "integration" of multifibre nerve activities. Double record of same PNA with t.c. = 10 s *(secound trace from above)* indicating the increase in mean activity and t.c. = 0.3 s (*bottom trace*) indicating the change in discharge pattern. During the apneustic pattern caused by N.i.s. cooling (*right*) mean PNA is nearly doubled whereas the relative increase in mean CSA amounts about 15%. The CSA discharge pattern is coordinated with the phrenic rhythm in a sliding manner both in the control record and during cooling. RSA is not perceptibly changed by the manipulation of respiratory rhythm. Note absence of any change in arterial blood pressure. (From Koepchen et al. 1981a)

This and analogue experiments in accordance with recent work of Trzebski and Kubin (1981) do not speak in favour of the possibility that projection of TRC output to cardiovascular efferents plays a decisive role in central coupling between the respiratory and cardiovascular transport systems in states of common activation. Consequently there remains as the main factor activation by common afferences, the effects of which partly bypass the respiration-related TRC.

Turning to the point of common afferents, we come back to the starting point, i.e., the question how ventral medullary chemosensitivity and drive have to be built into the cardiorespiratory control network. If ventral medullary drive (VMD) – to use a neutral term – would come from a uniform system we simply could handle is as one among other afferent inputs with the usual inhomogeneous diverging distribution within the network. In the meantime, a lot of experimental evidence has been accumulated against this assumption: Stimulation or elimination experiments at different sites of the ventral medullary surface produce different combinations of effects in various respiratory and cardiovascular parameters (e.g,., Koepchen et al. unpublished, Loeschcke 1980, Loeschcke et al. 1970, Schlaefke and Loeschcke 1967, Schlaefke and See 1980, Trebski et al. 1974). This applies even for different parameters of respiratory

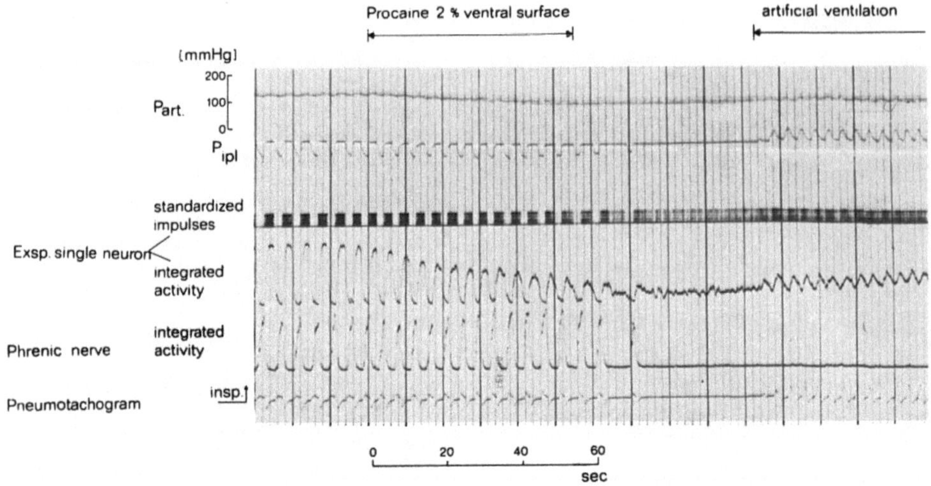

Fig. 5. Differentiated changes in phrenic nerve and expiratory neurone discharge parameters and blood pressure caused by addition of 2% procaine to continuous superfusion of ventral medullary surface by a dorsally inserted tube in an anaesthetized dog. Carotid sinus nerves cut. Application of procaine leads to decrease in blood pressue and expiratory single neurone activity, cessation of phrenic activity and respiration with quite different time courses. Decrease in expiratory neurone activity begins immediately after switching from control superfusion to procaine containing mock CSF, decrease in blood pressure begins some seconds later, whereas phrenic nerve activity continues for about one minute even with slightly enhanced amplitude before it stops suddenly. Note slight shift of expiratory level of interpleural pressure parallel with the decrease in expiratory neuronal activity.

innervation. One example is shown in Fig. 5 . Other relative time courses in the effects on different respiratory parameters than in the records of Fig. 5 were seen when the site of application of procaine was changed. Thus we can state that VMD is not uniform with respect to the contribution of its parts to the different parts of the network.

Another fundamental kind of non-uniformity is revealed by the fact that elimination of ventral strucutres (by procaine, cooling or coagulation) has not identical effects as removal of H^+ ions and does not produce an exact mirror-image of the CO_2-H^+ effect. One example of this discrepancy is manifested by two experiments: In Fig. 6 the effect of short-time ventral medullary cooling on phrenic nerve, renal and cervical sympathetic activity is represented. Together with phrenic activity and several cardiovascular parameters, renal nerve activity is distinctly reduced by the cooling. That means that the recorded renal nerve activity receives a considerable amount of drive which can be impaired by ventral medullary cooling. On the other hand, many of the same strains of the renal sympathetic which are inactivated by ventral cooling do not respond to changes in systemic pCO_2 over a wide range where marked answers of phrenic and cervical sympathetic activity occur (Koepchen et al. 1981a). It follows that there is a superficial medullary source of excitation which does not respond to the CO_2 - H^+ stimulus. Thus we have to discriminate between ventral medullary drive and ventral medullary chemosensitivity, or in other words between a chemosensitive and

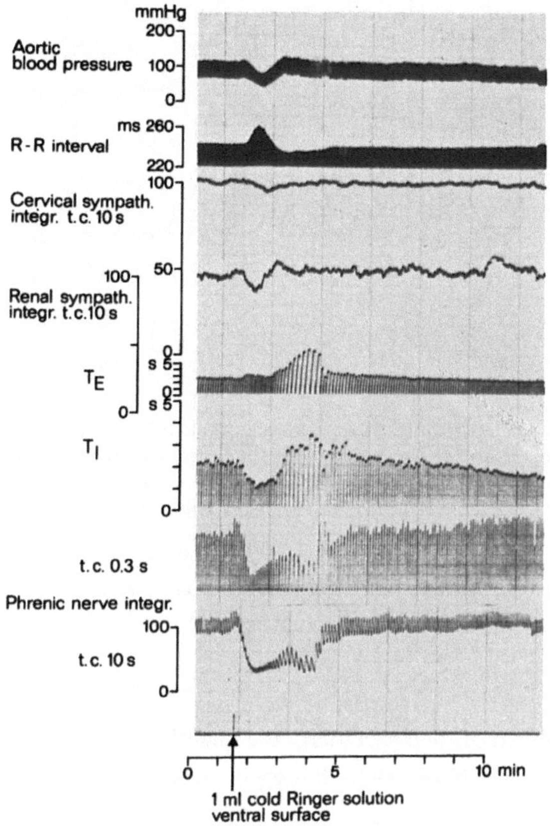

Fig. 6. Effects of short time cooling of ventral medullary surface by application of 1 ml cold Ringer's solution (5° C) through a curved tube inserted from dorsal side on various cardiovascular, sympathetic and respiratory parameters in an an aesthetized cat (experimental conditions see legend Fig. 4). General depressing effect on heart rate, blood pressure and renal (RSA) and cervical sympathetic activity (CSA) and profound changes in integrated phrenic burst amplitude and time parameters: RSA reacts much more distinctly than CSA. Time course of effects in blood pressure and heart rate are very similar to RSA. Note the much longer latency of recovery of phrenic discharge parameters: After initial decrease in T_I contemporarily with the sympathetic and cardiovascular changes there is a secondary increase in T_I and T_E outlasting for several minutes the other effects. (From Koepchen et al. 1981a)

a non-chemosensitive part of the ventral medullary source of drive. Both have in common the superficial site, the susceptibility to synaptic drugs (Feldberg 1980, Feldberg and Gürtzenstein 1976, Loeschcke 1980) and the properties to deliver drive to various parts of the network and to be afferent but not identical with the TRC.

In our analysis we have to ask in which manner the two parts have to be inserted into the network. To begin with the chemosensitive part, the question arises if the chemoreceptive elements are true receptors or synapses reacting to the H^+ concentration of ECF. Figure 7 illustrates that in both cases the response of the neurone z depends on a sufficient coaction between synaptic input from other sources and the H^+ concentration. Thus for this neurone and all dependent neurones the functional consequences are equivalent in both cases, and we can leave open this question without loss of understanding of the network. In any case we can handle the chemosensitive elements as one source like afferent inputs. One preferential pathway from the chemosensitive part reaches the TRC input – also its feedback part – other pathways bypass it as shown above.

As to the question of specific chemosensitive neurones in the medulla, we have to realize that all neurones in the network subordinate to neurone z in Fig. 7 likewise will appear to be chemosensitive, i.e., change their activity in response to changes

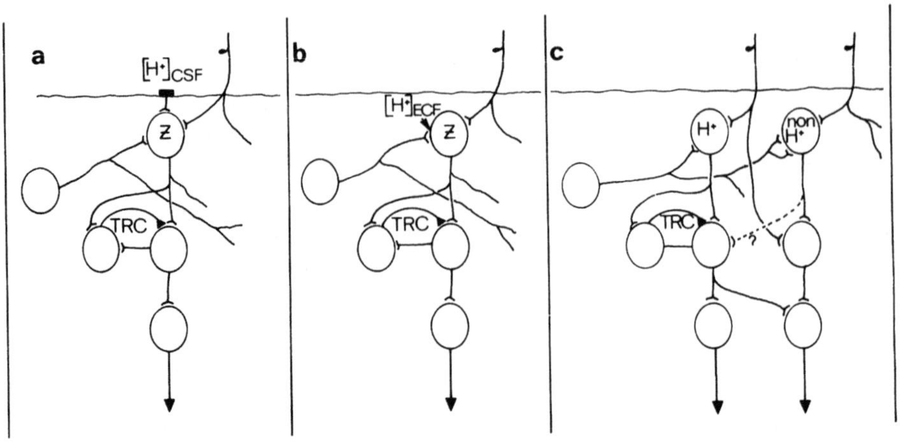

Fig. 7. Schemes of chemosensitive and non-chemosensitive superficial ventral medullary drives as source for the control network. The comparison between **a** and **b** illustrates that the consequences for the downstream stations of the network are equivalent in the case of a true receptor projecting on a neurone z (**a**) or a chemosensitive synapse at neurone z (**b**) because the net effect will depend on the sum of inputs to that neurone including the direct or indirect action of $[H^+]$. c illustrates the possibility that drives from chemosensitive and non-chemosensitive superficial structures can converge on the same output neurone, part of which bypass the TRC. The assumption of a chemosensitive drive reaching also the feedback part of TRC is derived from experiments with local weakening of the feedback function

in superficial $[H^+]_{ECF}$. Again the problem of their specificity is a quantitative and not a qualitative question, which can be answered only if more inputs are tested than only the chemosensitive one.

The non-chemosensitive source of ventral medullary drive differs from other parts of the reticular network by its accessibility from the ventral surface which of course is a morphological and experimental criterion, but does not imply functional peculiarity. Therefore we can expect that the activity in this part is fed by many different sources in an inhomogeneous manner as in other reticular areas of the lower brainstem with convergence reactions of single neurones (Koepchen et al. 1975). Thus in the network model we can handle the non-chemosensitve part as one among other portions of that network with diverging and converging inputs from peripheral and intracentral afferents and uneven distribution of its downstream actions.

In view of the profound changes in blood pressure and sympathetic activity caused by experimental interference at the ventral medullary structures, a nearly related topic ist the central origin of vasomotor tone. From the experiments mentioned above and earlier investigations at the ventral medullary surface (Lioy and Hanna 1978, Trzebski et al. 1974), it can be concluded that it originates from the chemosensitive as well as from the non-chemosensitive part of the VMD. It remains open, however, whether tonic sympathetic constrictor innervation is fed in addition also by other sources. Many indications speak in favour of this assumption. One additional source is revealed by the observation that in baroreceptor denervated animals cooling of the dorsal surface of medulla near the obex likewise leads to strong fall in blood pressure and sympathetic activity (Czachurski et al. 1979). One example where cardiovascular, sympathetic and

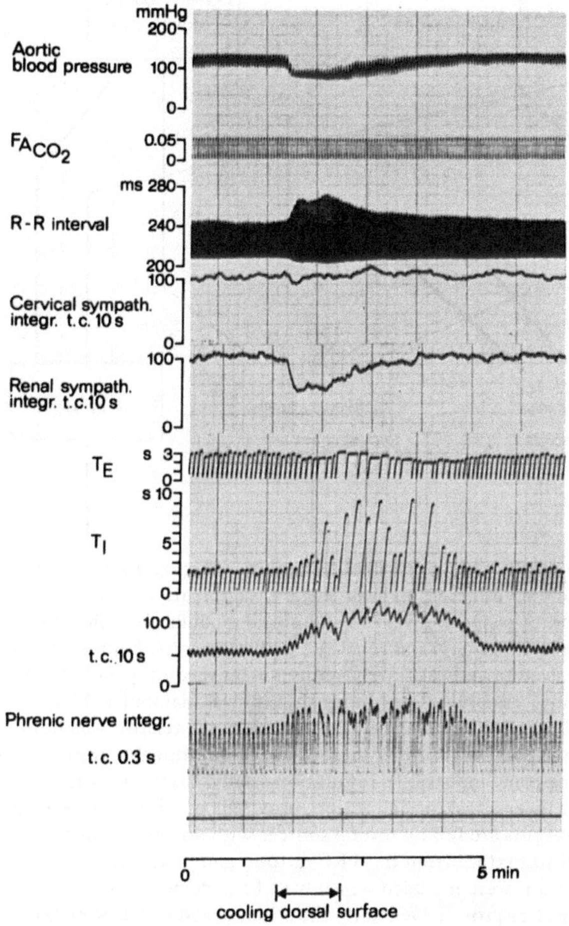

Fig. 8. Effect of bilateral dorsal medullary cooling on cardiovascular and respiratory outputs in the cat (experimental conditions see legend Fig. 4). Note the distinct decrease in sympathetic-cardiovascular parameters, with nearly no participation of cervical sympathetic activity, together with the striking increase in net phrenic activity. The typical effect of cooling of the ventrolateral NTS region is manifested by strong increase in T_I and its variability with minor changes in T_E. "Integrated" phrenic activity recorded with two different time constants of "integration" visualizes the strong increase in mean activity. (From Koepchen et al. 1981a)

respiratory effects of dorsal medullary cooling are recorded in a deafferented cat is shown in Fig. 8. Cold from the surface in this case reaches, with a little longer latency, also the region of the ventrolateral NTS and produces the same apneustic changes with net increase in phrenic activity occuring in case of isolated cooling of that region (see Fig. 4). This double effect is instructive because it demonstrates that – unlike the ventral medullary source – the dorsal medullary source is not an origin of excitation for the TRC. The possibility that the dorsal medullary source of sympathetic tone could be merely a relay station subordinate to the ventral source becomes unlikely by the observation of Amendt et al. (1979) that neurones from the dorsal as well as from the ventral region directly project on preganglionic neurones in the intermedio-lateral column of the spinal cord.

The main intension of the above general considerations on the relations between the classical "respiratory centre" and "cardiovascular centre" was to open ways for the concrete experimental analysis of the network unloaded from misleading questions concerning irradiation, specificity or non-specificity and labelling of single neurones as members of specific control systems. The practical experimental analysis has to

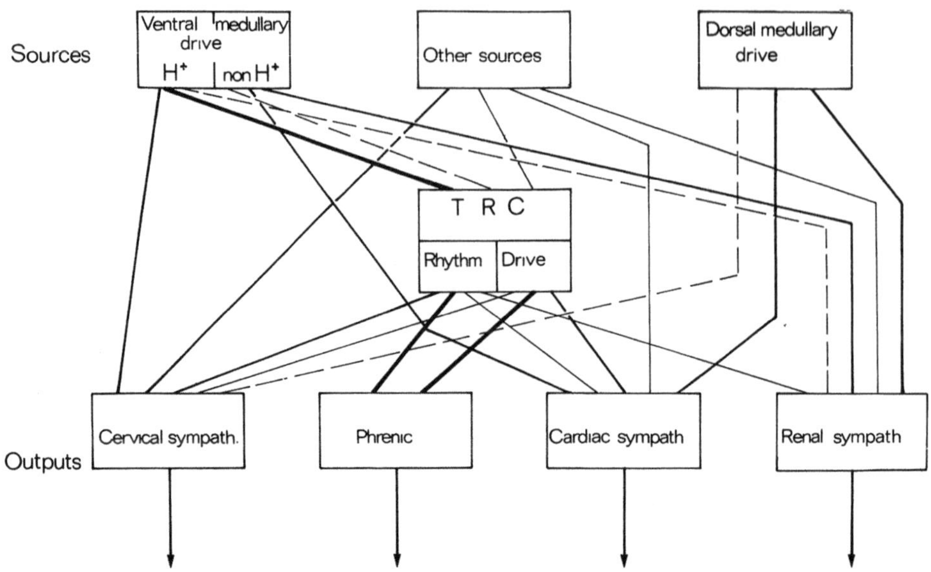

Fig. 9. Schematic representation of the results of an experimental analysis of several sources and outputs of the cardiorespiratory control network (examples of single experiments see Figs. 4,6,8). Sources were studied by variation of pCO_2 and ventral and dorsal medullary cooling, outputs by recording from phrenic, cervical sympathetic and renal sympathetic nerves. Source analysis of cardiac sympathetic fibres is derived from heart rate responses after vagotomy and therefore is confined to chronotropic sympathetic fibres only. Respiratory rhythm was manipulated by cooling the ventrolateral NTS region. *Strength of connecting lines* indicates the relative weight of the respective source for the different outputs. *Interrupted lines:* Weak or questionable connections. TRC: Tone-rhythm converter converting tonic inputs into respiration-related thythmic output. Rhythm means entraining action, Drive means superposing action, providing net drive. The scheme shows that the different sympathetic outputs are fed by several sources with different weights, receive diffifferent amount and type of admixture from the TRC output, and besides also drives bypassing the TRC. Only those connections were included which have been demonstrated experimentally, additional pathways are not disproved. Generally the scheme demonstrates marked inhomogeneity of the network despite the multiplicity of connections

quantify the relations between as many inputs as possible and as many outputs as possible, realizing that specificity is a quantitative and not a qualitative property. The BCR and the amount of its admixture to different outputs can be used as marker. An analysis of this kind has been started recently. The result was a first semi-quantitative table giving a survey on the central sources of excitation for the phrenic and some sympathetic and cardiovascular outputs (Koepchen et al. 1981a) Figure 9 presents a graphic compilation of the intracentral organization as derived from these experiments.

It is a general experience from such studies that the different efferent pathways for one system, especially the cardiovascular innervation, are highly differentiated and that parts of the cardiovascular outputs are more closely related to parts of the respiratory innervation than to other cardiovascular efferents. That means that the central network is not organized on the principle of our teleological classification according to control systems for respiration and circulation.

Finally, one important point should be emphasized: The confinement of the above considerations to the cardio-respiratory control and the respiration-related rhythm was due to the theme of the present symposium and should not be misunderstood in the sense of the existence of a specific common cardiorespiratory control network within the lower brainstem. The same general points of view developed on the basis of this example can be generalized in the same manner for other rhythms and other outputs of the network controlling skeletomotor innervation or general state of activity of the cortex. Also these outputs have to be included in the outlined kind of analysis. Thus this short review on the general problem of organization of "centres" and some related experimental approaches does by no means represent a conclusive settlement but a challenge to induce future experimental work.

Acknowledgements. Most experimental series underlying this analysis were supported by Deutsche Forschungsgemeinschaft and Stiftung Volkswagenwerk. The author expresses his thanks to Mrs. E. Gaebel for typing the manuscript, Mr. D. Winzer for drawing the schemes and Mr. P. Holzner for photographic preparation of the figures.

References

Adrian ED, Bronk DW, Phillips G (1932) Discharges in mammalian sympathetic nerves. J Physiol (London) 74:115–133

Amendt K, Czachurski J, Dembowsky K, Seller H (1979) Bulbospinal projections to the intermediolateral cell column; a neuroanatomical study. J Autonom Nerv Syst 1:103–117

Anrep GV, Pascual W, Rössler R (1936) Respiratory variations of the heart rate. I. The reflex mechanism of the respiratory arrhythmia. Proc R Soc London SerB 119:191–217

Barman SM, Gebber G (1976) Basis for synchronization of sympathetic and phrenic nerve discharges. Am J Physiol 231:1601–1607

Bradley GW, Euler von C, Martilla I, Roos B (1975) A model of the central and reflex inhibition of inspiration in the cat. Biol Cybernetics 19:105–116

Cohen MI, Gootman PM (1979) Periodicities in efferent discharge of splanchnic nerve of the cat. Am J Physiol 218:1091–1101

Czachurski J, Amendt K, Dembowsky K, Seller H (1979) The effect of reversible cold block of dorsomedial or ventrolateral regions in the brainstem on blood pressure, sympathetic activity and respiration. Pflügers Arch 379:R48

Daly M de B, Hazzledine JL, Ungar A (1967) The reflex effects of alternations in lung volume on systematic vascular resistance in the dog. J Physiol (London) 188:331–351

Feldberg W (1980) Cardiovascular effects of drugs acting on the ventral surface of the brain stem. In:Koepchen HP, Hilton SM, Trzebski A (eds) Central interaction between respiratory and cardiovascular control systems. Springer, Berlin Heidelberg New York

Feldberg W, Gürtzenstein PG (1976) Vasodepressor effects obtained by drugs applied to the ventral surface of the brain stem. J Physiol (London) 258:337–355

Frank C, Borchert J, Klüßendorf D, Koepchen HP (1976) Modes of interaction between vagal and CO_2 chemoreflex input at bulbar respiratory neurons. Bull Eur Physiopathol Resp 12:223

Golenhofen K, Hildebrandt G (1958) Die Beziehungen des Blutdruckrhythmus zu Atmung und peripherer Durchblutung. Pflügers Arch 267:27–45

Gürtzenstein PG, Hilton SM, Marshal JM, Timms R (1977) Experiments on the origin of vasomotor tone. J Physiol (London) 275:78–79

Hanna BD, Lioy F, Polosa C (1981) Role of carotid and central chemoreceptors in the CO_2 response of sympathetic preganglionic neurons. J Autonom Nerv Syst 3:421–435

Hering E (1869) Über Atembewegungen des Gefäßsystems. Sber Akad Wiss Wien Math Naturwiss Kl 2 Abt 60:829–856
Heymans C, Neil E (1958) Reflexogenic Areas of the Cardiovascular System. Churchill, London, p 271
Holst E v (1939) Die relative Koordination als Phänomen und als Methode zentralnervöser Funktionsanalyse. Ergeb Physiol 42:228–306
Hukuhara Jr T (1980) Spontaneous activity pattern and anatomical distribution of brain stem reticular neurones showing correlations with phrenic and renal sympathetic nerve activities. In: Koepchen HP, Hilton SM, Trzebski A (eds) Central interaction between respiratory and cardiovascular control systems. Springer, Berlin Heidelberg New York, p 21
Kao FF, Lahiri S, Wang C, Mei SS (1967) Ventilation and cardiac output in exercise: interaction of chemical and work stimuli. Circ Res 20, 21 (Suppl 1) 179–191
Koepchen HP (1962) Die Blutdruckrhythmik. Steinkopff Darmstadt, p 139
Koepchen HP, Thurau K (1959) Über die Entstehungsbedingungen der atemsynchronen Schwankungen des Vagustonus (respiratorische Arrhythmie). Pflügers Arch 269:10–30
Koepchen HP, Seller H, Polster J, Langhorst P (1968) Über die Feinvasomotorik der Muskelstrombahn und ihre Beziehung zur Ateminnervation. Pflügers Arch 302:285–299
Koepchen HP, Lazar H, Borchert J (1974) On the role of the nucleus infrasolitarius in the determination of respiratory periodicity. Proc IUPS, vol XI:81 (Abstr)
Koepchen HP, Langhorst P, Seller H (1975) The problem of identification of automatic neurons in the lower brain stem. Brain Res 87:373–393
Koepchen HP, Klüssendorf D, Sommer D (1981a) Neurophysiological background of central neural cardiovascular-respiratory coordination: basic remarks and experimental approach. J Autonom Nerv Syst 3:335–368
Koepchen HP, Lazar H, Klüssendorf D (1981b) Further experiments on the role of the nucl. infrasolitarius (N.i.s.) region in pontomedullary respiratory rhythmogenesis. Pflügers Arch 389:R53
Langhorst P, Lambertz M, Schulz G (1981) Central control and interactions affecting sympathetic and parasympathetic activity. J Autonom Nerv Syst 4:149–163
Lioy F, Hanna BD (1978) Cardiovascular effects of medullary chemoreceptors. Fed Proc Fed Am Soc Exp 37:744
Loeschcke HH (1980) Introduction to Session II: Ventral Surface of Medulla. In: Koepchen HP, Hilton SM, Trzebski A (eds): Central interaction between respiratory and cardiovascular control systems. Springer, Berlin Heidelberg New York, p 43
Loeschcke HH, Katsaros B (1959) Die Wirkung von in den Liquor cerebrospinalis eingebrachtem Ammoniumchlorid auf Atmung und Vasomotorik. Pflügers Arch 270:147–160
Loeschcke HH, Koepchen HP (1958a) Über das Verhalten der Atmung und des arteriellen Drucks bei Einbringen von Veratridin, Lobelin und Cyanid in den Liquor cerebrospinalis. Pflügers Arch 266:586–610
Loeschcke HH, Koepchen HP (1958b) Beeinflussung von Atmung und Vasomotorik durch Einbringen von Novocain in die Liquorräume. Pflügers Arch 266:611–627
Loeschcke HH, Koepchen HP (1958c) Versuch zur Lokalisation des Angriffsortes der Atmungs- und Kreislaufwirkung von Novacain im Liquor cerebrospinalis. Pflügers Arch 266:628–641
Loeschcke HH, Lattre de J, Schlaefke ME, Trouth CO (1970) Effects on respiration and circulation of electrically stimulating the ventral surface of the medulla oblongata. Respir Physiol 10:184–197
Mitchell RA, Loeschcke HH, Severinghaus JW, Richardson BW, Massion WH (1963) Regions of respiratory chemosensitivity on the surface of the medulla. Ann N Y Acad Sci 109:661–681
Preiss G, Polosa C (1977) The relation between end-tidal CO_2 and discharge patterns of sympathetic preganglionic neurons. Brain Res 122:255–267
Schlaefke M, Loeschcke HH (1967) Lokalisation eines an der Regulation von Atmung und Kreislauf beteiligten Gebietes an der ventralen Oberfläche der Medulla oblongata durch Kälteblokkade. Pflügers Arch 297:201–220
Schlaefke ME, See WR (1980) Ventral medullary surface stimulus response in relation to ventilatory and cardiovascular effects. In: Koepchen HP, Hilton SM, Trzebski A (eds) Central Interaction between Respiratory and Cardiovascular Control Systems. Springer, Berlin Heidelberg New York, p 56

Schwanghart F, Schröter R, Klüssendorf D, Koepchen HP (1974) The influence of novocaine block of superficial brain stem strucutres on respiratory and reticular neurons. In: Umbach W, Koepchen HP (eds) Central rhythmic and regulation. Hippokrates, Stuttgart, p 104

Schweitzer A (1937) DIe Irradiation autonomer Reflexe. Karger, Basel

Trzebski A, Kubin L (1981) Is the central inspiratory activity responsible for pCO_2-dependent drive of the sympathetic discharge? J Autonom Nerv Syst 3:401–420

Trzebski A, Zielinski A, Majcherczyk S, Lipski J, Szulczyk P (1974) Effect of chemical stimulation and depression of the medullary superficial areas on the respiratory motoneurones discharges, sympathetic activity and efferent control of carotid area receptors. In: Umbach W, Koepchen HP (eds) Central rhythmic and regulation. Hippokrates, Stuttgart, p 170

Target Sites for Anticholinesterases, Cholinolytics and Oximes on Ventral Medulla Oblongata

H. EDERY[1]

In recent years evidence has accumulated showing that many physiological functions including respiration, blood pressure, glucose regulation, vasopressin release may be greatly affected by chemical stimulation of selected sites on the ventral surface of medulla oblongata (reviews Feldberg 1976; Loeschcke 1980). It has also been shown that DFP (dyflos) applied in these sites elicits a profound vasodepressor response resistant to atropine and accompanied with respiratory disturbances (Edery and Guertzenstein 1974). The studies have now been extended to the powerful organophosphate (o.p) soman (S). It will be shown that the anticholinesterase topically or systemically administered provoked a long lasting centrally originating hypotension which could be reversed by selective protection of ventromedully chemosensitive areas with antimuscarinic drugs and some particular oximes. These areas were found to contain muscarinic receptors and cholinesterase. Partial results have been presented (Edery 1978).

Methods

Sixty cats of either sex (2.2–3.9 kg body weight) were anaesthetized with an iv. injection of 25 mg/kg sodium pentobarbitone supplemented by 5 mg/kg whenever required to maintain adequate surgical anaesthesia. Left femoral artery blood pressure (BP) was recorded with a pressure transducer. Cannulas were inserted into the right and left femoral veins for single injections and continuous infusions respectively. Occasionally 0.3 ml blood samples were obtained from the cannulated right femoral artery. The trachea was cannulated and respiration recorded with a thermistor probe. When required artificial ventilation was applied after vagotomy using an Assistor 644 Drager respirator which allowed monitoring of the reappearance of spontaneous respiration. In some experiments lead I, II and III ECG was recorded. Cholinesterase in 0.1 ml blood was determined by the method of Fleischer et al. (1955). Activity was expressed in μmol Ach split during 35 min incubation at 25°C.

To apply drugs on the ventral surface of the medulla oblongata the method of Feldberg and Guertzenstein (1972) was used. After removal of part of the occipital pars

1 Department of Physiology and Pharmacology, Sackler School of Medicine, Tel Aviv University, Ramat Aviv, Israel

Central Neurone Environment
(ed. by M.E. Schläfke, H.P. Koepchen and W.R. See)

basilaris and section of the dura mater, a special perspex cannula was positioned so that its lower openings remained on the chemosensitive areas. In some experiments the cannula was moved rostrally to apply it on the trapezoid bodies. The correct placement of the openings was checked at the end of each experiment by staining the areas with brom-phenol blue solution as previously described (Edery and Guertzenstein 1974). For topical application of solutions to the cerebral cortex two burr-holes, one in each parietal bone were drilled 7 mm lateral and 7 mm posterior of the median and coronary suture respectively. The dura was removed and a cylindrical perspex cup 1 cm long and of 5 mm internal diameter was lowered with a rack and pinion holder into each hole until the lower edge of the cup lightly but completely contacted the surface of the parietal cortex' gyrus lateralis. Silicone grease was used to seal bone edges. The cups were constantly kept filled with 0.1 mock cerebrospinal fluid (CSF) which was aspirated before each application of solutions. These were applied bilaterally onto the ventromedullary sites, trapezoid bodies or parietal cortex in a volume of 10 μl. For this purpose a suitable graduated syringe fitted to a polythene tube was used. The chemicals were allowed to act for 5 min and then removed by repeated alternated application and suction of 0.1 ml mock CSF. During these operations, care was taken not to touch the surface of ventral medulla to prevent hypertensive reactions (Edery 1975). To expose the IV ventricle a part of the vermis cerebellum was removed by suction after dissecting nape muscles, atlanto-occipital membrane and nibbling a piece of the occipital.

Application of solutions on the floor of IV ventricle was performed by means of circular filter paper pledgets of 5 mm diameter. The size was selected in accordance with previous measurements of cat's ventricular system (Levinger and Edery 1969), so that the pledget covered the entire central fovea of IV ventricle. 10 μl of the test solution was first dropped onto the pledget which was then gently deposited onto the fovea. Light pressure was applied to ensure good contact. The pledget was removed 5 min after application and the fovea rinsed with CSF. In a few experiments the topical application of chemicals was performed at two different sites in the same animal. In these cases the test drug was first applied to the parietal cortex and subsequently either to the IV ventricle or after turning the cat on its back, to the ventral medulla oblongata.

The maximum volume for single injections of S was 0.3 ml. In a series of experiments 20 μg/kg/min S was infused continuously (slow injector, B. Braun) at a rate of 0.275 ml/min during 1–1.5 h.

About 30 min before topical application or systemic administration of S, cats received 2 mg/kg atropine methyl nitrate i.v. supplemented by 0.5 mg/kg hourly. In a series of experiments this cholinolytic was replaced by atropine sulphate or benactizine in the same dose regime or by 0.5 mg/kg diazepam. The injection of cholinolytics had to be performed slowly to prevent or diminish a severe, though reversible fall of BP. At any rate, no chemical was tested until BP had completely recovered. Each drug effect described in the paper was replicated in at least four animals.

In a series of experiments eight cats of either sex were sacrificed with an overdose of sodium pentobarbitone and the medulla oblongata immediately excised and frozen. Superficial slices (0.5 mm thick) from four areas (see Fig. 1) of the ventral side were cut and muscarinic receptors ([^{3}H] QNB binding) and cholinesterase activity (Jōhnson and Russell 1975) determined in tissue homogenates.

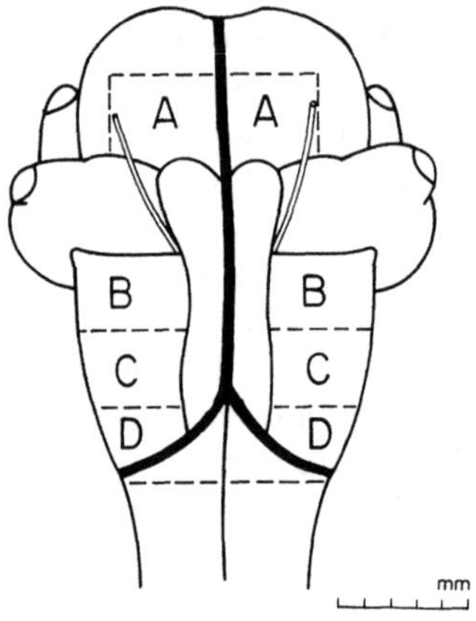

Fig. 1. Diagram of cat ventral surface of medulla oblongata showing the superficial areas dissected for determination of muscarinic receptors and cholinesterase activity

Materials

A stock solution of a pure specimen of pinacolyl-methylphosphonofluoridate (soman, S) was prepared in dry propylene glycol. The stock solution (10 mg/ml) was kept at – 10°C and aliquots diluted further before each experiment. Drugs used included atropine methylnitrate (Merck), atropine sulphate (Merck), benactizine hydrochloride (Aldrich), eserine salicylate (Sigma), diazepam (Roche) as well as oximes, namely N,N′-oxydimethylene-bis (pyridinium-4-aldoxime) dichloride (obidoxime), 1,1′-oxydimethylene bis (2,4′ dihydroximinomethylpyridinium) dichloride (HS-3), 1,1′oxydimethylene bis [(2-hydroximinomethyl-3′carboxamide) pyridinium] dichloride (HS-6) and 1,1′-oxydimethylene bis ([2-hydroxyiminomethyl-4′-carboxamide) pyridinium] dichloride (HI-6). Drug solutions for topical application were made up in mock CSF and for i.v. administration in saline. All weights refer to the salts.

Results

Effects of Soman, Cholinolytics and Oximes Applied to the Ventral Surface of Medulla Oblongata

Immediately or following a very short latency, after topical application of S at concentrations of 100–250 μg/ml on the chemosensitive areas of ventral medulla, the BP fell and in most cases continued to fall after removal of the o.p. from inside of the perspex applicator. The fall of BP was neither accompanied by bradycardia nor by major deviation of normal electrocardiographic pattern (Fig. 2). The extent and duration of the hypotensive response varied with the dose. It amounted to 20–90 mm Hg and lasted from 0.5 h to more than 3 h with the smallest and the highest dose respectively. In one case, a fall as large as 140 mm Hg was produced after application of 200 μg/ml S and in three cats a fall as great as 90 mm Hg was elicited with such small concen-

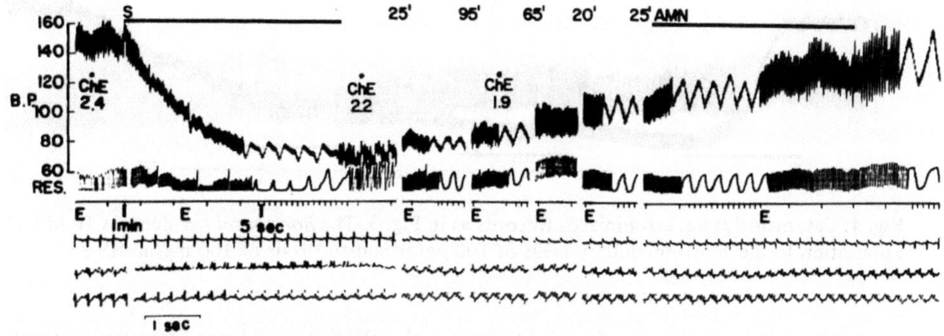

Fig. 2. Cat, male 2.9 kg, atropinized (2 mg/kg atropine methyl nitrate i.v. supplemented with 0.5 mg/kg hourly). Records from top downwards; blood pressure (BP scale mm Hg), respiration (RES) time (large division 1 min, small division 5 s) electrocardiogram (E) lead I, II, III and time (1 s). Breaks between panels indicate time intervals. The *horizontal bar* represents 10 μl bilateral application to the ventromedullary areas of 250 μg/ml soman (S) or 50 mg/ml atropine methyl nitrate (AMN). Blood cholinesterase (ChE) activity values at *dots*

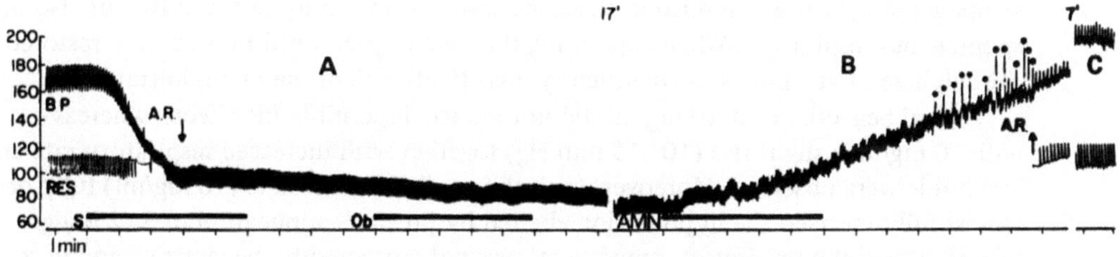

Fig. 3. Cat, male 2,8 kg, atropinized. Records of blood pressure (BP) respiration (RES) and *time* (1 min). Breaks between panels indicate time intervals. Downwards and upwards pointing *arrows* respectively indicate application and cessation of artificial respiration (A.R.). The *horizontal bar* represents 10 μl bilateral application to the ventromedullary areas of 100 μg/ml soman (S); 100 mg/ml obidoxime (Ob) or of 50 mg/ml atropine methyl nitrate (AMN). Between **A** and **B** there was an additional application of obidoxime. *Dots* on BP record indicates that spontaneous respiratory movements were detected by the artificial respirator

trations as 100 μg/ml that is 1 μg of S placed on each side of the ventral medulla. In all cases the BP gradually recovered and regained eventually its initial level. In some cases, however, the recovery was incomplete and BP remained constant at a level 10–20 mm Hg lower than the control. After recovery, hypotensive responses of similar magnitude and duration could be elicited in the same animal after repeating the application of a given dose of S.

Profound, albeit variable respiratory changes, accompanied the S-induced BP fall. In some experiments respiratory rate decreased whilst amplitude increased and in others, the opposite occurred or mixed changes were observed. In some cases, as illustrated in Fig. 3, apnoea appeared and artificial respiration had to be applied in order to continue the experiment.

The S-induced vasodepressor responses could neither be prevented by pretreating the animals with atropine sulphate instead of the methylnitrate, benactizine or diazepam

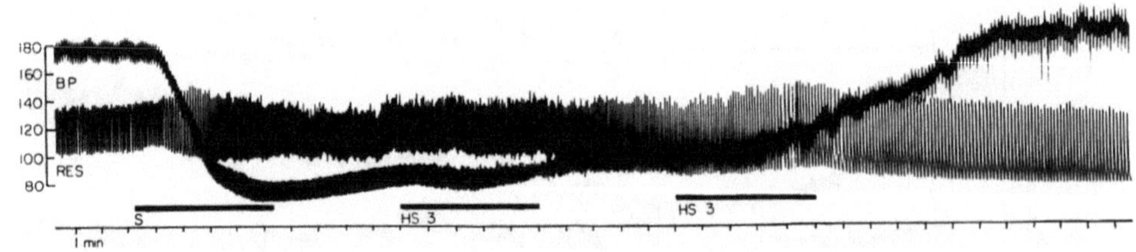

Fig. 4. Cat, male 4,0 kg, atropinized. Records as in Fig. 3. The *horizontal bar* denotes 10 μl bilateral application to the ventromedullary areas of 100 μg/ml soman (S) or of 100 mg/ml HS-3

nor by previous severance of vagi. Moreover, the BP fall could not be reversed after re-peated (at 20 min intervals) i.v. injections of 0.5 mg/kg of the above mentioned cholinolytics or of 1–2 mg/kg each of the oximes HS-3, HS-6, HI-6 or obidoxime. In contrast the BP was promptly restored when the cholinolytics or the oximes HS-6, HS-3 or HI-6 were topically applied to the chemosensitive areas of ventral medulla. Figure 2 illustrates the reversal of the long lasting S-provoked vasodepression by topically applied atropine methyl nitrate. Figure 3 shows that whilst topical obidoxime was totally ineffective, atropine methyl nitrate applied in the same way not only fully restored BP but also the spontaneous respiration. When comparing the time elapsed until BP was fully restored, the sulphate of atropine was consistently less effective than the methylnitrate.

Topical benactizine at 10 mg/ml did not induce discernible BP effects, whereas with 50 mg/ml a slight rise (10–15 mm Hg) together with increased respiratory rate and amplitude were observed. Moreover, topically applied benactizine (10 mg/ml) fully or almost fully reversed the hypotension elicited by previous application of S (250 μg/ml). However, the respiratory impairment was not corrected by benactizine and in some experiments even worsened.

HS-3, HS-6 or HI-6 when applied on the chemosensitive areas at a concentration of 100 mg/ml produced a slight rise of BP as previously had been shown for obidoxime (Edery and Guertzenstein 1974); but respiration remained unchanged. Figure 4 shows that BP which had fallen by topical S, started to recover already after the first topical application of HS-3 (100 mg/ml), and full recovery occurred following an additional dose of the oxime. Furthermore, the drug brought about a restoration of respiration which had been severely affected by the anticholinesterase. Similar results were obtained with HS-6 and HI-6. In all cases after two applications of these oximes (100 mg/ml) the BP which had been lowered by previous application of S (250 μg/ml) was entirely restored but the respiratory disturbances were corrected only by HI-6. This latter also reversed the hypotension induced by topical eserine (50 mg/ml). Topical obidoxime (100 mg/ml) did not antagonize either the hypotension or the respiratory changes elicited by topical S (100–250 μg/ml). On the contrary, in some cases S-induced vasodepression and dyspnoea worsened when obidoxime was applied three times at 10–15 min intervals.

Effects of Soman Applied to Parietal Cortex, Trapezoid Bodies and IV Ventricle

No alterations of BP or respiration occurred after topical application of S at concentrations up to 500 μg/ml to parietal cortex, trapezoid bodies or the IV ventricle (upper part Fig. 7). With 1 mg/ml S on the latter, a small fall of BP as well as reduction of respiratory amplitude and increased rate were observed. Both functions recovered fully 20 to 30 min after removal of the S pledget.

Effects of Intravenous Administration of Soman

No detrimental effects on BP and respiration occurred after a single injection of 5 to 10 μg/kg S into cats pre-treated with atropine either the methyl nitrate or the sulphate.

When S was administered continuously at a rate of 20 μg/kg/min into atropinized and artificially ventilated cats, blood pressure steadily declined. In some experiments, the fall was preceded by a transient rise (Fig. 6). The vasodepressor response was accompanied by fasciculations starting in neck and hindleg muscles and gradually spread all over the back and forelegs. The fasciculations disappeared within 30–45 min of commencing S injection. When BP had declined to its nadir, usually 15–35 min after initiation of S administration, a partial recovery could occur in spite of continuing o.p. injection. This is shown in Fig. 5. Nonetheless, in all cases the BP remained at a low steady level for 1–1.5 h after the end of S infusion, and then recovery began. Pari passu with the recovery, spontaneous respiration reappeared. Both BP and respiration fully regained their initial level and rhythm respectively within 2 h 15 min of cessation of S injection. If under such circumstances a second continuous infusion of S was instituted neither BP level nor spontaneous respiration were then affected as is illustrated in Fig. 5. When S administration was discontinued these functions remained unchanged for 45 min to 1 h 20 min.

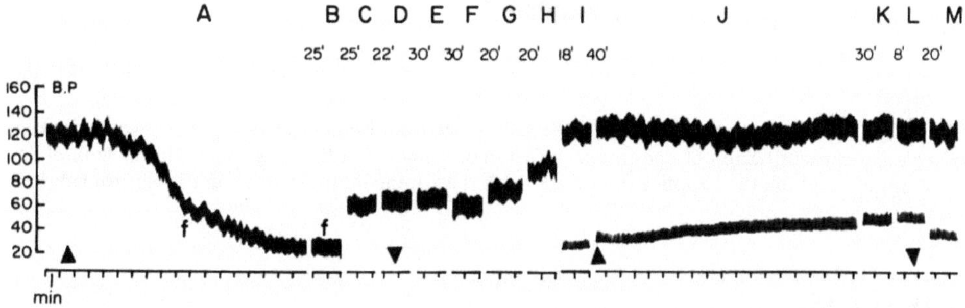

Fig. 5. Cat, male 3.2 kg, atropinized, artificially ventilated. Records of blood pressure (BP) and time (1 min). Upwards and downwards pointing *triangles* indicate commencement and cessation respectively of soman (S)–infusion (20 μg/kg min). Severe fasciculations occurred during the period marked f–f, letters A–M indicate panels. In panel I articicial ventilation was discontinued and respiration recorded

In these cases, as the animals seemed refractory to the anticholinesterase, the experiments were terminated with i.v. injection of an overdose of sodium pentobarbitone.

Antagonism by Topically Applied Atropine Methyl Nitrate, HS-3 and HS-6 of the Hypotension Elicited by Prolonged Intravenous Administration of S

a) Ventral Medulla Oblongata

Atropine methyl nitrate (50 mg/ml) applied to the chemosensitive areas of ventral medul la oblongata during the vasodepressor response elicited by continuous injection of S brought about a complete restoration of BP which subsequently maintained normal level throughout the experiment. The antagonistic effect was remarkable because in all cases a single application was sufficient, and considering that the massive administration of the poison was prolonged for 30–45 min after removal of the drug from the perspex applicator. Some of these findings are shown in Fig. 6.

The antagonistic action of HS-3 and HS-6 when applied to the ventral medulla was essentially similar to that of atropine methyl nitrate but less notable. Two applications of HS-3 or HS-6 (100 mg/ml) were always required to fully re-establish BP level. Nevertheless, once restored it remained unchanged not only until the end of S administration which was continued for 20–25 min after the second application of the oximes, but also for more than 1 h thereafter. The animals presented a steady normal BP at the time of sacrifice.

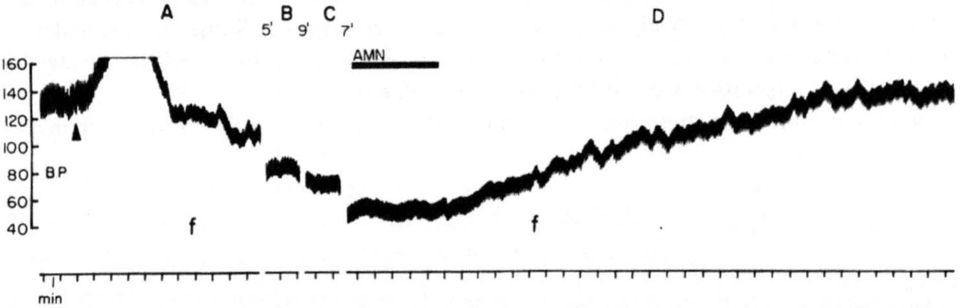

Fig. 6. Cat, female 2.7 kg, atropinized, artificially ventilated. Records as in Fig. 5. Upwards pointing *triangle* indicates initiation of continuous infusion of soman (S) (20 μg/kg/min). The *horizontal bar* denotes 10 μl bilateral application of 100 mg/ml atropine methylnitrate on the ventromedullary areas

b) IV Ventricle

Control pledgets soaked in mock CSF when applied to the IV ventricle fovea, neither influenced normal BP nor restored it when lowered as a result of continuous administration of S. When the pledget was soaked in 100 mg/ml either atropine methyl nitrate, HS-6 or HS-3 it was effective in antagonizing the vasodepression caused by the organo-

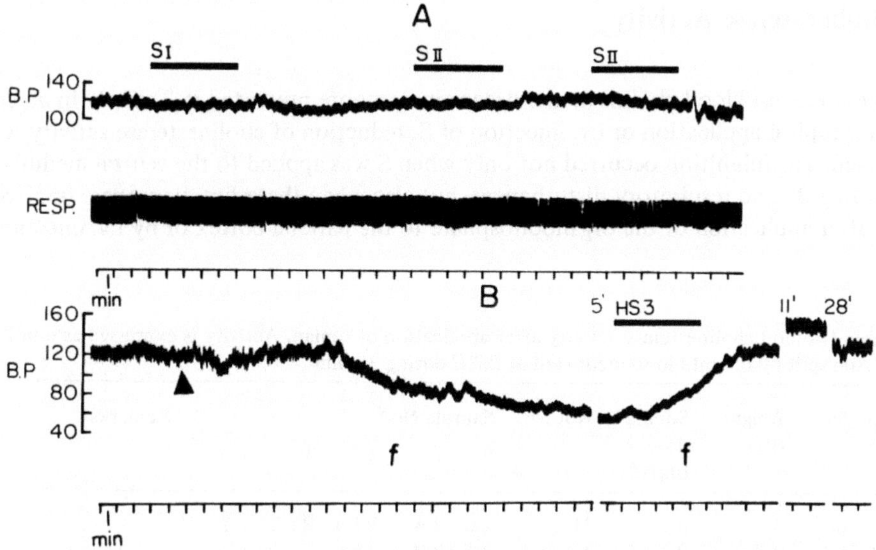

Fig. 7. Cat, male 3.8 kg, atropinized, artificially ventilated. **A** and **B** are panels of the same experiment. Records of blood pressure (BP), respiration (RES) and time (1 min). *Horizontal bar* represents 10 μl application on the IV ventricle central fovea of soman 100 μg/ml (S_I), 250 μg/ml (S_{II}), 500 μg/ml (S_{II}, 2nd mark) or of 100 mg/ml HS-3. Between **A** and **B** respiration recording was discontinued and artificial ventilation applied. Upwards pointing triangle indicates initiation of continuous infusion of soman (20 μg/kg/min). Severe fasciculations occured during the period f-f. Breaks in panel **B** indicate intervals

phosphate. The experiment of Fig. 7 (lower part) illustrates the rapid recovery of BP after application of an HS-3 pledget to the IV ventricle. In these experiments BP was maintained throughout administration of the remaining S, which continued for 40–45 min after the removal of the HS-3 pledget and also for 45 min to 1 h after completion of the infusion. After these periods the animals were sacrificed. The effects of atropine methyl nitrate (50 mg/ml) were very much like those of HS-3, viz. a single application sufficed to restore BP which had been lowered by prolonged administration of S. On the other hand, in similar conditions application of HS-6 (100 mg/ml) brought about only a partial recovery (30–40 mm Hg) of BP. No further improvement occurred when the application of the oxime was repeated thrice at 20 min intervals during the S-induced hypotensive response. As obidoxime did not antagonize the BP fall nor the respiratory disturbances elicited by S applied to ventral medulla, the oxime was not examined under the stringent conditions of continuous injection of the anticholinesterase.

c) Parietal Cortex

When applied to the parietal cortex, atropine methyl nitrate (50 mg/ml), HS-3 or HS-6 (100 mg/ml) did not modify the extent and course of the long lasting hypotensive response occurring after prolonged administration of S.

Cholinesterase Activity

The results of blood cholinesterase determinations are presented in Table 1. In all cases after topical application or i.v. injection of S, reduction of cholinesterase activity was ob served. The inhibition occurred not only when S was applied to the ventral medulla thus causing BP and respiratory disturbances, but also when these functions were unaffected as after application of the organophosphate to the parietal cortex or by i.v. injection.

Table 1. Blood cholinesterase activity after application of soman. Activity is expressed as μmol of Ach split by 0.1 ml blood incubated at 25°C during 35 min

Cat No.	Sex	Weight (kg)	Soman[a] dose (μg)	Route[b]	Sample No.[c] 0	1	2	3	4	Remarks
1	m	3	0.2	M	2.5	2.4	R1.4	R1.1	1.7	
2	f	2.7	2	M	2.5	2.0	2.0			
3	m	3.5	2	M	3.4	2.8				
4	m	3.5	2	M	2.8	2.5	R2.3			
5	m	3.0	2	M	3.0	2.2				
6	m	2.9	5	M	2.4	2.2			1.9	
7	m	3.0	5	M	2.0	1.6				
8	m	2.5	5	M	2.8	2.3	2.2			
9	f	2.6	5	M	3.2	3.0				
10	f	2.4	5	M	2.9	2.4	2.4	R2.1	2.1	
11	m	3.5	5	M	2.6	2.5	2.4		2.3	
12	m	3.4	5	M	2.0	1.9	R1.4	1.3	1.3	
13	m	2.8	5	M	1.6	0.8				
14	f	2.6	5	M	2.7	2.5		R2.2	R1.8	
15	m	3.1	5	M	2.7	2.1	2.1	2.1		
16	f	3.2	2	C	2.1	1.8	R0.85			Repeated dose was 20 μg
17	m	3.9	2	C	3.3	3.1	R2.6	R2.4		Repeated doses were applied on M
18	m	2.8	5	C	2.7	1.8	1.8			
19	f	2.1	0.2	i.v.	3.0	2.4	R2.4	2.4		
20	f	2.4	0.2	i.v.	2.6	2.4	R2.2			Repeated dose was 1 μg
21	f	2.2	2	i.v.	3.0	2.4	R2.1	R1.6	R1.4	The last application was on M

[a]Numbers refer to the total amount applied.
[b]M: chemosensitive areas of ventral medulla: C: parietal cortex; i.v.: intravenous injection.
[c]: Sample No.0 (control) was withdrawn 5 min before, whereas samples 1–4 were obtained at 5–10, 20–30, 50–90 and 100–120 min respectively after application of soman. R, indicates that application of soman was repeated 10 min before withdrawal of the sample, though the time interval refer to the first application.

Muscarinic Receptors and Cholinesterase Activity on the Surface of Ventral Medulla

Table 2 shows that concentration of muscarinic receptors and cholinesterase activity gradually decreased (analysis of variance) from pons in caudal direction to other ventromedullary sites. In addition there was a positive correlation between receptor concentration and enzyme activity in all the sites.

Discussion

The present experiments showed that S-induced long lasting hypotension, conspicuously resistant to systemically administered cholinolytics and oximes, originated centrally and was most probably triggered when the o.p. interacted with the superficial neurons (Petrovicky 1968, Trouth et al. 1973, Shahar and Edery 1976) populating the chemosensitive areas of ventral medulla oblongata. It is difficult to avoid such a conclusion considering two main facts. First, the striking fall of blood pressure occurring when a minute amount of the o.p. was topically applied on the ventromedullary but not on other brain areas. Secondly, the prompt reversal of the S-induced hypotension, effected even after massive systemic administration, when antagonistic drugs were applied on the ventromedullary sites. It was difficult to estimate the actual concentration of the anticholinesterase effective on the neural tissue but surely it must have been considerably smaller than that applied, as a substantial part must have passed via the efficiently absorbing local vessels (Lipscomb and Boyarsky 1972) to the general circulation. This was clearly indicated by inhibition of blood cholinesterase.

The S-provoked hypotensive response should be attributed to the peripheral vasodilation consecutive to diminished central sympathetic vasomotor tone. Participation of cholinergic vasodilator pathways could be excluded as they were blocked by pretreatment with atropine. Consistent with our view, de Neef and Porsius (1981) recently reported that paraoxon decreased the peripheral resistance in cats treated with cholinolytics. Furthermore, Preston and Heath (1972) showed that prolonged administration of S in atropinized rabbits caused vasodilation. The cat however appeared more sensitive than the rabbit as S infused at a rate of 25 μg/kg/min in this species elicited a barely perceptible hypotension, whereas it was invariably severe in cats. The initial rise, frequently preceding the fall of blood pressure, should be attributed to the release of adrenaline (Edery 1969) leading to increased peripheral resistance. The original observation of Preston and Heath (1972) that blood pressure did not recover within about three hours of cessation of S infusion, in contrast to sarin and DFP, was confirmed. However, in the present experiments after that time blood pressure did regain its initial level together with spontaneous breathing and subsequent tolerance to repeated administration of the o.p. Tolerance to the respiratory impairment induced by S has already been reported in rats (Meeter and Wolthuis 1968), and guinea pigs (Adams et al. 1972) and is attributed to adaptation of the respiratory centre occurring independently of cholinesterase reactivation. Surely, the latter could not have occurred in our experiments.

It could then be speculated that after the administration of the first dose of S, most of the critical targets including nervous cholinesterase must have been occupied or inactivated by the anticholinesterase, thus rendering them insensitive to a repeated dose.

In biochemical terms, it seems reasonable to expect that the primary action of S at the superficial ventromedullary neurons must be cholinesterase inactivation and consequently synaptic accumulation of acetylcholine. Incidentally, direct and indirect evidence indicate that these neurons are cholinergic and cholinoceptive. Nevertheless, alternative or additional mechanisms besides cholinesterase inhibition should be considered. For instance, S has shown to interfere with tissue respiration and to inhibit enzymes of the intermediary metabolism including succinic dehydrogenase (Jovic et al. 1971). This enzyme has been visualized histochemically in the cytoplasma of the superficial ventromedullary neurons (Edery and Gaaton, unpublished experiments). It has also been reported (Bullock et al. 1977), that S as well as paraoxon and DFP inhibited the cholinesterase of *electrophorus electroplaque* and also combined with the acetylcholine post synaptic receptor. However, S at difference with other anticholinesterases, besides rapidly "aging" the cholinesterase, causes a two-phase depolarization resulting in excitation.

Similarly, S in the rat isolated diaphragm preparation depolarized the muscle cell membrane, an effect occurring independently of cholinesterase inhibition (Abraham and Edery 1977). It could then be speculated that a similar mechanism of excitation could be operating at the superficial ventromedullary cell's membrane. Assuming a direct excitatory effect of S would be more compatible with the fact that repeated topical application resulted always in hypotensive responses. In contrast, the respiratory disturbances were rather variable thus suggesting independence from vascular changes. They could have resulted from asynchronic stimulation of S of the superficial ventral neurons. In regard to this, it should be recalled that the location of these cells coincide with the areas from which Trouth et al. (1973) applying electrical point to point stimulation, elicited either inspiratory, expiratory or mixed responses. It could then be assumed that some neurons, regularly eliciting inspiration, shift to expiration under the influence of S. This curious effect has been reported (Bay et al. 1973) to occur in neurons of the medullary reticular formation of atropinized cats treated with S.

The rapid reversal of S hypotension by atropine when topically applied on ventral medulla as well as the IV ventricle contrasted with the ineffectiveness after its i.v. injection. The variance could be readily attributed to larger concentrations attained at these critical sites when the drug was locally applied. The rise of blood pressure occurring after topical benactizine, HS-3, HS-6 or HI-6, suggests that these drugs like atropine (Guertzenstein 1973) could have blocked the inhibitory influence exerted by superficial ventral neurons on structures controlling sympathetic vasomotor tone (Schlaefke et al. 1979). They might thus have increased their activity resulting in augmented peripheral resistance and concomitantly rising blood pressure.

In the present experiments obidoxime not only failed to reverse but sometimes even accentuated S-induced vasodepression. A similar failure has been reported during S-induced block in rat isolated diaphragm preparation (Wolthuis and Cohen 1967). Although several hypotheses have been proposed, the mechanism of the interaction is not fully understood.

To explain the antagonism against S by HS-3, HS-6 and HI-6 on a biochemical level a number of actions attributed to oximes should be considered. Those include (a) reac-

Table 2. Muscarinic receptors concentration and cholinesterase activity in superficial slices of ventral medulla. Values represent mean ± S.E.M. of eight cats

Area (See Fig. 1)	Muscarinic receptors ([^{3}H] QNB pmol/mg protein)	Cholinesterase activity (μmol Ach split/min/mg protein)
A	0.415 ± 0.055	0.106 ± 0.011
B	0.113 ± 0.015	0.054 ± 0.004
C	0.088 ± 0.009	0.046 ± 0.004
D	0.078 ± 0.055	0.039 ± 0.002

tivation of inhibited cholinesterase; (b) inactivation of the o.p. by direct reaction; (c) atropine-like effect, i.e. competition for the cholinergic receptor. Possibilities (a) and (b) seem unlikely as the cholinesterase undergoes rapid "aging" after S and because the interaction of this latter with oximes has been shown to proceed very slowly in vitro thus unlikely to occur in vivo. Postulate (c) could be possible as muscarinic receptors are now evident in the superficial layer of ventromedullary areas (Table 2) and Amitai et al. (1980) demonstrated that these oximes blocked muscarinic receptors in mouse brain. Blockade of muscarinic receptors could also explain the reversal by HI-6 of the hypotension induced by carbamate anticholinesterase eserine.

Finally, it is believed that the ventromedullary chemosensitive areas represent selective target sites for anticholinesterases to produce severe vasodepression. Conceivably, these or their homologues sites in species, other than cats, should be protected with adequate concentrations of effective cholinolytics and oximes to restore the BP in anticholinesterase poisoning.

Acknowledgements. My thanks are given to Dr. Eva Haber, Dr. H. Gazit, Mrs. G. Porath and Mr. N. Oz for participation in some experiments.

References

Abraham S, Edery H (1977) Presynaptic effects of soman in rat isolated diaphragm. Isr J Med Sci 13:1142–1143

Adams EK, Yamamura HI, O'Leary JF (1972) Spontaneous recovery of central respiratory function after an irreversible AChE inhibitor. Eur J Pharmacol 20:377–380

Amitai E, Kloog Y, Balderman D, Sokolovsky M (1980) The interaction of bis-pyridinium oximes with mouse brain muscarinic receptor. Biochem Pharmacol 29:483–488

Bay E, Adams NL, von Bredow KD, Nelson JD (1973) Respiratory phase shift of pattern in the medullary reticular formation after soman in the cat. Brain Res 60:526–532

Bullock JO, Farquharson DA, Hoskin FCG (1977) Soman and receptor-ligand interaction in electrophorus electroplaques. Biochem Pharmacol 26:377–343

Edery H (1969) Release of pharmacologically active substances during alkylphosphate intoxication. Abstr 4th Int Pharmacol Cong 184–185

Edery H (1975) Barosensitive areas on the ventral medulla oblongata. Abstr 6th Int Pharmacol Cong 325

Edery H (1978) The origin of soman central vasodepression and its reversal by cholinolytics and oximes. In: Plaa GL, Duncan WAM (eds) Proceedings of the first international congress on toxicology. Academic Press, London, New York, p 452
Edery H, Guertzenstein PG (1974) A central vasodepressor effect of dyflos. Br J Pharmacol 50: 481–487
Feldberg W (1976) The ventral surface of the brain stem: a scarcely explored region of pharmacological sensitivity. Neuroscience 1:427–441
Feldberg W, Guertzenstein PG (1972) A vasodepressor effect of pentobarbitone sodium. J Physiol (London) 244:83–103
Fleischer JH, Pope EJ, Spear SF (1955) Determination of red blood cell cholinesterase activity in whole blood. Arch Ind Hlth 11:332–337
Guertzenstein PG (1973) Blood pressure effects obtained by drugs applied to the ventral surface of the brain stem. J Physiol (London) 229:395–408
Johnson CD, Russell RL (1975) A rapid, simple radiometric assay for cholinesterase, suitable for multiple determinations. Anal Biochem 64:229–238
Jovic R, Bachelard HS, Clark AG, Nicholas PC (1971) Effects of soman and DFP in in vivo and in vitro on cerebral metabolism in the rat. Biochem Pharmacol 20:519–527
Levinger IM, Edery H (1969) Casts of the cat cerebroventricular system. Brain Res 11:294–304
Lipscomb WT, Boyarsky LL (1972) Neurophysiological investigations of medullary chemosensitive areas of respiration. Respir Physiol 16:362–376
Loeschcke HH (1980) Chemical alterations of cerebrospinal fluid acting on respiratory and circulatory control systems. In: Wood JH (ed) Neurobiology of cerebrospinal fluid. Plenum Press, New York London, p 29
Meeter E, Wolthuis OL (1968) The spontaneous recovery of respiration and neuromuscular transmission in the rat after anticholinesterase poisoning. Eur J Pharmacol 2:377–386
Neef de JH, Porsius AJ (1981) Central effects of paraoxon on haemodynamics in the cat. Naunyn Schmiedebergs Arch Pharmakol 317:168–172
Petrovicky R (1968) Über die Glia marginalis und oberflächliche Nervenzellen im Hirnstamm der Katze. Z Anat Entw 127:221–231
Preston E, Heath C (1972) Atropine-insensitive vasodilatation and hypotension in the organophosphate-poisoned rabbit. Arch Int Pharmacodyn Ther 200:231–244
Schlaefke ME, Kille JF, Loeschcke HH (1979) Elimination of central chemosensitivity by coagulation of a bilateral area on the ventral medullary surface in awake cats. Pflügers Arch 378:231–241
Shahar A, Edery H (1976) Large multipolar nerve cells in the ventral medulla oblongata of the cat. Proc 6th Eur Cong Electron Micr II:614–616
Trouth CO, Loeschcke HH, Berndt J (1973) Topography of the respiratory responses to electrical stimulation in the medulla oblongata. Pflügers Arch 339:153–170
Wolthuis OL, Cohen EM (1967) The effects of P_2S, TMB_4 and LuH_6 on the rat phrenic nerve diaphragm preparation treated with soman or tabun. Biochem Pharmacol 16:361–367

Histochemical Studies on the Central Chemosensitive Fields of the Medulla Oblongata

R. DERMIETZEL, I. WILLENBERG, and A.LEIBSTEIN[1]

The ventral surface of the medulla oblongata contains a network of superficially located neurons which are believed to represent the neuronal substrate of central chemosensitivity of respiration.

While most of the recent morphological studies were focused primarily on the reconstruction and identification of the different superficially located nuclei of the formatio reticularis, probably involved in the chemosensitive drive, the intention of our previous investigations aimed at a better characterization of the neurons by elucidating the transmitter characteristics of the nerve cells in question. This histochemical approach was initiated by recent pharmaco-phsyiological studies of Dev and Loeschcke (1979) and Fukuda and Loeschcke (1979) which clearly indicated that a cholinergic mechanism is involved in the mediation of central chemosensitivity. The primary intention of this study was to identify the *cholinergic neuronal* matrix in the superficial layers of the medulla oblongata.

Our different histochemical series were designed as follows:

a) We wanted to evaluate the overall cell distribution in the different chemosensitive areas by means of serial sectioning and subsequent cresylviolet-staining.

b) A discrimination of the cholinergic neurons from the non-cholinergic cells was obtained by applying the acetylcholinesterase (ACHE) technique of Karnovsky and Roots (1964). Light microscopical and electron microscopical experiments were both performed during this phase of our experiments.

c) To differentiate the cholinergic cells we made use of a highly sensitive technique, which allows the identification of nicotinergic receptors, the α-bungarotoxin-HRP technique. This method, originally introduced by Lentz and Chester (1977), was refined in our laboratory to allow light microscopical observation of nicotinergic cells in serial sections.

[1]Institute of Anatomy, Universitätsklinikum der Gesamthochschule Essen, D–4300 Essen, Fed. Rep. of Germany

Central Neurone Environment
(ed. by M.E. Schläfke, H.P. Koepchen and W.R. See)

Results

In the caudal area (area L) of cats and an analogue location in rats medium sized nerve cells (20–25 μm) were frequently obtained. They occurred either single or in

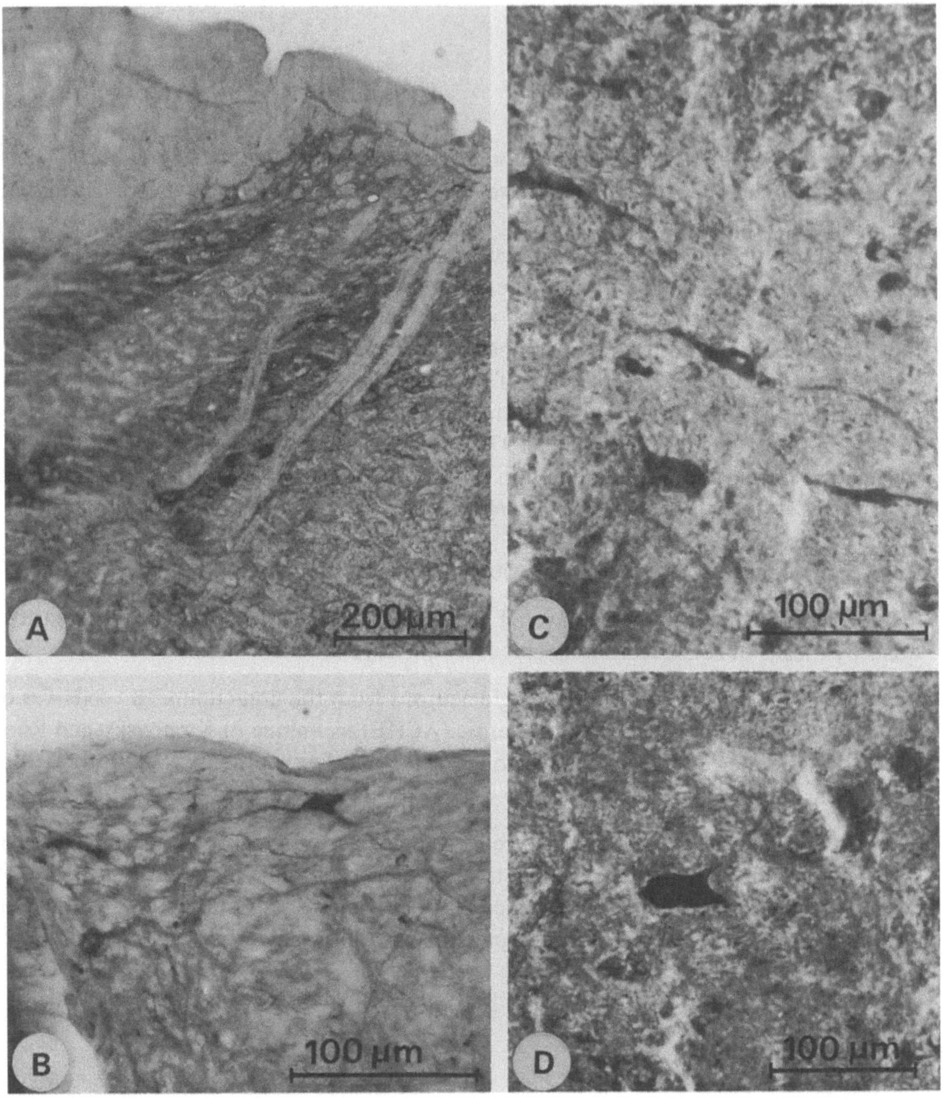

Fig. 1. A, B. Acetylcholinesterase (ACHE) positive neurons. **A** Small groups of neurons occur in the fascicles of the hypoglossal roots of the caudal area (area L). **B** Superficially located ACHE positive cells from the intermediate area (area S). **C, D** α-bungarotoxin-HRP positive cells indicating the presence of nicotinergic receptor sites (NRS). C NRS-positive cells from the caudal area. The dendrites of the cells show some reaction products. D Single NRS-positive cell from the intermediate area

groups of a few cells. Such groups were detected preferably in the fascicles of the hypoglossal roots (Fig. 1A). The pattern of these cholinergic nerve cells changed in the intermediate areas and in part in the rostral area (M). Here, we found complexes of small sized cells (10–15 μm) containing a dense reaction product. Some of these cells were localized very superficially in the intermediate area, no more than 5–10 μm below the surface. This was not the case in the rostral area (Fig. 1B).

Electron microscopical studies of ACHE positive cells allowed a precise localization of the reaction product. It was found in the synaptic clefts of axo-dendritic contacts and in different parts of the Golgi apparatus of the neurons (Fig. 2A, B).

Quantitative mapping of the cholinergic cells indicated that only a certain part of the whole cell population of the neurons in the superficial layers of the medulla are cholinergic in nature. To obtain the portion of cholinergic cells serial sectionings of four medullae of cats and four medullae of rats were performed. Every third section of these series was stained with cresylviolet and the subsequent section reacted for ACHE. By this procedure a direct visualization of the ACHE-positive proportion of cells was possible. Figure 3 gives the ratio of ACHE positive cells on the basis of the whole cell population. For a further discrimination of the cholinergic cells we made use of the α-

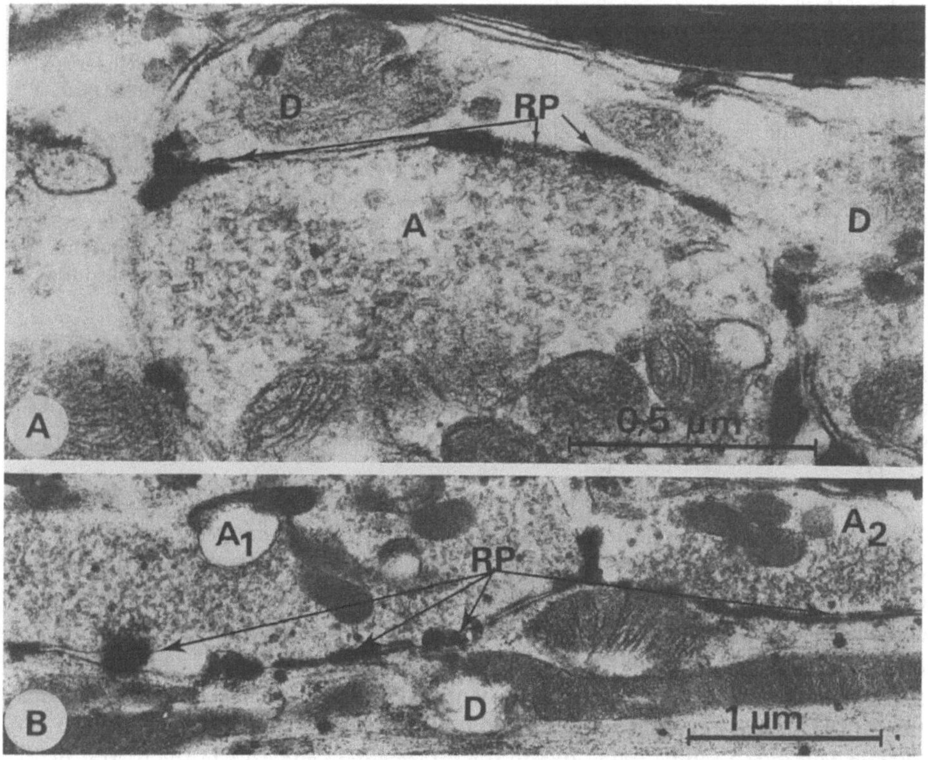

Fig. 2. A, B. Electron micrograph of ACHE positive axo-dendritic synapses from the intermediate area. The reaction product occurs as an electrondense precipitate within the synaptic cleft. D, dendrite; RP, reaction product; A, axon

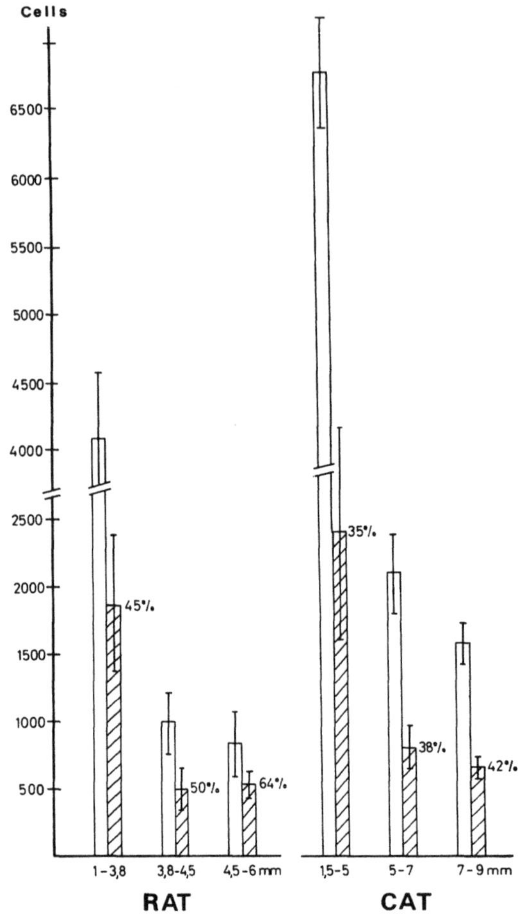

Fig. 3. Histogram depicting the percentage of ACHE positive cells on the basis of the whole cell population. Counting of the cells was performed in an area of the ventral zones of the medulla which includes the following nuclei (listed from caudal to rostral): nuc. reticularis ventralis med. obl., nuc. reticularis lateralis magnocellularis, nuc. reticularis parvocellularis (4.5–6 mm, rat; 7–9 mm, cat), nuc. paragigantocellularis lateralis, nuc. reticularis gigantocellularis pars α (1–3.8 mm, and 3.8–4.5 mm, rat; 1.5–5 mm, and 5–7 mm, cat). The zero point refers to the foramen coecum of the ventral aspect of the medulla oblongata

bungarotoxin-HRP technique. This technique allows the detection of nicotinergic receptor sites. By conjugating the snake venom α-bungarotoxin with horseradish peroxidase (HRP) it is possible to visualize nicotinergic cells light microscopically (Fig. 1C, D).

As is known, from in vitro pharmaco-physiological studies (Fukuda and Loeschcke 1979), nicotine applied on the ventral surface of the medulla creates a remarkable excitation of chemosensitive neurons. Hence, a detection of these receptor sites seems of considerable importance for scrutinizing the chemosensitive units. α-bungarotoxin conjugated with HRP gives a brownish reaction product when reacted with diaminobenzidine. The specificity of the reaction was ascertained by control experiments with α-bungarotoxin and nicotine as preadsorbents. By means of this technique quantification of the nicotinergic cells proved possible. It turned out that about 80% of the cholinergic cells in the caudal field, 70% in the intermediate and 14% in the rostral field represent nicotinergic neurons (the data refer only to rats and therefore the notations "caudal", "intermediate", and "rostral" field are to be taken as a mere analogue to the cat nomenclature).

What About the Rest of the Cell Population?

Recent experiments performed by Loewy and McKellar (1981) indicate that part of the superficial cells of the caudal area in rats are serotinergic neurons. These experiments are in good agreement with our own observations. Screening of the chemosensitive fields for catecholamine and serotonine fluorescence showed single neurons in the nucleus interfascicularis n. hypoglossi of the rat, the side where Loewy and McKellar (1981) described their serotinergic cells. Neither in the intermediate nor in the rostral area a significant amount of fluorescent cells was found. An obvious concentration of catecholamine containing cells was evident only in the superficial portion of the nucleus reticularis lateralis magnocellularis. This nucleus has, however, so far not been found to be involved in the chemosensitive drive. To check whether other transmitters than acetylcholine or catecholamines occur in the neurons of the chemosensitive areas we started immunohistochemical screening for neuropeptides. Though our experiments are very preliminary at the present stage we feel justified to give the following list of peptigergic neuronal presentation in the chemosensitive fields (Table 1). With the exception of endorphins we lack any physiological proof that the other neuropeptides are involved in the control of central chemosensitivity. Opiate receptors, however, have been localized in brain stem respiratory areas, recently. According to Pokorski et al. (1981) endogeneous opiates may contribute to respiratory control. Fentanyl, a morpphine-like substance, depresses ventilation and the opiate antagonist naloxone stimulates ventilation when applied to the intermediate chemosensitive area (S).

Though we have no definite immunohistochemical proof, our results point in the same direction.

Table 1. List of peptigergic neuronal presentation in the chemosensitive fields

Neuropeptides	Vasopressin	Somatostatin	Substance P	β-Endorphin
Representation in ch. f.[a]	Ø Ø	Single cell units	Fibers	Single cell units
Distance from the f.c.[b]		5,5–4 mm (L)	5,5–3 mm (L, S)	5,8–4 mm (L)

[a]ch.f., chemosensitive fields; [b]f.c., foramen coecum.

Conclusion

Our histochemical results clearly indicate that cholinergic neurons are well represented in the chemosensitive zones on the ventral surface of the medulla oblongata. This is in agreement with pharmaco-physiological findings which proved the presence of a cholinergic action in the mediation of the chemosensitive drive. Furthermore, a certain part of the cholinergic neurons are of the nicotinergic type. Catecholamines containing neurons occur only occasionally. Immunohistochemical mapping for neuropeptides revealed the presence of substance-P reactive fibers, somatostatin-like and β-endorphin-like neuronal reactivity. The functional significance of these findings has to be elucidated in further pharmaco-physiological experiments.

Acknowledgements. R. Dermietzel was supported by a grant from the Deutsche Forschungsgemeinschaft (SFB 114, Bionach).

References

Dev NB, Loeschcke HH (1979) Topography of the respiratory and circulatory responses to Acetylcholine and Nicotine on the ventral surface of the medulla oblongata. Pflügers Arch 379:19–27

Fukuda Y, Loeschcke HH (1979) A cholinergic mechanism involved in the neuronal excitation by H^+ in the respiratory chemosensitive structures of the ventral medulla oblongata of rats in vitro. Pflügers Arch 379:125–135

Karnovsky MJ, Roots L (1964) A "direct-coloring" thiocholine method for cholinesterase. J Histochem Cytochem 12:219–221

Lentz TL, Chester J (1977) Localization of acetylcholine receptors in central synapses. J Cell Biol 75:258–267

Loewy AD, McKellar S (1981) Serotonergic projections from the ventral medulla to the intermediolateral cell column in the rat. Brain Res 211:146–152

Pokorski M, Grieb P, Wideman J (1981) Opiate system influences central respiratory chemosensors. Brain Res 211:221–226

Round Table Discussion

Three main topics were discusssed at the round table: 1. Action of central chemosensitivity on sympathetic efferents, 2. convergence of various inputs on central chemosensitivity, 3. neurophysiological mechanisms of respiratory rhythmogenesis.

Chemosensitivity and the Sympathetic Discharge

Trzebski reminds of his communication at the Heidelberg symposium 1980 where he demonstrated CO_2 response in cervical sympathetic activity below the CO_2 threshold for rhythmic phrenic activity. He adds some new findings on hyperventilated cats where addition of 7% CO_2 caused a blood pressure increase before the start of phrenic activity. Evoked phrenic potentials were recorded following stimulation of dorsal or ventral respiratory brain stem regions. The stimulation effect from the dorsal respiratory complex was reduced by addition of CO_2 in the apnoeic range. A possible interpretation could be that E neurones which are still active in a range of CO_2 where I neurones are silent inhibit the I neurone complex. This would be another indication that the CO_2 response in the sympathetic efferents is not mediated by I neurones. Schlaefke agrees, but adds that both systems may have the input in common. Koepchen agrees with the conclusion of Trzebski. He adds information from own experiments where the difference in action of CO_2 on both systems was demonstrated also in the suprathreshold range for phrenic activity. In apneustic states provoked by cooling of the ventrolateral NTS region the net increase of phrenic activity by far exceeded the increase in sympathetic activity demonstrating that a great part of the CO_2 sensitivity seen in the cervical sympathetic does not come from the output of the respiratory tone-rhythm converter, but bypasses it.

Kiwull-Schöne asks Trzebski if he had chemodenervated his animals. According to their own observations, a lack of CO_2 response in subthreshold phrenic activity after hyperventilation was prolonged during hypoxia after peripheral chemodenervation. Trzebski responds that he has not chemodenervated his animals. He points to his experiments together with Polosa where they saw a CO_2 response in blood pressure and sympathetic activity during hypocapnic silence of phrenic activity already with 1.5% inspiratory CO_2. This would be another confirmation that both systems have different CO_2 thresholds. Cunningham reminds of the observation of Mitchell that respiratory neurones are hyperpolarized by CO_2.

Cohen cites his work with Gootman on rhythmic discharges in sympathetic nerves where three frequency ranges were observed: 3/s, 10/s, and a lower one which could be locked with the respiratory rhythm. In his experiments the 10/s rhythm was favoured

Central Neurone Environment
(ed. by M.E. Schläfke, H.P. Koepchen and W.R. See)

with increasing CO_2. He proposes systematic study of sympathetic rhythms under CO_2. Polosa suggests that the increased rhythmicity under CO_2 could be caused by recruitment of sub-threshold sympathetic neurones. Cohen points to the fact that not only mean sympathetic activity but also the degree of rhythmicity, i.e. of synchronization of the neurones was enhanced. See adds the finding that blocking the intermediate area at the ventral medullary surface causes disappearance of respiratory rhythm in splanchnic and cervical sympathetic activity. Koepchen emphasizes that also reduction of basic activity of the sympathetic can reduce the rhythmicity in the discharge. This can be observed if sympathetic activity is reduced more strongly than that of the respiratory oscillator, e.g. by baroreceptor excitation. In this case, respiratory rhythmicity disappears in the reduced sympathetic discharge already in a range where phrenic activity is hardly reduced.

Convergence of Various Inputs with Central Chemosensitivity

Cunningham (chairman) turns the discussion to the relation between chemosensitive and non-chemosensitive drives from the ventral medulla. Schlaefke remarks to that point that the function of central chemosensitivity would be dependent on a basic level of excitation from non-chemosensitive sources. She reminds of the analogy of the drive from peripheral chemoreceptors under hyperoxia. Cunningham comes back to the approaches of Grodins and other earlier workers who tried to explain in a quantitative manner the whole ventilatory response as the sum of the single responses and cites the conclusion of Lloyd that a summation of the partial effects would exceed the actually observed ventilation. He points to the general viewpoint of redundancy of afferent information and the selective function of the central nervous system. He says that the scheme outlined by Prof. Loeschcke at the blackboard could be a good model of this selective action and would be able to overcome these difficulties of over-summation of afferent inputs (Fig. 1).

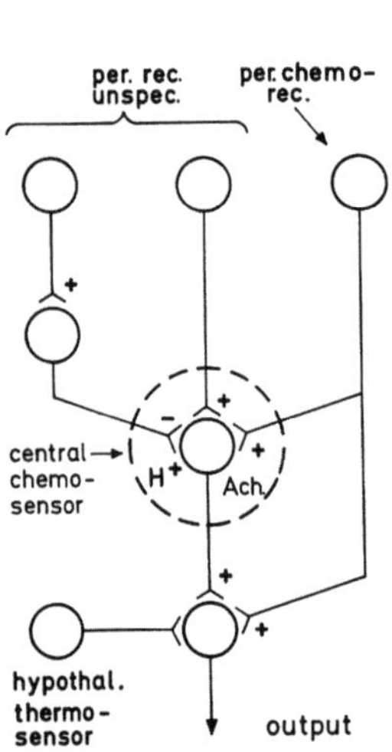

Fig. 1. H.H. Loeschcke's outline on the blackboard summarizing his review on *Central Chemosensitivity*, presented at the Symposium. The central chemosensor is formed by H^+-sensitive cholinergic synapses of ventral medullary neurones, connecting e.g. with afferents from various peripheral unspecific receptor systems and with collaterals from peripheral chemoreceptor fibers. The central chemosensor-output converges together with direct peripheral chemoreceptor pathways and with the hypothalamic thermosensor on the integrating respiratory neuronal network, the latter producing the respiratory center output

Von Euler points to the differentiation between additive and multiplicative interactions. In the case of multiplicative interaction a decrease of chemical drives results in the decrease also of the action of other drives. This would reduce the danger of hyperventilation in contrast to the case of additive interaction. Javaheri asks if the drives of exercise and CO_2 would act on ventilation in an additive manner. Cunningham says that this is the case in man. Von Euler reports that he found a parallel shift of CO_2 response curves, i.e. additive action of exercise during locomotor activity in thalamic cats.

Koepchen tells about studies of his group on renal sympathetic activity where in some strands the discharge was not dependent on CO_2 but could be strongly reduced by ventral medullary cooling. This would mean that there must be a non-chemosensitive drive coming from the ventral medulla. He raises the question if such a non-chemosensitive drive would differ from other intramedullary, non-chemosensitive drives by any other criterion except its site near the ventral surface. Loeschcke responds that the superficial site would be no special criterion. There would be specific inputs into the system and specific outputs, and the specificity of the network in between would be due to the anatomical pathways.

Honda reports that patients after bilateral dissection of the carotid bodies show ventilation in the normal range but reduced CO_2 sensitivity. Schlaefke thinks that the reduced CO_2 sensitivity in this case could be the general result of reduced activity in the whole system. In confirmation of Loeschcke's remark, she points to the role of the nucleus paragigantocellularis which may receive collaterals from many afferent systems and would be of special importance not only for the control of the respiratory system, but also for that of the sympathetic efferents and regulation of body temperature. In addition, this part of the ventral medulla would be responsible also for the interaction between these systems. Koepchen reminds of his studies together with Langhorst and Seller, where they found extensive vegetative-somatic convergence on single neurones in the medulla oblongata. In this respect, the more ventral parts of the reticular network, like the nucleus paragigantocellularis, would not basically differ from other parts of the reticular formation of the lower brainstem. It would have to be studied systematically if there are differences in the convergence of different afferents between different parts of the medullary reticular nuclei. Schlaefke reminds of earlier discussions with Koepchen where she raised the question why, in spite of the existence of so many unspecific, non-feedback respiratory drives, the elimination of central chemosensitivity leads to respiratory arrest. At that time, they agreed that this would be a quantitative problem, which has to be analysed now in more detail.

Edery asks where the corresponding areas would be located in humans and non-human primates since in these species the whole surface would be covered by the pyramids. Dermietzel responds that they had studied the medullas of very young humans. Also in man the whole surface would not be covered by the pyramids and there could be seen for instance the olives and the hypoglossus roots. The region of the chemosensitive structures in humans would correspond with the nucleus arcuatus. Schlaefke confirms in response to the question of Cunningham that the cell deficits in children suffering from respiratory insufficiency are located in the same region. Von Euler asks whether the nucleus arcuatus in man would be analogue to the nucleus paragigantocellularis lateralis or to the nucleus reticularis lateralis. Dermietzel responds that the nucleus reticularis lateralis would be located more laterally, whereas the nucleus arcuatus lies superficially at the same location where the nucleus paragigantocellularis in cats reaches the ventral surface. Ahmad asks for the site of the nucleus arcuatus in rat, and Dermietzel says that the structures in rat are similar as in man, whereas in cats an analogue accumulation of cells at that site is absent. There are tremendous species differences.

Petersen compares the well-coordinated motor act of respiratory movements with laughing which as a final effect can be likewise evoked by many different afferent stimuli. If his interpretation of the contributions to this symposium would be correct, the afferents for the respiratory motor act would converge upstream the entrance of the chemoreceptor afferents, so that all of them would be modulated in common by the influence of chemosensitivity. In contrast, the NTS and retroambigual regions would be downstream of the site of convergence which is presumably in the nucleus paragigantocellularis. In this connection, von Euler cites two recent anatomical studies by Andrezik et al. 1981 in the Journal of Comparative Neurology where it is demonstrated that the nucleus paragigantocellularis receives many collaterals from ascending and descending pathways and in addition inputs from pontine and medullary nuclei besides very strong connections between both sites. It would be remarkable that there is also a strong input from the raphe nuclei, so that part of the nucleus paragigantocellularis belongs to the serotonergic system. See confirms that they have found convergence of various afferents from peripheral nerves on cells in the intermediate area, which in addition are activated by increase of P_{CO_2} or H^+ ion concentration. Schlaefke warns of the generalization that everthing converges in the ventral medulla and especially in the intermediate area. Convergence has been proved for afferents from working muscles. Recent studies together with Kumazawa have shown that also afferents from polymodal receptors are converging there. Other afferents, however, would bypass the intermediate area as e.g. acoustic afferents which likewise are able to stimulate respiration and which are even more effective after elimination of central chemosensitivity. The latter phenomenon can be interpreted as a stabilizing factor in the system. In contrast, other afferents have a destabilizing action as e.g. those from the hypothalamus. Koepchen emphasizes that the evidence of upstream convergence does not exclude the possibility that additional convergence takes place downstream, so e.g. much convergence from different sources on sympathetic output neurones was demonstrated similar as in respiratory motor neurones. Therefore the same combination of excitation or inhibition from different sources found in the output does not necessarily occur in an upstream superimposed central neurone in the medulla.

Von Euler asks Schlaefke to which input she did refer speaking of exercise afferents. Would these be the afferents from working muscles or from higher structures like basal ganglia or nucleus subthalamicus producing locomotor activity together with respiratory drive? He believes that both inputs play a role and plans experiments with cooling of the ventral medullary surface during centrally induced locomotor activity. This kind of activity acts together with the CO_2 stimulus in an additive manner. The input from working muscles produces another kind of interaction. In this context Cunningham reminds of the difference of opinion where in the range of low P_{CO_2} the peripheral drive from working muscles is effective according to Asmussen and is not according to Déjours. Von Euler feels that both mechanisms could exist.

Albers raises the point that the hypothalamic thermoregulation does not necessarily disturb the respiratory regulation. He interprets the panting as an invention of evolution in arctic animals in order to avoid wetting of the fur. It would be useful if panting started immediately at the beginning of muscular work to provide the dissipating of additional heat produced by the exercise. Cunningham asks if the panting at the start of muscular work would occur also in cold environment. Albers responds that in his studies on sledge dogs in Canada panting started already within seconds after beginning of the run. Cohen remarks to that point that in the respiratory response to muscular work one has to differentiate between hyperventilation and panting. Irrespective of whether the exercise drive would come from central or peripheral sources, it would be a tonic input to the respiratory oscillator. The fact that blocking the chemoreceptor area would eliminate the exercise drive but not the response to hypothalamic stimulation, would not allow conclusions about the site of convergence. This could be just a quantitative difference. Von Euler agrees to a large extent. The subthalamic stimulation for the production of locomotion would be difficult and artificial and has to be localized very precisely to avoid side-effects. The episodes of spontaneous locomotor movements in the thalamic cat are more suitable for analysis. He believes in Kao's evidence of afferent drives from working muscles and thinks that the convergence from central and peripheral drives activated by muscular exercise could be of different nature.

Petersen asks if the bronchomotor reflexes have been studied during cooling of the intermediate area, especially if the carotid body bronchomotor reflex would be preserved. Schlaefke confirms.

Neurophysiological Mechanisms of Respiratory Rhythmogenesis

Richter gives a survey of those themes in the symposium which in his view have been the most important with respect to the rhythmogenesis of respiration. He stresses von Euler's comparison between the respiratory motor act and other motor rhythms and emphasizes that the old model of Burns and Salmoiraghi should be overcome, which is based on only two symmetrical phases of the respiratory cycle with the corresponding two neuronal populations. Instead one has to assume three respiratory phases. The third phase of the respiratory cycle should be introduced which he himself would call post-ramp phase, whereas von Euler had the opinion that this would be the first phase of expiration because the functionally important event during this phase is the movement of air out of the lungs. Centrally this is the phase directly after the end of the ramp where a new inspiration cannot yet start. It would be important to clear up further the function of the respiratory pattern generator including the off-switch mechanism on a neuronal basis. In this connection, the contribution of von Euler concerning the role of the nucleus paragigantocellularis and by Marek on the effect of peripheral afferents on inspiratory and expiratory activity and the finding of Cohen upon the correlation between respiratory neuronal discharges had to be discussed. A further important point of view would be the question of a primary pacemaker of the oscillation raised by Hukuhara with his analysis of the stability of the discharge pattern of respiratory neurones.

Von Euler remarks that the knowledge on the ramp generator and its site would be very small inspite of extensive investigations and that the possibility should be considered that non-spiking interactions between neurones could be responsible which cannot be detected by extracellular recordings. Referring to Marek's experiments, Cohen says that the stimulation experiments on chemoreceptor afferents as well as his own earlier intracentral stimulation in the pontine region demonstrate that the excitability in the inspiratory and in the expiratory population increase during the burst. This is manifested already by the fact of a ramp-like discharge which, according to the popular theory, is caused by some sort of re-excitation. Here the correlation analysis of the discharge of neighbouring neurones could bring further insight into the mechanisms. He cites the thesis of Maden who found in correlation analyses of adjacent neurones in the NTS region correlation peaks centred around zero with a high likelihood of firing synchronously within half a millisecond of each other. This led to the hypothesis that there are dendro-dendritic interactions. This would agree with the finding of Richter that the dendrites of neighbouring neurones are very closely intermingled. Thus the ramp-generation could be a non-synaptic mechanism. Von Euler adds the remark that the pre-excitation and even a balance between re-excitation and re-inhibition could be caused by non-spiking processes. The evidence would be on the way that the R_α cells and perhaps even the R_β cells are not parts of the ramp-generator but outpout neurones. Referring to the results of Maden, Richter stresses that the delays between the cell discharges of 0.2 to 0.5 ms are compatible with a conventional chemical dendro-dendritic interaction. He reports own recent investigations together with Kalia, where they found an R_β neurone with a fairly small cell body, with a large dendritic tree extending over the whole section of the brain stem and including the whole reticular formation, so that these must be highly integrating neurones. There would be a lot of chances to have dendro-dendritic synapses with interactions through local variations of transmitter release without spiking. Marek shows a slide from his experiments with stimulation of chemoreceptor afferents at different times of the phrenic bursts. The increase of rate of rise and maximal amplitude of phrenic activity was found to be the higher the later the time of stimulation was with respect to the phrenic discharge. The maximal effect was obtained at 90% of inspiratory time. Thereafter a delay of the time of stimulation of only 10–20 ms is sufficient to abolish the phrenic response on the same stimulus, in spite of the fact that the integrated phrenic amplitude at that time was still 80–90% of its maximum control value. Richter tries to give a neurophysiological interpretation of these findings. He thinks that the excitatory stimulations during the first time of inspiration are acting on the early inspiratory neurones and that these neurones decide on the further course of inspiration and also on the off-switch mechanism. He refers to the finding of Merrill that these neurones project onto all parts of the brainstem. The action of the stimulus on these neurones also produces a time-shift of the off-switch. In his intracellular recordings of inspiratory neurones he saw that lateron there was still an increased input resistance and in a small set of neurones a depolarized level of membrane potential. This would result in a further increase of stimulus effect. At the end of the ramp, when the stimulation effect is already diminished, there is a decrease in the input resistance because the inhibition by the off-switch begins to appear. Thereafter comes the phase 1of expiration. At that time, stimulation activates, in addition to bulbospinal inspiratory neurones, also those neurones which fire during that time, and these neurones have some inhibitory effect on bulbospinal respiratory neurones, too. In this period, bulbospinal inspiratory neurones show both a declining pattern of excitation and a very weak inhibition resulting in a fading of the stimulation effect.

Subject Index

MIX
Papier aus verantwortungsvollen Quellen
Paper from responsible sources
FSC® C105338

If you have any concerns about our products,
you can contact us on
ProductSafety@springernature.com

In case Publisher is established outside the EU,
the EU authorized representative is:
Springer Nature Customer Service Center GmbH
Europaplatz 3, 69115 Heidelberg, Germany

Printed by Libri Plureos GmbH
in Hamburg, Germany